Physician to the West

Physician to the West

SELECTED WRITINGS OF
DANIEL DRAKE
ON SCIENCE & SOCIETY

EDITED WITH INTRODUCTIONS BY
Henry D. Shapiro & Zane L. Miller

The University Press of Kentucky

Standard Book Number 8131-1197-8
Library of Congress Catalog Card Number 73-94071

Frontispiece: Daniel Drake, aet. 65,
from *Pioneer Life in Kentucky*

Copyright © 1970 by
THE UNIVERSITY PRESS OF KENTUCKY
Lexington, Kentucky 40506

A cooperative publishing agency serving Berea College, Centre College, Eastern Kentucky University, Kentucky State College, Morehead State University, Murray State University, University of Kentucky, University of Louisville, and Western Kentucky University.

Contents

Preface	vii
Daniel Drake: The Scientist as Citizen, by Henry D. Shapiro	xi
Daniel Drake, the City, and the American System, by Zane L. Miller	xxiii
Chronology	xxxv
Some Account of the Epidemic Diseases Which Prevail at Mays-Lick in Kentucky (1808)	1
Notices concerning Cincinnati (1810)	5
Anniversary Address to the School of Literature and the Arts (1814)	57
Natural and Statistical View, or Picture of Cincinnati and the Miami Country (1815)	66
Valedictory to the Class at Lexington in the Transylvania University (1818)	125
Anniversary Discourse on the State and Prospects of the Western Museum Society (1820)	131
Inaugural Discourse on Medical Education (1820)	151
Introductory Lecture for the Second Session of the Medical College of Ohio (1821)	168
From the Western Journal of the Medical and Physical Sciences *(1828–1849)*	183

The People's Doctors; A Review by 'The People's Friend' (1830)	195
Oration on the Causes, Evils, and Preventives of Intemperance (1831)	203
Remarks on the Importance of Promoting Literary and Social Concert in the Valley of the Mississippi (1833)	223
Discourse on the History, Character, and Prospects of the West (1834)	240
Discourse on the Philosophy of Discipline (1834)	260
Rail-road from the Banks of the Ohio River to the Tide Waters of the Carolinas and Georgia (1835–1836)	285
Introductory Lecture on the Means of Promoting the Moral and Intellectual Improvement of the Students and Physicians of the Mississippi Valley (1844)	295
Introductory Lecture, Thirtieth Session of the Medical College of Ohio (1849)	315
Systematic Treatise on the Principal Diseases of the Interior Valley of North America (1850)	329
A Bibliography of the Writings of Daniel Drake	381

Preface

BECAUSE DANIEL DRAKE concerned himself with virtually all aspects of American life during the first half of the nineteenth century, this volume attempts to be more than a collected edition of his writings on science and society. It is intended to be a selection representative of Drake's varied interests during a busy career in public life, conceived as a tribute to the "founding father" of the University of Cincinnati and its College of Medicine on the One Hundred Fiftieth Anniversary of their establishment. More important, however, this volume is intended as an initial attempt at assessment of Drake's career in the context of the problems which beset American science and American society during his lifetime.

Daniel Drake was among the most prominent physicians of his age, known nationally as well as in the Ohio Valley region. A graduate of the University of Pennsylvania Medical College, he was a member of the first medical faculty organized west of the Appalachians, at Transylvania University in Lexington, and later Dean of the Medical Department. He founded the Medical College of Ohio in Cincinnati, in 1819, and the Medical Department of Cincinnati College, in 1835. During the later years of his life he was a member of the distinguished faculty gathered at the Louisville Medical Institute. For twenty years Drake edited the influential *Western Journal of the Medical and Physical Sciences* (Cincinnati) and its successor, the *Western Journal of Medicine and Surgery* (Louisville). His early writings on the climate, geology, and natural history of the Ohio Valley were pioneering attempts to assess the habitability of the trans-Appalachian West and to examine in general the conditions under which civilization can flourish. This work, and his efforts on behalf of the Western Museum Society of Cincinnati, earned him membership in the American Philosophical Society, the American Antiquarian Society, the Philadelphia Academy of Natural Sciences, and the Wernerian Natural History Society of Edinburgh. When the National Association for the Advancement of Science was organized in Washington, D.C., he was elected to membership as a matter of course. In the early 1830s he wrote extensively on cholera, and began in earnest the study of the endemic and epidemic diseases of the West which was to yield his magnum opus, *A Systematic Treatise . . . on the Principal Diseases of the Interior Valley of North America,* widely hailed as "the most valuable and important original production . . . that has yet appeared from the pen of any of our physicians."[1]

Medical doctor, educator, scientist, Drake was also active in local affairs, and in local politics where he identified himself with National Republican, and later Whig factions. As early as 1810 he asserted the desirability of internal improvements as essential to the social and moral, as well as the economic growth of his region and his nation. Drake was an active promoter of canals and railroads. He worked for the establishment of manufactories in the West. He was instrumental in the organization of schools and colleges, hospitals and clinics, libraries, a museum, and literary, scientific, and professional societies without number. He served as an officer of the Cincinnati Branch of the Bank of the United States, a trustee of the Cincinnati City Corporation, the Cincinnati Lancasterian Seminary, the Cincinnati College, and Miami University in Oxford, Ohio. He spoke often and wrote profusely on behalf of projects for social betterment, including Henry Clay's candidacy for president of the United States, Texan independence, the manumission of slaves and their transportation to Africa, the establishment of institutional ties between sections of the nation to insure the preservation of the Union, educational institutions for mechanics, and the cause of temperance. For a time he ran a combination drug- and dry-goods store in Cincinnati, where the first soda-water was manufactured west of the Alleghenies, in 1816.

What Daniel Drake did was unique, but the way he did it was not, and it is this which has most interested us in him. We have tried to see his career as a whole, and to emphasize what we regard as the critical intellectual commitments which underlay and unified his diverse activities. The first of these, which we have called Baconianism following the lead which Drake himself provides, involved an assumption about the utility of observation and induction based on "natural" classification as the means by which knowledge of the universe might be obtained. The second, which we have called Whiggery in an effort to distinguish it from the elitism of the Federalists and the romanticism of contemporary Jacksonianism, involved a conviction that such knowledge might form the basis for the conscious construction of a viable social order, and that in the absence of such knowledge a viable social order could not indeed exist. If man did not recognize the restrictions which environment placed on his freedom, and seek his own ends in the context of this knowledge, he would become the absolutely unfree victim of his environment. Never

---

[1] *American Journal of the Medical Sciences,* n.s. 20 (July 1850): 109–20. For other reviews see Emmett Field Horine, M.D., *Daniel Drake (1785–1852): Pioneer Physician of the Midwest* (Philadelphia, 1952), pp. 363 ff.

merely abstract philosophical positions, for Drake these were guides to action as over the course of almost half a century he sought to deal with the dominant problems of his time and place. We have seen him as a consequence as a kind of representative figure and have sought, through an examination of his life and work, to understand better the period in which it occurred.

Many persons have contributed to the preparation of this volume. Mr. Richard J. Haupt, Director of the Cincinnati Historical Society, and Mr. Thomas Rees, Librarian of the University of Cincinnati Medical Center Libraries, assisted us in locating Drake material and have graciously permitted us to publish items from their collections. Dr. Whitfield J. Bell, Jr., of the American Philosophical Society Library, Dr. John Blake of the National Library of Medicine, Mr. Samuel T. Suratt, at the time Archivist of the Smithsonian Institution, Miss Roemol Henry of the Transylvania College Library, and Miss Joan Titley of the University of Louisville Medical Library offered advice and encouragement most welcome, and made the facilities of their respective libraries available for our use. The staff of the Rare Books Division of the Margaret I. King Library at the University of Kentucky guided us through the Horine Collection of Drakeana. Librarians at the Cincinnati General Hospital Library, the Cincinnati Public Library, the Lloyd Library, and the Pennsylvania Historical Society Library were uniformly courteous and helpful. Mrs. Frances Forman at the Cincinnati Historical Society Library checked citations and as usual provided assistance beyond the call of duty. Mr. Ted Eversole assisted in the preparation of the bibliography and read proof on the entire volume.

A grant from the Sesquicentennial Committee of the University of Cincinnati aided us in the preparation of the manuscript. We wish to acknowledge our gratitude for the financial and moral support thus provided, and for the assistance of Dean Clifford G. Grulee, Jr., of the College of Medicine, Dr. Charles Aring, Vice-President Frank Purdy, and Provost for Academic Affairs Thomas N. Bonner. Our colleagues, our friends, and especially our wives have suffered our enthusiasms and must not be unremembered.

The Medical Foundation of the Cincinnati Academy of Medicine generously provided a subsidy to aid in the publication of this volume, which would otherwise have been a thinner and, in our opinion, a less satisfactory tribute to Drake and to the University and the Medical College which he founded. For this assistance in particular we are grateful.

NOTE ON THE TEXTS

We have attempted here to present a selection from Drake's voluminous writings which will illuminate both the man and his times. Drake specialists may well disagree with our choice of items for inclusion, as with our sense of Drake and his place in history during the first half of the nineteenth century. To them, and to students of the period more generally, we can only plead the limitations of space and our desire to make available in a modern edition material which to us seems at once representative of Drake's varied interests and characteristic of his attitudes toward the pursuit of science and the conduct of society. This is the principle which has guided us not only in the selection of the pieces for this volume but in our determination to restrict editorial commentary to the brief notes which precede each item. Drake can speak for himself, and it has been our desire to permit him to do so, in the terms and in the language of his time, not ours. All footnotes are, as a consequence, Drake's own, as are (we hope) any inconsistencies in spelling, although we have corrected obvious typographical errors. Deleted passages are indicated in the conventional manner and consist in all cases of material represented elsewhere in this collection, or, in a few instances, of material which functioned rhetorically rather than substantively in the original. Items I, II, III, VI, VII, X, XI, XII, XIII, XIV, XV are from the collections of the Cincinnati Historical Society and are reproduced by permission; items IV, V, VIII, IX, XVI, XVII, XVIII are from the collections of the Cincinnati General Hospital Library and are reproduced by permission.

Daniel Drake: The Scientist as Citizen

DANIEL DRAKE is remembered by historians for what he did, not for what he was. Because he prepared a monumental survey of the natural and social history of the Mississippi Valley as related to the occurrence of disease west of the Alleghenies, his scientific career has been seen as a prelude to and a preparation for the publication of this *Systematic Treatise, Historical, Etiological, and Practical, on the Principal Diseases of the Interior Valley of North America, as They Appear in the Caucasion, African, Indian and Esquimaux Varieties of Its Population.* Because he was actively engaged in the organization and operation of a number of scientific and educational institutions in the new West of the early nineteenth century, his public career has been seen as a series of incidents relating to the histories of the several academies, colleges, museums, medical schools, hospitals, professional and scientific associations, and moral improvement societies with which he was connected, or at best to the history of the development of social organization in the Ohio Valley. His varied activities have never been seen in relation to each other, however, and only the imposed patterns of personal success or personal failure have been recognized as giving order to his life.

Daniel Drake's career had an order and a logic of its own, generated by his personal engagement with the problems of science and society in America and by the dynamics of his private attempts to deal with these problems as public issues. As physician and citizen, he was caught in an ambiguity bred of chronological and geographic distance. An American colonist in the cities of the Ohio Valley in the nineteenth century, like those English colonists in the cities of the Eastern seaboard at an earlier time, he was at once heir to history and reenacter of history. As a student and practitioner of medicine and science, he looked backward to the absolute certainty of Enlightenment rationalism while hastening by his own efforts the rejection of rationalism and the emergence of an empiricism dependent upon the immanence of Nature's revelation of her ways. As a citizen of Cincinnati, Lexington, and Louisville, he looked backward to the stable social and cultural order of the past and of the

East, while his own efforts to reproduce that society and that culture in the West hastened not only the disintegration of the old order but the emergence of a new order, in which the West stood alongside the East as a discrete but equivalent section of the nation. Like Benjamin Franklin, with whom he has often been compared, Daniel Drake looked at once towards the East and towards the West, towards the past and towards the present, and moved freely between the two worlds. But where Franklin was at home in both, Drake was at home in neither.

Viewed in this way, the details of Drake's life, and the success or failure of his varied activities cease to be the exclusive focus of our interest. Drake himself ceases to be merely the achieving individual, isolated in time and space, and becomes instead a kind of representative figure struggling to resolve for himself, in the context of the newly settled regions of the trans-Allegheny West, the tensions which dominated American intellectual life during the first half of the nineteenth century. In the same way, Drake's activities cease to be merely events in a unique biography, but become symptomatic acts of America's search for the institutional and conceptual means of resolving those tensions, which were her inheritance from the failing rationalism of the American Enlightenment. Drake's inability to resolve these tensions to his own satisfaction, and his personal ambiguity in the face of the changes which were occurring in American science and American society, especially after 1820, do not decrease his importance as a representative figure. Both his perception of the dilemma and his response to it were characteristic. He opted for a kind of Baconian empiricism as a replacement for rationalism and as an alternative to that contemporary idealism which seemed to deny the very possibility of science. He opted for the Whiggery of Adams and Clay as a replacement for the elitist mercantilism of Federalism and as an alternative to the laissez faire of the radical Jacksonian democrats. He found in the institutions which Baconianism and Whiggery generated essential mechanisms for mediating between a discredited past and an impossible present.

Daniel Drake was born in 1785 near Bound Brook, New Jersey, the second child of Isaac Drake and Elizabeth Shotwell.[1] In 1788 the Drakes joined several neighboring families in a migration to the Ohio country, and young Daniel grew up near Mays Lick, Kentucky, south of Maysville (then called Limestone) on the Ohio River. The woods, he remembered in his old age, were his first teacher. Drake received a more conventional education as well,

---

[1] The standard modern biography of Drake is Emmett Field Horine, M.D., *Daniel Drake (1785–1852): Pioneer Physician of the Midwest* (Philadelphia, 1961); a fuller account, although lacking scholarly apparatus, is Edward D. Mansfield, *Memoirs of the Life and Services of Daniel Drake, M.D., Physician, Professor, and Author* . . . (Cincinnati, 1855).

however, for life in frontier Kentucky among transplanted easterners was more stable and more civilized than the romanticism of retrospection would admit. By the time he was fifteen Daniel had determined on a career in business and had made arrangements to become apprenticed to a saddler in Lexington. His father, however, had chosen medicine for his son's profession. In December 1800 Isaac Drake brought young Daniel to Cincinnati and placed him in the care of Dr. William Goforth, who promised to provide board and lodging for four years and to teach the lad the essentials of physic, midwifery, and surgery. At the conclusion of his apprenticeship, Drake decided to continue his education in medicine at the University of Pennsylvania, at that time the most prestigious of the three colleges of medicine in the United States. Despite his own lack of funds and Dr. Goforth's offer of partnership, in the autumn of 1805 he set out on horseback for Philadelphia.

Daniel Drake studied medicine at the University of Pennsylvania when that institution was dominated by Benjamin Rush, whose systemic etiology had been an almost perfect transfer to medical theory of the principles of Newtonian mechanics and the analogous sociological nominalism of Hobbes, Locke, and Montesquieu. Rush's intellectual achievement was in many ways the capstone of the American enlightenment, and it won him an international reputation as "the American Sydenham," not so much for his methods, which were those of rationalist deduction, but because of his practical dominance of medical theory in the United States. Drake attended Rush's lectures during his stay in Philadelphia, corresponded with him as with a revered teacher upon his return to the Ohio Valley, and in general took him as a model to be emulated. What was most viable in Drake's image of Rush was not the scientific content of the latter's teaching, however, but his presence and his prestige as the intellectual and moral leader of the profession. It was this which Drake sought for himself, often at the expense of the more practical and immediate matter of earning a living from the practice of medicine. Like Rush, Drake concentrated his efforts on teaching and on the conduct of an extensive consulting practice by correspondence with physicians in other areas of the country. Like Rush, Drake sought to lead his colleagues and direct his students toward the establishment of professional standards in medical education and in the practice of medicine, and like Rush he was not unwilling to have the texts of his addresses on these and other subjects published for the use of the profession. Just as Rush had seen it as his duty to instruct his community and the nation during the great yellow fever epidemic of 1793, so Drake saw it as his duty to instruct his community and the nation during the cholera epidemics of 1832 and 1833. And just as Rush had concluded his career with the preparation of a systematic treatise on pathology, so Drake sought to conclude his career with the preparation of a systematic treatise on pathology.

Drake's *Systematic Treatise* was no systematic treatise at all, however. It was not an attempt to erect a viable nosology or to develop a theory of the nature and transmission of disease. It was rather the record of forty years' systematic observation of the natural history of Western America, systematically presented. It was the raw data for analysis of the pathology of the interior valley of North America and of the relationship of environment and etiology. It was descriptive science, and as science it was influenced by Benjamin Rush only to the degree that Rush had repeated as his own the injunction of Hippocrates, that physicians should observe the conditions of a place in order to understand the diseases which were likely to occur there.[2] Drake's *Systematic Treatise* was descriptive science, and that in fact was all it could be, for as a scientist Drake had been influenced less by the rationalism of Rush than by the empiricism of Rush's intellectual opposite, the professor of materia medica, natural history, and botany at the University of Pennsylvania, Benjamin Smith Barton.

Although eclipsed in the eyes of historians by his more famous colleague, Barton's reputation as teacher and as scientist was secure in 1805 when Drake first went to Philadelphia, and he was in fact elected to Rush's chair of the Institutes and the Theory and Practice of Medicine upon the latter's death in 1813.[3] Indeed, Barton may prove to have been the more influential of the two. Educated in the medical colleges of Philadelphia, Edinburgh, and Göttingen, Barton brought to his teaching the more informal training as a scientist which he had received from his uncle, the astronomer David Rittenhouse, and a Baconian enthusiasm for the increase of knowledge rare in that era of secular confidence. Largely at his uncle's instance, Barton was appointed professor of botany and natural history in 1789, while he was still abroad completing his education. During the next twenty-five years of teaching and writing, he is said to have elevated the study of these subjects in America to the level of systematic science, while earning for himself the appellation "father of American materia medica." He was a vigorous exponent of the need for careful and systematic observation as the first step in understanding a knowable universe, and his own works, the published records of his observations of Nature's ways,

---

[2] Cf. Benjamin Rush, "Valedictory Charge. June 5th, 1805," *Philadelphia Medical and Physical Journal* 2, part 1 (December 1805): 131–35.

[3] William S. Middleton, "Benjamin Smith Barton," *Annals of Medical History*, n.s. 8 (November 1936): 417–91; Jeannette E. Graustein, "The Eminent Benjamin Smith Barton," *Pennsylvania Magazine of History and Biography* 85 (October 1961): 423–38.

bear witness to his impatience with the vogue of classifying and system-building a priori which dominated contemporary American science. Throughout his life he urged colleagues and students alike to adopt the methods of descriptive science in order to interrogate "Nature, who is ever willing to answer, and to be understood."[4] Between 1804 and 1809 Barton edited the *Philadelphia Medical and Physical Journal* as a vehicle for the publication of "original communications relative to all the branches of Medicine, Natural History, and Physical Geography."[5]

The extent of personal relationship between Drake and Barton is unfortunately unknown, but the impress of Barton's teachings on Drake's own career as a scientist, and more particularly on his early writings, is clear.[6] It was for Barton's *Medical and Physical Journal* that Drake prepared in 1807 his first medical essay, "Some Account of the Epidemic Diseases which Prevail at Mays-Lick, in Kentucky," and for which the "calendarium flora" that eventually grew into his first book, *Notices concerning Cincinnati,* in 1810 after the *Journal* ceased publication, had originally been intended. Although it is not certain that Drake did "take a ticket" for Barton's lectures on materia medica and natural history during the winter of 1805–1806,[7] he certainly attended some of them, and in any case the two men met socially on a number of occasions, perhaps through the offices of Drake's classmate and fellowlodger, W. P. C. Barton, nephew and future biographer of the naturalist. Drake is known to have obtained from his father a piece of copper excavated from an Indian mound near Mays Lick for presentation to Barton, who may be

---

[4] "Dedication," *Phila. Med. and Phys. J.* 1, part 2 (Feb. 27, 1805): iv.

[5] "General Plan of the Philadelphia Medical and Physical Journal," ibid. 1, part 1 (Nov. 13, 1804): v–vii.

[6] Drake's interest in the ethnography, geology, mineralogy, meteorology, and botany of the West, if not directly traceable to Barton's suggestions, is in striking accord with the latter's ideas of the proper subjects for investigation in natural history, cf. Benjamin Smith Barton, *A Discourse on Some of the Principal Desiderata in Natural History, and on the Best Means of Promoting the Study of This Science, in the United-States* (Philadelphia, 1807). Drake's commitment to empiricism and to the necessity for scientific societies are strangely foreshadowed in this discourse, a kind of summary of the positions on the proper conduct of science which Barton had espoused for two decades. The only Drake-Barton correspondence that has been located is now in private hands, and was not available for the purposes of this study. I am indebted to Dr. Whitfield J. Bell, Jr., for calling its existence to my attention.

[7] Mansfield says Drake's funds were sufficient to permit him to enroll only in the lectures of Rush, Woodhouse, Wistar, and Physic, presumably on the basis of personal conversation with Drake or correspondence now lost (*Memoirs . . . of Daniel Drake, M.D.*, p. 67); Horine asserts that Drake did enroll in Barton's lectures, on the basis of Drake's own reminiscences and on the presence of notes from Barton's lectures in Drake's MS "Medical Diary or Common place Book" (*Daniel Drake . . . Pioneer Physician*, p. 80). On the one hand, these notes may be the result of personal conversations with Barton rather than formal classroom lectures, an hypothesis to which their placement at the end of the diary lends credence; on the other hand, Barton did from time to time permit impecunious students to attend his formal lectures without payment of fees (cf. Middleton, "Benjamin Smith Barton," p. 487).

presumed to have taken the occasion to encourage the younger man to take an active interest in the "antiquities" of the Ohio Valley upon his return to the West.

Drake did in fact excavate a number of Indian mounds in the vicinity of Cincinnati and gather reports of the excavations of other sites in Ohio, apparently with the primary intention of testing (or proving) Barton's hypotheses concerning the characteristics of the prehistoric inhabitants of Ohio.[8] In the summer of 1806 Drake filled blank pages in his commonplace book with notes of his observations of the Ohio catfish (*Silurus*), although whether he intended thus to begin a catalogue of Ohio River fish as recommended by Barton is not clear.[9] It was during these early years after his return from Philadelphia, however, that he began the systematic collection of data on the climate and geology of the Ohio Valley region along lines laid down by Barton, which was to yield his principal scientific publications prior to the *Systematic Treatise* and form the basis for his increasing interest in the influence of physical environment on all aspects of man's life.[10]

The science of Benjamin Rush, characteristic of the late eighteenth century in America, sought to utilize the deductive reasoning of rationalism to discover the organizing principles of a static universe. It found reliance upon observation unmixed with cogitation inadequate to the goals of philosophy and emphasized the need for leaps of the imagination, by which the true nature of Nature's order might be perceived, and the reduction of this perception into laws achieved. It was the kind of science which might be carried on successfully by isolated men of genius, and it had as its goal the erection of descriptive systems capable of accounting for the stable relationships between

---

[8] "Dr. Barton supposes that the remains of antiquity discovered throughout the continent of North America and more particularly in the Western parts, are the work of a Nation or Nations of people which are not now completely extinct but which have degenerated into the present aboriginal Indian tribes from the influence of war, pestilence, etc., etc. . . . . These curious remains, which have been found in different tumuli (and particularly in one at Cincinnati) [seem] to have been some for ornament and some to have been a part of their apparatus for worship," Drake, "Medical Diary or Common place Book," MS in National Library of Medicine, Bethesda, Maryland, pp. 123–24; cf. *Natural and Statistical View, or Picture of Cincinnati and the Miami Country* (Cincinnati, 1815), p. 218. Alexander Wilson, the Philadelphia ornithologist, examined Drake's "collection of curiosities" during the spring of 1810, "Extract of a Letter from Alexander Wilson," *The Port-Folio*, 3d ser., 3 (June 1810): 507.

[9] "Medical Diary or Common place Book," pp. 167–69; cf. Barton, *Discourse on Some of the Principal Desiderata in Natural History*, p. 27.

[10] Drake's short-lived interest in mineralogy, again one of Barton's "desiderata," may have been influenced as well by the repeated injunctions of Alexander Wilson as to "the duty of every citizen, at all times . . . to contribute his mite to promote his nation's independence" by discovering the useful plants and minerals native to the United States ["The Naturalist. No. III," *The Port-Folio*, 3d ser., 2 (August 1809): 119–23; "The Naturalist. No. V," ibid. (November 1809), pp. 426–29]. Drake met Wilson in Philadelphia during the winter of 1805–1806 ("Medical Diary," p. 165) and again in Cincinnati in 1810 (*supra*, n. 8), and appears to have been a regular reader of *The Port-Folio* during these years (cf. *Natural and Statistical View, or Picture of Cincinnati and the Miami Country*, p. ix).

entities defined in terms of their essential forms. The science of Benjamin Smith Barton and the early nineteenth century, on the other hand, depended not upon ratiocination but upon the inherent order of the universe to yield a knowledge of the processes of Nature to the practitioners of empiricism. It found the products of rationalist science inadequate or incorrect when compared with observed reality, and sought to reform the practice of science by restoring to it the reliance upon observation and the principles of induction with which it was said to have had its modern beginnings. It called itself Baconian after the author of the *Novum Organum,* whose works, it seems, were less read than talked about during this period. What is at issue is not the accuracy of the name, however, but what that designation meant: the substitution of "acute observation, accurate comparison, judicious arrangement, and logical induction" for the metaphysical speculation which had characterized rationalist science.[11] Thus where nineteenth-century idealism saw in the failures of rationalism a fundamental error concerning the possibility of knowledge itself, Baconian science saw only a failure of technique. Not study of esoteric doctrine but only simple observation was required, of a kind which any man might perform. For the Baconians, the scientist became the impartial vehicle for Nature's revelation of her laws, prophet instead of priest.

This shift involved more than merely the changing mood of American science in the early nineteenth century. The pursuit of science was being democratized, even as the focus of scientific activity—at least in natural history, medicine, and the geologic and meteorologic sciences—was moving toward investigation of dynamic or developmental processes. A new generation, armed with a rediscovered conviction of the universal availability of knowledge to the human understanding and with the techniques of inductive reasoning, sought admission to the scientific community (and hence to the nation's intellectual elite) at a moment in history when their services were most needed. Studies of the distribution of plants and animals, of mineral deposits and geologic forms, of climatic conditions, and of the occurrence of endemic and epidemic diseases required systematic observation carried on in different places and often over long periods of time. The necessity for correlating such observations in order to permit the pattern of natural processes to appear required the development of a new spirit of scientific cooperation and new agencies for the conduct of cooperative science. No longer merely the

---

[11] The quotation is from Drake, "Introductory Lecture for the Second Session of the Medical College of Ohio," November, 1821, MS in the Cincinnati General Hospital Library, and reprinted *infra.* On the revival of Baconianism in the United States, see George H. Daniels, *American Science in the Age of Jackson* (New York, 1968), and, e.g. "Remarks on the Writings of Lord Bacon, Translated from the Lectures of M. Garat, Professor of Metaphysics in the Normal Schools of France," *The Port-Folio,* 3d ser., 2 (December 1809): 514-18.

associations of men of genius but the institutional mechanisms by which Nature's laws became known, these scientific agencies welcomed the contributions of "amateurs," coordinated their activities, and reinforced their zeal by providing them with credentials to the scientific community at large.[12]

It is in the context of these shifts in scientific method, in the problems to which scientists addressed themselves, and in the sociology of science in America that Daniel Drake's career must be considered. His fascination with the possibility of generalization based upon observation is apparent in the "Medical Diary or Common place Book" which he kept during his first visit to the Philadelphia Medical College in 1805–1806, and certainly in the medical and physical researches in which he engaged after his return to the West. The problems in which he became interested—the climate and geology of the Ohio Valley, the course of epidemic diseases in time and space, the impact of environment on man and on society—like those he recommended for investigation by others, were characteristically directed at an elucidation of the dynamic processes of nature and were of a kind moreover for which systematic observation was sufficient. Even his early writings—"Some Account of the Epidemic Diseases which have Prevailed at Mays-Lick," the attack on Volney's *View of the Soil and Climate of the United States, Notices concerning Cincinnati, Natural and Statistical View, or Picture of Cincinnati and the Miami Country*, the "Letter to Correa de Serra" concerning the geology of the Ohio Valley—give evidence of his commitment to the inductive method and of his recognition of the dangers of rationalist generalization derived from abstract principles. After the start of his teaching career, moreover, he took every opportunity to recommend the inductive method both to students and colleagues, as in the passage briefly quoted above from his "Introductory Lecture for the Second Session of the Medical College of Ohio." In his "Introductory Lecture on the Necessity and Value of Professional Industry" at Transylvania in 1823 he advised his hearers: "Observations should be well made, for conclusions true to nature can never flow from false premises. The most distinguished physicians of every age, from Hippocrates to the existing galaxy of European and American practitioners, have surpassed their contemporaries in the accuracy of their observations, rather than in the depth and ingenuity of their reasonings. And it may be safely affirmed, that a faculty for

---

[12] Cf. Henry D. Shapiro, "Daniel Drake's *Sensorium Commune* and the Organization of the Second American Enlightenment," *Bulletin of the Cincinnati Historical Society* 27 (Spring 1969): 43–52. No better example may be found of the encouragement of "amateurs" and the coordination of their observations by these new agencies for scientific cooperation during this period than in the MS records of the early meteorological surveys conducted by the U.S. Patent Office and the Smithsonian Institution (Smithsonian Institution Archives, "Meteorological Correspondence," etc.)

patient, acute, and enlightened observation, is more important in the practice of medicine, than an excursive imagination or uncommon strength of understanding."[13]

Systematic observation, natural classification, and inductive generalization were never merely a matter of convenience for Drake, but the essential means by which Nature's ways might become revealed. If his own contributions to science were ultimately inconclusive, the fault lay with the method, not its practitioner. In the *Systematic Treatise* (1850), for example, Drake carried the inductive method to its logical conclusion by offering some seven hundred pages of carefully digested data on the geology, the climate, and the sociology of the Mississippi Valley as a background to an examination of its pathology. Drake erected no theories and founded no "schools" because he mistrusted theories and "schools," not because he tried and failed. Throughout his life he heeded the advice of Benjamin Smith Barton, to "hasten the fall of some of the present Systems . . . [and] by walking in the rose-spread paths of Truth, with Nature as your guide, . . . prepare the way for those more permanent systems which Providence *has* given man the ability to found."[14] Drake published the records of his observations as evidence by which the adequacy of current theory might be judged and as the basic data out of which those "more permanent systems" might emerge, and his contributions were regarded as being of sufficient importance in their own time to warrant his election to the leading scientific societies of the nation, including the American Philosophical Society, the American Antiquarian Society, the American Geological Society, the Philadelphia Academy of Natural Sciences, and the National Institute for the Advancement of Science.

In light of Drake's Baconianism, those vigorous efforts to establish educational and scientific institutions in the Ohio Valley which are so striking an aspect of his career must be seen as the result of more than just a desire to reproduce the civilization of the East in the West. This was part his purpose to be sure, and much of Drake's early work on the natural and social history of the West was in fact a defense of the region as an area for human habitation and a demonstration of the feasibility of establishing the agencies of sophisticated cultural life on the frontier.

It was not merely the absence of the conventional pattern of institutions but the necessity for certain kinds of institutions which especially concerned Drake. In his *Anniversary Address, Delivered to the School of Literature and*

---

[13] *An Introductory Lecture, on the Necessity and Value of Professional Industry* (Lexington, 1823), p. 22.
[14] Benjamin Smith Barton, ["Dedication,"] *Phila. Med. and Phys. J.* 2, part 1 (December 1805): v.

the Arts in 1814, for example, he acknowledged that newly settled areas could not hope to compete with the Eastern sections of the nation or with Europe on equal terms. What was the West's disadvantage was also her advantage, however, for the lack of an established social and intellectual elite which made institutions of the older style impracticable made the real democratization of knowledge a possibility, while the absence of "licensed authority" left the mind "free to expand, according to its original constitution."[15] Drake did not therefore conclude that institutions were unnecessary for the fulfillment of man's potentialities. Quite the contrary, he regarded the educational and scientific institutions which he supported as necessary mechanisms for the guidance of the understanding towards apprehension of Nature's laws, and hence towards the freedom of action which such understanding made possible. As a scientist, moreover, Drake saw in these agencies for the cooperative investigation of Nature's order an essential means for systematic observation. The journals, scientific and medical societies, natural history museums, libraries, academies and colleges, botanical gardens, and hospitals which he organized during a long and busy life were thus designed to serve as vehicles through which Nature's laws would become revealed. At issue was no simple re-creation of institutions desirable for their own sake. Upon their operation depended the success of the inductive method, and upon the success of the inductive method depended the possibility of science itself and hence of man's claim to a special status in the natural order of the universe. Drake's efforts in this regard were then not only characteristic of his generation in science, which sought to make Baconianism work, but symptomatic also of that larger effort to preserve through Baconianism the nondeterminist naturalism of the Enlightenment in the face of challenges from both a naturalist and an idealist determinism.

By his fortieth year, Drake's convictions concerning the possibilities of science in the West and the proper conduct of the practice of medicine were fully mature. He had little to say on these subjects after 1825 that he had not said before, if not so well as he would in later years. The critical need for systematic observation, both as the basis for scientific knowledge and as a normal function of the practicing physician, the desirability of broad education in literature and the sciences, the interrelatedness of the disciplines, and the necessity for diligence and honorable behavior in all fields of endeavor were his regular themes to medical and lay groups alike in the numerous public addresses he was invited to give during the last decades of his life. His professional activities also reveal no sharp break from the pattern he had

---

[15] *Anniversary Address, Delivered to the School of Literature and the Arts* (Cincinnati, [1814]), pp. 5–6, 8.

established in Cincinnati during the teens and early twenties. He continued to teach, he remained active in support of professional and scientific societies, and he pursued his investigations into the consequences of the physical environment of the Mississippi Valley on the development of social institutions and on the pathology of the region, whether his residence was in Cincinnati, Lexington, or Louisville. The single new venture of this later period, his editorship of the *Western Journal of the Medical and Physical Sciences,* was in fact no more than a logical continuation of his earlier commitment to the establishment of viable agencies for scientific communication.

After 1825, however, Drake began to address himself more and more to the problems of social and political organization which were becoming so dominant in the American consciousness during the early ante-bellum period. Although his focus was usually upon the West and its relationship with the East and with the nation as a whole, it seems correct to say that Drake's principal concern was with democracy itself, and the anarchic spirit of antinomianism which flowed from it. Its manifestations were all about him: the decline of manners and morals in the West; intemperance; the disrespect of students for their teachers, of young people for their parents, of "the people" for their leaders, of states and even sections of the nation for the nation as a whole; the medical and pseudo-scientific fads which opposed "natural" knowledge and "natural" cures to the "artificial" systems of the professionals; the proliferation of colleges and medical schools, especially in the West, where unteachables were taught by the unlearned; the electoral triumph of Andrew Jackson, itself a symbol for the rest. From the early 1820s in Lexington when he associated himself with Henry Clay, Drake was a National Republican in politics and a Whig in ideology. He supported Clay for the presidency in 1824 and subsequently defended him against charges of "corrupt bargain" during the dispute over the electoral returns from that contest. He lampooned Jackson and the cult of Jackson in *The People's Doctors: A Review.* Above all, he saw in the construction of institutions of social control the essential mechanism for the fulfillment of America's promise as heir to Enlightenment liberalism.

Whiggery to Drake was scientific. Unlike the idealism of the transcendentalists and the determinist naturalism of the laissez faire Jacksonians, it implied the knowability of a natural universe of which man was a part but with which he was not identical, and the necessity for such knowledge if man was to be more than the victim of his environment. Although analogous to a great machine, the world in which man lived was not self-regulating, at least so far as man himself was concerned. Social theories which depended upon the "natural" actions of "natural" men were as a consequence necessarily false. Just as the temperate man had to combat with his will and his good sense a natural desire for stimulants, counterbalancing one faculty with another, so the temperate society had to combat with its will and good sense its natural desire for disunion and anarchy, counterbalancing one faction against another and thus obtaining that equilibrium which was health. Just as the maintenance of temperance among men depended upon the effect of the body's external environment on its operations if will and good sense were to be effective, so the maintenance of equilibrium in society depended upon an understanding of the sociology of the body politic and of the effect of its external environment on its operations.

Whiggery regarded this understanding as available to all citizens and necessary to their proper conduct, hence necessary also to the health of the body politic. Whig social theory, unlike Federalist mercantilism, was not to be the result of the ratiocination of a few but rather the obvious conclusions of any man's empirical assessment of the nature of man and society in America. Whig institutions, in the same way, were not to be the arbitrary creations of an elite seeking to achieve an abstract end defined without reference to the social and political realities with which they were to deal, but the scientific creations of a society determined to resolve the tensions within itself.

The shift in Drake's career was, then, primarily one of focus. His later interest in the solution of social problems, and his explicit commitment to Whiggery after the mid-1820s may thus be seen as a continuation and a broadening of his earlier interests and commitments as a scientist and physician. As early as 1810, in his essays on the geology and climate of Cincinnati and in the *Notices concerning Cincinnati,* he began the assessment of the consequences of physical environment on social organization and social development which evolved into his more sophisticated analyses of the 1830s and 1840s. At the same time, as a Baconian he developed an active interest in the organization and operation of scientific and educational institutions which influenced his more pointedly Whiggish efforts of later years. Indeed for Drake the pursuit of science was always a practical activity, for what was true was necessarily functional. Science was thus the handmaiden of humanity, providing that knowledge which was essential not only for man's happiness or prosperity but for his freedom, and hence for the attainment of his proper estate.

Henry D. Shapiro

Daniel Drake, the City, and the American System

AMONG THE MAJOR achievements of the people who settled the basin of the Mississippi during the late eighteenth and early nineteenth centuries was the establishment of a series of towns. These urban communities fostered and encouraged flux and intensified the pace of the economic, social, and cultural development of the region. At the same time the internal structure, form, and composition of the towns themselves constantly shifted.[1] Response to this chronic instability varied, but many of the most successful, cultivated, and sophisticated residents, though exhilarated by the potential and actual growth of their cities, distrusted the direction and pace of change and sought to regulate what they regarded as its disruptive and destructive power. To a significant extent their uneasiness in the face of virtually unprecedented vitality was produced by a conflict of experience and theory, for much of the impetus behind their quest for order grew out of an inherited intellectual amalgam of science and religion which asserted that man lived in an orderly universe which ought to be harmonious and predictable. In large part, it was Daniel Drake's preoccupation with the process and consequences of urbanization and development in a vast unsettled region, his persistent search for stability, and his stubborn attempt to explain an admittedly unique experience through the application of old ideas which make him an important figure in American history.

Several circumstances drew him to the study of urbanization and development in the West. First, he was a member of a scientific community wedded to the methods, principles and assumptions of Bacon, Newton, Locke, and Linnaeus. As such, he felt it was his obligation systematically to observe and gather data about the operation of nature in his portion of God's orderly and well regulated universe and to help men bring themselves and their institutions into accord with that order. Second, he occupied, he believed, a unique vantage point to make significant contributions to man's study of nature, for

---

[1] Richard C. Wade, *The Urban Frontier: Pioneer Life in Early Pittsburgh, Cincinnati, Lexington, Louisville and St. Louis* (Chicago, 1964).

he lived in a largely unexplored and uncharted territory and in the midst of a diverse population drawn from civilizations which already stood high in the hierarchy of social and intellectual development. Here he might observe facts and witness the functioning of laws of nature and society never before seen by man. He might, then, through science, bring man closer to fathoming God's design so that disorder and pain and distress—the Divine punishment for disobeying the laws of existence—might be avoided.

Drake's inability to contain his curiosity emerged early. His first major publication, *Notices concerning Cincinnati* (1810), grew out of a plan to make a "calendarium flora" and publish it in a medical journal as an exhibit of the "progress" of the vegetation of the place. But his scientific preconceptions led him farther afield. As he put it, "the physical sciences are so intimately connected, that the narrow limits [originally] prescribed, have been overstepped. . . ."[2] What emerged was a treatise on the diseases of the city, on the factors which Drake felt might have some influence in producing them, and a call for more research.

The volume falls into five chapters on topography, geology, climate, the condition of the town, and diseases. The first two are almost wholly descriptive. The third, on climate, is more pointed. It includes a section arguing that he had found "interesting indications" of the influence of climate on vegetables and the lower orders of animals. Here Drake also attacked the notion that the interior was significantly warmer than Philadelphia, the idea that great cities were always several degrees warmer than the countryside, and the widespread assumption among Cincinnatians that their own climate had changed as the result of fifteen years of town growth. But what was really needed before these and other questions could be settled, Drake argued, was a history not only of the climate of the East and West coasts and the Allegheny and western mountains, but also of the "broad shallow valley, or vast platform comprehended between them."[3]

The next chapter, on the condition of the town, Drake included because the "well established fact, that customs, manners and habits exert a decided modifying influence on diseases" made it necessary "to exhibit a concise statement, the items of which, abstractly considered, are insignificant, but taken aggregately, appear to be of too much moment not to be considered."[4] He drew several conclusions about the condition and the health of the town. Bathing was less common than it ought to be, tobacco was too much used by males from the age of ten, and there were "not a few . . . who daily but not

[2] Daniel Drake, *Notices concerning Cincinnati* (Cincinnati, 1810), p. [3].
[3] Ibid., p. 11.
[4] Ibid., p. 29.

quietly become intoxicated, and no *very* inconsiderable number [who] . . . fall victims to that habit." The town's ventilation, moreover, was defective. It had few alleys, no common worthy of the name, and too few houses with porches. And the grid, laid out north and west, deflected the transit of the prevailing southwestern wind through the city.

But Drake was also careful to note the principal obstacle to scientific analysis and generalization. Cincinnati's population was recently arrived and was drawn from almost every state in the Union and from virtually every country of western Europe. Such a population, "must necessarily exhibit much *physical,* as well as moral diversity. The climate and soil have not yet introduced an uniform constitution of body; nor customs, manners and laws an uniform moral character."[5]

Consequently, in the final chapter on diseases, Drake felt constrained to do no more than indicate the principal endemic diseases "and their supposed sources."[6] The chief matter for discussion, in fact, was the febrile diseases and typhus which had plagued the community in recent months. Drake contended that there were "sickly states of the atmosphere" which acted as a predisposing cause of these disorders. But the "exciting cause" was the "miasma or noxious exhalation" rising from brickyards, cemeteries, butcheries, and tanneries. The danger of severe air pollution, however, was not yet acute because of the efficiency of trees in intercepting the poisonous gases rising from the bottom land west of the city and destroying their virulence. "The forest," Drake noted, "should be considered in the light of a rampart against a perpetual enemy, and preserved in the most sacred manner."[7]

The little volume was a success. Though intended only for his "medical and scientific friends," the book attracted a broader audience, principally among travelers seeking information about the city and its hinterland. As a result, in 1813 Drake began a more extended and "less professional work, of a similar kind," which he called a *Natural and Statistical View, or Picture of Cincinnati and the Miami Country.*[8]

Here, as in the 1810 piece, Drake prepared an introduction explaining his objective and approach. In it he touched on the significance of the disproportion of geography to history in the West. He noted that his account of a village in the woods was inherently different from the history of a populous city and would be read for a different purpose. The city histories, of recent vogue east of the mountains, exhibited the progress of improvements and expatiated on

---

[5] Ibid., p. 30.
[6] Ibid., p. 32.
[7] Ibid., pp. 32, 35.
[8] Daniel Drake, *Natural and Statistical View, or Picture of Cincinnati and the Miami Country* (Cincinnati, 1815), p. v.

the arts in a way to inspire imitation or adoption. But his work, by necessity, dwelled upon natural objects and advantages which might encourage immigration to the new country. It contained, then, a larger proportion of natural history than books about eastern cities.

Drake also emphasized the relationship between urbanization and regional development. Since the *"relations* of a town with the surrounding country, are an essential part of its history," he wrote, the city could not be properly understood unless both were studied. To underscore the point, he opened the book with a map of the Miami country and the adjacent part of Kentucky "so as to exhibit an entire view of the tracts dependent on Cincinnati as their emporium." The *Natural and Statistical View,* therefore, comprises the first systematic discussion of what is now the Cincinnati metropolitan region.[9]

The rest of the book, though laden with statistics and detached facts, was studded with significant observations underlining this point. The price of land, he found, varied according to its proximity to important towns or cities, and the most valuable was located within three miles of Cincinnati. It was the construction of two large breweries in Cincinnati which led to the growth of barley in the region, not the reverse. The residents of Indiana depended almost entirely on Cincinnati for foreign imports because little mercantile capital was available in Lawrenceburg and there was no town on the Great Miami River to serve as a "depot of merchandize for that region." A rise in food prices in Cincinnati markets over the past few years indicated that the urban population was growing at a more rapid rate than the rural, and suggested an increase in the number of grazing farms, the establishment of larger dairy operations, and the cultivation of more extensive gardens in the Cincinnati area. Manufacturing in Cincinnati, though increasing, still lagged because the town was settled before the surrounding country, and commerce and land development left little capital to invest in factories. But it was still impossible, he admitted, to "give a faithful portraiture" of society because the constant influx of diverse elements blurred the social distinctions which characterized old communities and slowed the molding effect of the physical, legal, and social environment.[10]

In the concluding pages of the book Drake broadened his vision geographically and chronologically to look into the urban future of the Ohio and Mississippi valleys. The region, he felt, was destined to be the site of not one great metropolitan center, but of a series of magnificent cities. Among them, Cincinnati would outstrip Pittsburgh and Louisville and become the metropolis of the Ohio.

[9] Ibid., pp. v–viii.
[10] Ibid., pp. 32, 57, 142, 166–68, 170.

This was, to Drake, a self-evident proposition. Pittsburgh had flourished in the past because it had been at once a commercial entrepot and manufacturing center—the "Birmingham of America"—for the towns below. But the flow was changing, and as the drain of Cincinnati capital upstream lessened, the opportunity for investment in manufacturing and internal improvements which would free Cincinnati increased. Pittsburgh, moreover, had no rich agricultural hinterland to develop, and her coal and iron could be floated down river. Louisville was also vulnerable. Its fortunes rested on a natural advantage readily surmounted by the construction of a canal around the falls of the Ohio. Once that was done, it remained only to connect Cincinnati to the Great Lakes by canal and to tighten its hold on the Miami country by extensive road construction. By creating such a diverse base and widespread transport connections, Drake concluded, his townsmen could not "fail to realize their most glowing anticipation of greatness."[11] Since success in the clash of urban imperialisms was not mechanistically determined by geography but by what men did with their geography, the glowing anticipations of greatness, bolstered by Drake's scientific enquiries into the natural history of the region, amounted to self-fulfilling prophecies. By 1830 Cincinnati was, in fact, the Queen City of the West.

Drake, meanwhile, was doing all he could to make his vision a reality. While pursuing his medical practice and scientific studies, he opened and helped manage a combination general store and pharmacy and served on the board of the National Bank at Cincinnati. He also belonged to the group of entrepreneurs which established the Cincinnati Manufacturing Company, an ambitious enterprise housed in a nine-story building which contained machinery for the production of, among other things, flour, wool, cotton, and flax seed oil.[12]

This combination of diverse economic activity and persistent investigation of Western development led Drake away from his old but vague commitment to Jeffersonian policies and into Whiggery and an alliance with Henry Clay. In 1819 he attacked the widespread notion that the West was destined for the foreseeable future to be agricultural and commercial and that it could not enjoy the benefits of an adequate domestic market for home manufactures until the forests were subdued and agriculture established. His experience suggested that it was a delusion to believe manufactures would not flourish "until we are an old people" or that the contiguous forests somehow discouraged manufacturing. He preferred to put part of the rising generation in workshops and encourage others to cultivate trees. The sugar maple, Drake believed, deserved

[11] Ibid., pp. 226–32.
[12] Ibid., p. 137.

special solicitude, for its cultivation would relieve the city of purchasing sugar in Louisiana, thus providing cash to pay off commercial debts in Philadelphia and freeing additional capital for local investment.[13]

Drake questioned the notion that there was a relationship between the age of a community and a given stage of economic development, but he was less confident that high culture could flourish in a new city. Culture, he felt, like science, was cumulative; it was built on the achievements of the past and grew out of the total environment. It took an ancient, or at least an old city, with a relatively homogeneous population, to produce an urbane civilization.

That did not, however, relegate the intellectual—the scientist and philosopher—to oblivion in the new towns of the West. Quite the contrary. The urban frontier provided positive and "peculiar aids" to the "development of understanding." Here the community was not old, divided, and specialized, but new, diverse, and united. Here the facts relative to all the arts and sciences could be readily assembled and correctly classified so that those objects associated by natural affinities would rise in beauty and order just as the grand architect of nature and society intended. This was possible because here, in a new country, resided people who were mature in the scale of civilization, peoples readily schooled in the method and philosophy of Bacon, Locke, Newton and Linnaeus.

Western cities, moreover, possessed the means for creating institutions to disseminate the correct scientific methodology. Although the region lacked great individual patrons of the arts and sciences, its undeveloped condition made private and public prosperity inseparable, and knowledge was the common basis of both. Since knowledge was the key to wealth and power—to the creation of an independent and economically diversified society—western urbanites should be willing, even in a depression, to join associations and contribute money for the establishment of museums, schools and universities, just as they joined together to build manufacturing, commercial, and financial institutions, and transportation facilities. In western towns, according to Drake, there was no conflict between the intellectual and the practical man. Here religion, reason, and circumstance combined to point the way to amity and concord among philosopher, philanthropist, and entrepreneur. The cities could be true scientific laboratories, observation posts on a strategic frontier of knowledge in God's empire.[14]

---

[13] Daniel Drake, "Observations on the Means and Importance of Preserving Fruit and Forest Trees," *Western Spy,* Oct. 23, 1819, pp. 2–3.
[14] Daniel Drake, *Anniversary Address, Delivered to the School of Literature and the Arts at Cincinnati, Nov. 23, 1814* (Cincinnati, n.d.).

It was not surprising, then, that Drake, who regarded himself neither as a politician nor a statesman but as a scientist—a naturalist who applied geographic and geological observations to patriotic and social duty—should speak out during the South Carolina secession crisis of the early 1830s. His studies of the role of cities in the development of western culture provided the foundation on which he built a theory of urban federalism. Admitting that the Constitution established a complex system likely but not necessarily fated to decay, he asked what could be done to avert its decomposition.

He began by attacking the old idea that a federal system would not work in an extensive geographic area. This had been true in the past, but the discoveries of modern science—physical, mechanical, political, and moral—now made such a system possible. In Drake's perspective the country formed an arch held together by the weight of the West on top. New York City and New Orleans and their contiguous spheres of influence were the pedestals from which the superstructure of the arch—the Hudson, Niagara, and Great Lakes maritime basin to the east and the lower Mississippi Valley to the south—reached into the West and absorbed vital nourishment. In this view, South Carolina was geographically the most isolated part of the Union. Nonetheless, so long as the West remained in place, the Union was safe.

Thus, in Drake's analysis, the West was the keystone which ought to hold the federal arch together, but it was not functioning as such. The heart of the crisis was not in the South, but in the West. Although it constituted a great physical region in which the very soil, water, and vegetable and animal life migrated harmoniously from part to part, the West was defective as a human ecological region. Still characterized by diversity and change, Western culture was not old enough for the operation of natural laws to weld it into a unit. And since it was federal and state policy to encourage intercourse among regions wherever nature favored it, there was a chance that the polar attractions of New York and New Orleans might pull the West and the Union apart. History—the accumulation of institutions and customs over time—was not being made fast enough in the West to overcome geography and mobility.

Drake proposed to speed up the process, to make social science the handmaiden of history. If the divisive pull of the cultural influence of New York and New Orleans were to be offset, the West had to become and remain united. He called for the creation of a social and literary communion which would amalgamate the West's diverse and migrant populations into one social compound. Since "communities, like forests, grow rigid by time," he suggested that western cities take the lead in molding uniform customs and manners by establishing an educational system, with common curricula and books, stretching from the elementary grades through the university level. The effort would be coordinated through periodic conventions to be held in the cities of the West.[15]

But the resolution of the crisis did not depend exclusively on the success of such grandiose schemes. Before regional unity could be assured, urbanites had to mitigate the disruptive and divisive influences within their communities. Drake, in short, was becoming more and more concerned with the internal structure, organization, and operation of the city. This interest, of course, was not new. It dated back at least to *Notices concerning Cincinnati* (1810), and from then through the 1820s he spent a great deal of time and effort helping to establish voluntary business, educational, philanthropic, and social institutions. These were designed to harmonize the elements of city life and to elicit the order which, his intellectual assumptions told him, were implicit in the very existence of cities and necessary for their perpetuation. But after 1825, in an atmosphere pervaded by a heightening sense of crisis created by sporadic race riots, antiabolitionist violence, and outbreaks of nativist agitation, Drake's attention turned from the creation of order to the deterrence of disorder. Anarchy, he warned in 1834, threatened to "deluge our pleasant places and rush in desolation along our streets."[16]

One prolific source of this danger, Drake felt, was intemperance. It was not, however, peculiar to cities. "Man is endowed, by the Creator," he told a Philadelphia temperance audience, "with certain desires, the regulated gratification of which is necessary to existence." Among these was the propensity for stimulants, such as coffee, tea, wine, beer, and liquor, and intemperance was nothing more or less than an excessive propensity for these stimulants. But civilized life demanded a sustained and diversified excitement of all man's organs. Therefore, within the advanced civilization of the United States, the abnormal desire for stimulants was a chronic problem which tended to concentrate in the great cities where the organization of society was most complicated and the pace and pressure of life most intense. But that was not all. Cities also contained a concentration of secondary sources of the abnormal propensity for stimulation. In that category Drake placed the habitual drinking of ardent spirits (constant use of potent stimulants heightened the natural desire), dinner and supper parties, gambling, the theater, idleness (more of a problem for the rich since they did not have to work for food and clothing),[17]

---

[15] Daniel Drake, *Remarks on the Importance of Promoting Literary and Social Concert, in the Valley of the Mississippi, as a Means of Elevating Its Character, and Perpetuating the Union* (Louisville, 1833); *Discourse on the History, Character, and Prospects of the West* (Cincinnati, 1834).

[16] Daniel Drake, *A Discourse on the Philosophy of Discipline* (Cincinnati, 1834), p. 34.

[17] Drake believed man was a naturally indolent animal which had to be disciplined or coerced to work. Necessity, in the case of the poor, functioned quite well, and was hence a beneficent force.

Sabbath breaking, and fashion. Though all of these did not necessarily exacerbate the desire for stimulation, they nonetheless provided the opportunity for the indulgence of an otherwise normal and necessary natural desire.

Drake was particularly concerned with the influence of fashion. Its power was rooted in the principle of human nature which made man an imitative animal and involved the respect for public opinion. Although it was not limited to any special class, community, or state of society, Drake was convinced by his observations that the influence of fashion was most operative in cities and among the highest circles of society. This was a dangerous situation because it was precisely that urbane group which set the tone of national life. "The great men of the land," Drake stressed, "should look to *their* example. . . . Those who fill high offices—the distinguished and learned professions—the aristocracy of wealth—the men of our chief cities—the community of self-styled gentlemen—the *magi* who wield the baton of fashion . . . ; these are they who govern the destinies of the multitude. . . . These are the men, among whom the reformation should commence!"[18]

Nothing less than national survival, Drake argued, hinged on the progress of the temperance cause. Intemperance led to rejection of religion, to broils, assassinations, duels, neglect of family and business, and to insanity. Nations need not, he believed, like men, pass through the cycle of life and death. But they would if they gave in to their vices and follies. If America took temperance, intelligence, and religion as its motto, he predicted, it would defy the revolutions which had overtaken other nations and endure from generation to generation.

Intemperance, then, was one source of disorder which endangered the cities and through them the perpetuation of the Union. But there were other sources of discord. Drake had always viewed the social diversity of the West with ambiguity. It contradicted his assumption of ecological balance, his conviction that in nature and society, despite the mobility of men, animals, vegetables, and even physical matter, everything had its appropriate geographic locale and that each movement set off a countermovement which tended to produce equilibrium and stability. That, however, could only occur if things that did not belong to the emerging new system were either transformed or eliminated.[19]

By the 1830s immigrants seemed to many to constitute an element which did not fit into the American system. Drake was sure, however, that in Cincinnati immigration could continue and the problem would be resolved.

---

[18] Daniel Drake, *An Oration on the Intemperance of Cities* (Philadelphia, 1831), pp. 27–28.
[19] The idea of a new society coming into existence bothered Drake, for he thought of civilization in terms of static levels or stages. He had no clear way of explaining how one stage of society moved or was transformed into another. He was never sure if the West would produce a unique America or would eventually become another Europe in a different place.

But it would not happen naturally. Social and national peculiarities persisted and gave society a diversified character. Although that produced collisions of minds and thus new ideas and original plans of social improvement, "[s]uch a community is not a compound, but an unconsolidated mass; and to acquire uniformity, it must be subjected to the crucible of amalgamation. The schoolhouse is that crucible."[20]

Yet the ultimate goal was not a new race. The object was to avoid the establishment of a permanent and distinct caste of foreigners. He wanted to build a community, not a confederation of clans. Although natives and newcomers would exert a mutual influence on each other, the former, since they were indigenous to the region—somehow a natural product—formed the central governing element. They would prevail, and Germans, Irish, Scots, and English would be absorbed and assimilated into the native culture after giving it a character of freshness and vigor. If all our efforts, he contended, were directed toward quickening this operation, the process could continue indefinitely. The melting pot could be a permanent feature of Drake's community.

Black emigrants to Cincinnati presented an entirely different problem. Drake's systematic scientific observations led him to the conclusion that Negroes could never rise to equality with whites in an environment favorable to and therefore dominated by whites. For this Drake could give no clear biological or environmental reason. It was, to him, merely another of those self-evident facts of nature, readily observed in a variety of places and hence true. Drake was satisfied with this fatalistic attitude toward ultimate causation because it fit his preconceptions about a divinely designed and orderly universe which men might observe and study but might never, at least very soon, totally fathom. Indeed, man could never wholly understand his world until he had been exposed to all its facts and functionings and applied the correct scientific method in unraveling their meaning and significance.

Drake was neither fatalistic nor sanguine about the place of blacks in the American system. To be sure, he argued that the condition of Negroes under slavery was meliorating and that the direction of the institution was toward gradual and voluntary emancipation. So long as the North left slavery alone these trends would continue. Hence the real crisis of race in America was in the North, not the South, and it centered about free blacks, not slaves.

At the heart of the problem was the tendency of freed blacks to colonize in northern cities. Bearing the visible physical and behavioral marks of slavery,

---

[20] Daniel Drake, "[Remarks] On the Education of Emigrants," *Transactions of the Fifth Annual Meeting of the Western Literary Institute and College of Professional Teachers* [1835] (Cincinnati, 1836), p. 81.

they stoked the enthusiasm of abolitionists who were, in the final analysis, responsible for the steady barrage of criticism of the South which made southern whites so anxious about the perpetuation of their system and increasingly inflexible. To make matters worse, most whites in northern cities found it difficult to mingle with black emigrants in workshops, poorhouses, hospitals, schools, and churches. "We hold them," Drake noted, "at arm's length . . . , and, according to the instinct . . . of an immense majority . . . they are, and should be kept, a distinct and subordinate *caste*," a separate clan within an area which contained the potential and aspiration for real community.[21] Not surprisingly, then, mobs periodically rose to drive the blacks from the cities. Negroes did not fit in the American system. The crisis could only worsen as more and more blacks were emancipated. Abrupt and total abolition, in Drake's view, would be catastrophic.

Drake proposed a final solution. All free Negroes in the South and emancipated slaves should be barred from traveling to, visiting, or settling in the North. They should, instead, be deported to Africa, preferably Liberia. This scheme, Drake held, would be good for blacks, whites, and the Union. It would tend to quiet the abolitionist attack on slavery, reassure the South and reduce its pressure for more slave territory, and pave the way for the continued softening of the institution of slavery and for gradual voluntary emancipation.

Equally important, deportation would be good for the blacks and fit into God's design for the world. Only by being a free man under a democratic and Christian government could Negroes acquire the characteristics, traits, and habits of free men. That is what America did for the Irish. It transformed them by permitting them to participate fully in a free society. The blacks, confined by prejudice to the status of a caste, could not do that in America, but they could in Liberia. The Negro middle class, Drake argued, should take the lead in promoting this end. In Liberia they could help build a black America —a black melting pot—in a climate favorable to the fullest development of their race. To Drake, race, nationality, and region were virtually identical. If he read God's design correctly, Utopia would arrive when all mankind was classified and ranked into races and each was placed, physically, in its appropriate geographical setting.

"Separation is, then," Drake argued, "the only effectual remedy." If this were not accomplished, he predicted, the physical proximity of the races in the cities would produce racial mixing and increasing numbers of mulattoes fired with high aspirations. They would, however, inevitably be placed with the blacks, resent it, and join with a minority of whites to demand not only the

[21] *Dr. Daniel Drake's Letters on Slavery to Dr. John C. Warren of Boston* [reprinted from *The National Intelligencer,* April 3, 5, 7, 1851] (New York, 1940), p. 34.

abolition of slavery but also liberty and revenge. Then "rebellions, amounting to local civil wars, will carry dismay throughout our cities, and drench their streets with blood."[22]

Drake was a prisoner of his intellectual era. He inherited a scientific method and set of assumptions about the nature of the world which prevented him from reconciling himself to change as the natural order of things. This tension between his intellectual heritage and his experience produced a mania for institution building, a desire to spread a regulating layer of institutions across the broad and rapidly changing geography of Western cities and countryside. It encouraged him to participate in the affairs of his city, region, and country. A self-styled scientist and philosopher, he is now remembered as a founder and builder of educational and medical institutions, philanthropic, professional, business and social organizations, urban community and regional community, not as the author of a two-volume, 1400-page work on the diseases of the Mississippi Valley; and appropriately so, for institutions frequently outlive their intellectual origins and provide us an avenue of change as well as a tie to the past. They can, if respected and used but not worshiped, keep alive the hope that we may escape our "prison of history."[23]

ZANE L. MILLER

[22] Ibid., p. 36.
[23] The phrase is Daniel Boorstin's.

Chronology

1785	born, near Bound Brook, N.J., to Isaac Drake and Elizabeth Shotwell, October 20.
1788	Drake family migrates to Kentucky, settling at Mays Lick.
1800	apprenticed to Dr. William Goforth, Cincinnati physician, December.
1804	enters into partnership with Dr. Goforth, May 18.
1805–06	attends medical classes at University of Pennsylvania, November–March.
1806	returns to Kentucky, begins practice of medicine at Mays Lick, April.
1807	takes over Goforth's medical practice in Cincinnati, April 10. marries Harriet Sisson, niece of Jared Mansfield, surveyor-general of Northwest Territory, December 20.
1808	Harriet Drake born, October 24 (d. September 20, 1809).
1810	"Daniel Drake & Co." established (with brother Benjamin) for sale of drugs and medicines, groceries, paints, surgical instruments, stationery, and books, April–May. *Notices concerning Cincinnati,* part I, published May–June.
1811	Charles Daniel Drake born, April 11 (d. April 1, 1892). *Notices concerning Cincinnati,* part II, published May.
1811–12	builds three-story brick house at 429 East Third Street, Cincinnati.
1812	organizes First District Medical Society in Cincinnati.
1813	elected trustee of Cincinnati City Corporation, April 5. John Mansfield Drake born, July 1 (d. February 5, 1816). announces preparation of *Natural and Statistical View . . . of Cincinnati and the Miami Country,* solicits subscribers at $1.00 each, August–September. organizes "School of Literature and the Arts" with Josiah Meigs and Peyton Short Symmes, October. organizes Cincinnati Manufacturing Company, November.
1814	organizes Cincinnati Lancasterian Seminary, with Edmund Harrison and Joshua L. Wilson, April (instruction begun April 17, 1815).

1815	"Daniel Drake & Co." sold. The Drakes leave for Philadelphia, October.
1815–16	attends medical classes at University of Pennsylvania, November–March; M.D. degree awarded May 7.
1816	*Natural and Statistical View . . . of Cincinnati and the Miami Country* published, February 16 (copyright date 1815).
	"Isaac Drake & Co." established (with father and brother Benjamin) for sale of medicines, paints, dyes, patent medicines, groceries, stationery, hardware, dry goods, and "artificial mineral waters."
	returns to Cincinnati, resumes practice of medicine, June.
1817	takes Dr. Goforth into partnership, January 1 (Goforth dies May 12).
	elected to medical faculty of Transylvania University, January 7.
	takes Dr. Coleman Rogers into partnership after death of Goforth, July 9; they announce facilities for medical instruction of students in "any number that may apply," July 11.
1817–18	professor of materia medica and botany, Medical Department of Transylvania University, November–March; resigns, March 24, is accused of seeking to wreck the Medical Department.
1818	appointed a director of the Office of Discount and Deposit, Cincinnati branch, Bank of the United States, February (until 1820).
	elected to membership in American Philosophical Society and American Antiquarian Society, April.
	delivers subscription lectures on botany at Lancasterian Seminary Hall, May.
	advertises inauguration of formal lectures on medicine at Lancasterian Seminary Hall beginning November: Rev. Elijah Slack, chemistry and pharmacy; Coleman Rogers, surgery and anatomy; Drake, botany and materia medica, physiology, and practice of physic, May 27.
	organizes "Western Museum Society" with William Steele, Rev. Elijah Slack, Jesse T. Embry, James Findlay, September.
	organizes Cincinnati Society for the Promotion of Agriculture, Manufactures and Domestic Economy, October.
	seeks charter from Ohio General Assembly for a Medical College of Ohio to be located in Cincinnati and a Cincinnati College to succeed the Cincinnati Lancasterian Seminary, December; charters granted, January, 1819.
1819–22	professor of the institutes and the practice of medicine, and president of the College, Medical College of Ohio; instruction not begun until November 11, 1820.

1819	appointed a trustee of the Cincinnati College, March. Harriet Echo Drake born, July 19 (d. September 9, 1864).
1819–20	lectures on mineralogy and geology at Western Museum Society, December–April. hard feelings within the Cincinnati Medical Society (organized 1818) over organization of Medical College of Ohio. Drake and Slack resign, December, organize The Medico-Chirurgical Society of Cincinnati, January.
1820	Coleman Rogers challenges Drake to a duel, January 5. Drake does not accept. Drake and Slack vote to dismiss Rogers from the Medical College of Ohio faculty, January 14. Drake and Rogers dissolve partnership, February 23.
1820–21	seeks charter from Ohio General Assembly for a Commercial Hospital and Lunatic Asylum at Cincinnati, to be staffed by Medical College of Ohio faculty and used for the instruction of students, December; charter granted, January.
1821–22	hard feelings within the Medical College of Ohio faculty; Drake dismissed as professor and president by vote of his colleagues, March 7.
1822	announces preparation of a "Treatise on the Diseases of the Western Country," April.
1823	elected to medical faculty of Transylvania University, February; accepts appointment, May.
1823–27	professor of materia medica and medical botany, Medical Department of Transylvania University, November–March, 1823–24; professor of the theory and practice of medicine, and dean of the medical faculty, November 1824 to March 1827.
1825	death of Mrs. Drake, September 30. Drake's 40th birthday, October 20.
1827	resigns from Transylvania University medical faculty; returns to Cincinnati, March. Joins Dr. Guy W. Wright in establishing *The Western Medical and Physical Journal*. visits Philadelphia and Washington, investigates techniques for eye care and operation of eye infirmaries, March–May. establishes Cincinnati Eye Infirmary, July; incorporated by the Ohio General Assembly, December.
1828	delivers first discourse on intemperance, March.
1828–38	edits *The Western Journal of the Medical and Physical Sciences,* founded April 1828, as a continuation of and competitor to *The Western Medical and Physical Journal;* publication suspended after vol. XII, no. 1.
1830–31	professor of the theory and practice of medicine, Jefferson Medical College, Philadelphia, November–February.
1831	organizes Medical Department of Miami University, to be located in Cincinnati, February. Legal action by Trustees of Medical College of Ohio to prevent institution of instruc-

	tion (May) results in compromise by which the faculties are merged in a reorganized Medical College of Ohio.
1831–32	professor of clinical medicine, Medical College of Ohio. resigns, January 19, 1832, is accused of trying to wreck the college. Pamphlet war ensues.
1832	publishes *Practical Treatise* on the history, prevention and treatment of cholera, July. Warns of cholera in Cincinnati, October. Pamphlet war ensues.
1832–33	lectures on anatomy and physiology at Ohio Mechanics Institute, Cincinnati, December–April.
1833	attends Literary Convention of Kentucky, held in Lexington; urges stronger regional associations as a means of preserving the Union; is accused of secessionist tendencies, November.
1834	helps organize Hamilton County Temperance Society, January.
1835	organizes Medical Department of Cincinnati College, May–June.
	urges establishment of rail connections between the Ohio River and the Carolinas, August; subsequently organizes The Louisville, Cincinnati and Charleston Rail-road Co., October, 1836 (construction never begun).
1835–39	professor of the theory and practice of medicine, and dean of the medical faculty, Medical Department of Cincinnati College. Instruction suspended by action of the Trustees, August, 1839.
1836	chairman of Cincinnati "Friends of Texas" supporting Texan struggle for independence from Mexico, April.
1838	principal speaker at Cincinnati Semicentennial Celebration, December.
1839	elected to faculty of Louisville Medical Institute, August.
1839–49	professor of clinical medicine and pathological anatomy, Louisville Medical Institute.
1840–49	edits *The Western Journal of Medicine and Surgery,* with Lunceford P. Yandell and others.
1841	organizes Physiological Temperance Society of the Louisville Medical Institute.
1844–45	begins composition of *Systematic Treatise.*
1845	Drake's 60th birthday, October 20. Begins composition of reminiscential letters to his children, December.
1849	resigns from Louisville Medical Institute, March; returns to Cincinnati. MS of *Systematic Treatise,* vol. I, sent to printer.
1849–50	professor of special pathology, practice, and clinical medicine, Medical College of Ohio. Resigns at end of session.
1850	*Systematic Treatise* published, April.
	elected "permanent member" of American Medical Association, May.

1850–51	writes on slavery.
1850–52	professor of pathology and the practice of medicine, Louisville Medical Institute.
1851	organizes Cincinnati Medical Library Association.
1852	resigns from Louisville Medical Institute, February; accepts appointment as professor of the theory and practice of medicine, Medical College of Ohio.
1852	dies, November 5.

Some Account of the Epidemic Diseases Which Prevail at Mays-Lick, in Kentucky*

(1808)

The *Philadelphia Medical and Physical Journal* was published between 1804 and 1809 by Drake's former teacher, Benjamin Smith Barton, professor of materia medica, botany, and natural history at the University of Pennsylvania. Designed to serve as a vehicle for the publication of "original communications relative to all the branches of Medicine, Natural History, and Physical Geography," it was among the first of those new scientific journals of the nineteenth century that encouraged the development of descriptive rather than ratiocinative science and that emphasized the investigation of American rather than European problems. It seems indeed to have been a model for many of them, including Drake's own *Western Journal of the Medical and Physical Sciences*. In general, the *Philadelphia Journal* reflected Barton's own primary interest in natural history rather than in more specialized medical subjects, and was properly a physicians' rather than a medical journal. Many of its articles appear to have been solicited by the editor.

Drake's brief essay, submitted in 1807, was prepared after his removal to Cincinnati where he took over the medical practice of his first teacher, Dr. William Goforth. Its careful compilation of data, its attempt to isolate the unusual and hence critical factors, and its refusal to confuse generalization based on observation with causal hypothesis is characteristic of Drake's later work in science, and especially in epidemiology, and may be said to have formed the basic methodological program of the empiricism to which he was committed.

* "Some Account of the Epidemic Diseases which prevail at Mays-Lick, in Kentucky. In a letter to the Editor, from Dr. Daniel Drake," *Philadelphia Medical and Physical Journal*, III, part I ([March ?] 1808), 85–90.

To fill up this sheet, I will copy from my common-place-book some observations on the topography and diseases of that part of Kentucky in which I lived, after my return from Philadelphia, till about three months ago. The village in which I lived is 12 miles from the Ohio. It is remote from any marsh, pond, or considerable stream of water; the land is fertile and rolling; the springs, though numerous, are most of them transient. The inhabitants of this little place, and the surrounding country, had, for many years, enjoyed a high degree of health. An epidemic, till last year, was almost unknown to them; but it formed a sad reverse. A fever of the typhous or typhoid kind, attended with bilious symptoms, prevailed in every house in the village, and in many in its vicinity. I shall not attempt a history of it, but will merely give you a few of the results of my observations on it.

1. The majority were attacked between the 1st and 10th or 15th of October, but several both before and afterwards.

2. Sometimes typhous symptoms appeared at the commencement, but in most cases it was at the beginning somewhat inflammatory.

3. In November and December it was attended with more typhous symptoms than in September and October.

4. Bilious symptoms were present in almost every instance. In some cases, large quantities of bile were discharged.

5. It was certainly not infectious, for visitors did not take it, and yet, in two families, it gradually attacked almost every member of them.

6. The tongue, in almost every case, was covered with numerous small papillae, which were more obvious to the sight than touch. They occurred whether the tongue were dry or moist, blackish or whitish. They also occurred in every case of indisposition which I witnessed that autumn, from whatever cause.

7. Either during the formation, progress, decline, or convalescence of this fever, a diarrhoea uniformly occurred.

8. Pains in the extremities were very common. They were sometimes periodical. They generally occurred towards the decline of the fever, and, in almost every case, indicated a favourable termination. They discovered that the sensibility of the system was not exhausted. In a case that terminated fatally, no pain attended through the whole course of

the fever, neither could any be excited by blisters and sinapisms. When these pains were violent, they were most effectually relieved by blisters over the part affected, and by sweating.

9. Boils and other abscesses were extremely common; they were favourable appearances. They generally occurred about the termination of the disease.

10. When a free determination to the skin, either spontaneously, or by the use of sudorific medicines, took place, the disease generally terminated favourably.

11. In most cases, quotidian intermissions, or remissions, were observable.

12. Occasional chills were not uncommon; and, as they indicated the existence of considerable sensibility, they were a favourable symptom.

13. When strong emetic, cathartic, and sudorific medicines were exhibited, and operated freely at the commencement, they generally destroyed the fever.

14. I bled freely in two or three instances, when there seemed to be considerable inflammatory diathesis; but, as they proved to be among the most dangerous cases that occurred, I left it off.

15. The general plan of treatment, and one which I partly derived from Dr. Duke, a respectable and old practitioner, was to exhibit emetics and cathartics freely at the commencement, and at any subsequent period when they seemed necessary; to exhibit diaphoretic medicines at every period of the disease; to apply blisters and sinapisms during the whole course of the fever, but more especially towards the latter stages; and to exhibit stimulants and tonics freely, after the transient inflammatory symptoms of the commencement were abated.

The cause of this fever I shall not attempt to assign; but will mention those circumstances which were attendant upon it.

1. The summer and autumn were remarkably dry. Almost every spring was exhausted. The wheat, &c., ripened nearly two weeks earlier than usual; and whole fields of corn were destroyed. Almost every different kind of tree defoliated much earlier than usual; and the leaves of some were dried up without assuming those beautiful colours that precede their fall.*

---

* This was more especially the case with the Pau-pow (Annona glabra?). The leaves became dry, and curled up without assuming that light yellow colour which precedes their fall, generally. This drying up uniformly commenced at the apex of the leaf; but the yellow colour generally commences at the base, and, in most cases, on one side of the petiole.

2. In proportion to the number of showers which fell, we had very little lightning and thunder.

3. There was, I think, more east-wind than usual.

4. Several different species of insects were uncommonly numerous.

a. The army worm (your Phalaea migratoria?).†

b. A green worm, about the same size, which committed great ravages upon the leaves of the Hackberry-tree.

c. Small insects not much unlike, but much larger than *Pediculi;* with a tuft of white filaments from 3 to 6 inches in length, rising out of the superior posterior part of their bodies. These insects I saw exclusively on the limbs of the Beach-tree. I saw them in no other state than the one I have mentioned. It was about the middle of September.

d. A worm which destroys the unripe ears of Indian corn. This worm is seen every summer, but was uncommonly numerous and destructive last summer. In some fields scarcely an ear was unaffected.

I do not pretend to see any connection, after the manner of cause and effect, between these facts and our little epidemic, but, as I cannot assign any cause for it, and as these occurrences were all contemporary, I thought them worth mentioning. Evils often seem gregarious.

I am yours, &c.,

Daniel Drake.

Cincinnati, Ohio, July 22d, 1807.

† This worm has not appeared during the present year.

*Notices concerning Cincinnati**

(1810)

Between 1807 and 1810 Daniel Drake was busily engaged in the construction of a career. He practiced medicine. He associated himself with Cincinnati's intellectual and scientific community, formally in the Cincinnati Lyceum and the Cincinnati Library Association, informally at the home of Captain Jared Mansfield, surveyor-general of the Northwest Territory, where a group of local literati regularly gathered. He met and married Mansfield's niece, Harriet Sisson. He excavated a large Indian mound on the site of what is now downtown Cincinnati and collected information concerning excavations of prehistoric encampments elsewhere in Ohio. He kept records of temperature, rainfall, and wind direction in order to describe the climate of Cincinnati, and he sought comparative data from meteorological observers in the East. He botanized with an eye to the preparation of a catalogue of medicinally useful plants native to the West and to the comparison of the flora of the West with that of the Atlantic states and Europe. He gathered information on the existence of useful minerals and on the geography and geology of the Ohio Valley. During the winter of 1809–1810 he put together his observations in a little book which he called *Notices concerning Cincinnati*.

Entirely lacking the boosting quality of his *Natural and Statistical View, or Picture of Cincinnati and the Miami Country* (1815), Drake's *Notices* was an ecological survey of the Cincinnati area. Whether the preparation of a "calendarium flora" led him to an examination of climate and climate to an examination of endemic diseases, or vice versa, or whether he sought to combine the emphases of both his teachers, Benjamin Rush and Benjamin Barton, is not clear. In fact the separate sections of the book stand as virtually discrete entities. It was here, however, that Drake developed his conception of the critical importance of environment in the development of the patterns of man's life.

It was a primary goal of science to describe this relationship, according to Drake, and of the scientist to point to the consequences for man's happiness of the particular circumstances in which he had placed himself. This was a theme of Drake's writings throughout his career, whether his focus was on the necessity for the establishment of

---

* *Notices concerning Cincinnati. By Daniel Drake.* Cincinnati: Printed for the Author, at the Press of John W. Browne & Co. 1810.

social and cultural institutions in the new cities of the West, on the consequences of topography and climate on the prevalence of endemic and epidemic diseases, or on the dangers of "fashion" as a determinant of personal and social behavior. His concept of environment was no simple-minded one, however. As a physician accepting the conventional dualism of contemporary biomedical theory, Drake knew that man's body was the first of the series of concentric environments in which he lived, and the one with which he must come to terms first, as a conscious, sentient, knowledgeable being. Although he would later lay great stress on physiology as "the science of human nature" and hence an essential clue to understanding man's behavior, he rejected out of hand the naturalist determinism of those who regarded human nature as necessarily immutable. Science for Drake was descriptive, not predictive, a useful and beloved servant, a guide to human action; with the aid of science man could overcome his innate desires through an exercise of will as easily as he could overcome geographic barriers to social intercourse by the construction of a railroad or a canal.

PREFATORY REMARKS

Such a Calendarium Flora as would exhibit the progress of vegetation at this place, and answer for insertion in the Medical and Physical Journal, or some other Magazine, and nothing more, was at first intended. But the physical sciences are so intimately connected, that the narrow limits then prescribed, have been overstepped, and the addition of notices respecting our soil, climate and diseases, now renders the floral calendar the most inconsiderable part.

Even, however, in its extended form, no great degree of importance is attached to this humble prodromus; and the original design, of sending it to the editor of that Journal, would still be adhered to, did not the writer wish to distribute several copies of it among medical and other friends, as an acknowledgment for similar information communicated to him. To them he addresses it without hesitation, and only requests that they make a careful distinction, between what is given as fact, and what as hypothesis, or deduction: The latter *may* be correct, the former can scarcely be incorrect.

It only remains to observe, that the *manner* in which it is executed may probably be considered too elaborate and methodical for such an immature production; but in addition to the existance of insuperable obstacles to the attainment of an easy and elegant style, the writer labored to be concise, and conciseness requires method.

Cincinnati, Ohio, May 1, 1810.

I. TOPOGRAPHY

Cincinnati, situate on the northern bank of the river Ohio, in a bend of gentle curvature, is in 39 deg. 7 min. N. Lat. and about 84 deg. 30 min. W. Longitude.

Its site is not equally elevated. A slip of land, called the BOTTOM (most of which is inundated by extraordinary freshes, though the whole is elevated several feet above the ordinary high water mark) commences at Deer-creek, the eastern boundry of the town, and stretches down the river, gradually becoming wider and lower. It slopes northwardly to the average distance of 800 feet, where it is terminated by a bank or glacis, denominated the HILL, which is generally of steep ascent, and from 30 to 50 feet in height. In addition to this, there is a gentle aclivity for 6 or 700 feet farther back, which is succeeded by a slight inclination of surface, northwardly, for something more than half a mile, when the hills, or real uplands commence. These benches of land extend northwestwardly, (the upper one continually widening) nearly two miles, and are lost in the interval grounds of Mill-creek. The whole, form an area of between 2 and 3 square miles; which however comprehends but little more than a moiety of the expansion which the valley of the Ohio has at this place. For on the southern side, both above and below the mouth of Licking river, are extensive elevated bottoms.

The hills surrounding this alluvial tract, form an imperfectly rhomboidal figure. They are between 2 and 300 feet high, but the angle, under which they are seen from a central situation, is only of a few degrees. Those to the S.W. and N.E. at such a station, make the greatest, and nearly an equal angle; those to the S.E. and N.W. also make angles nearly equal. The Ohio enters at the eastern angle of this figure, and after bending considerably to the south, passes out at the western; Licking enters through the southern, and Mill-creek through the northern angle.—Deer-creek, an inconsiderable stream, enters through the northern side. The Ohio, both up and down, affords a limited view, and its valley forms no considerable inlet to the E. and W. winds. The valley of Licking affords an entrance to the S. wind, that of Mill-creek to the N.W. and that of Deer-creek, (a partial one) to the N.E. The other winds blow over the hills that lie in their respective courses.

The Ohio is 535 yards wide from bank to bank, but at low water much narrower: no extensive bars exist, however, near the town. Licking river, which joins the Ohio at right angles, opposite the town, is about 80 yards wide at its mouth. Mill-creek is large enough for mills, and has wide alluvions; which, near its junction with the Ohio, are annually overflown. Its general course is from N.E. to S.W. and it joins the Ohio at a right angle.

Ascending from these valleys, the aspects and character of the surrounding country are various. On the southern, or Kentucky side of the Ohio, the land is hilly, and the interval grounds narrow; on this side the land is more level, and the interval spaces wider. These spaces are covered with large sycamores (platanus occidentalis) hackberries (celtis occidentalis) poplars (liriodendron tulipifera) the beech (fagus Americanus) the buckeyes (esculus) hickories and walnuts (juglans) honeylocusts (gleditsia triacanthos) 2 or 3 oaks (quercus) paupow (annona glabra) grape-vines (vitis ferotina) 2 species of ash (fraxinus) sugar trees (acer saccharinum) black locust (robinia pseudacacia) and most of the other 40 or 50 trees and shrubs, which compose the *Arbustum Terrae Fertilis* of the western country. While the shores of the creeks and rivers are embellished by willows (salix) cotton trees, (populus deltoid) and red maples, (acer rubrum.)

The uplands produce either the trees already mentioned, or the numerous species and varieties of oak, or beech, or the whole blended together, according to their differences in fertility and moisture.

No barrens, prairies, or pine lands are to be found near the town.

II. GEOLOGY

The internal structure, of the site of our town, demonstrates, that it is wholly "made ground," and that water has been the immediate agent. On the upper division, or Hill, the soil near the eastern and western extremities is better, but in the middle it is extremely thin, exhibiting every where the loam over which it is spread. This loam, which constitutes the second stratum, is from 4 to 8 feet in depth. It presents but few varieties, affording besides the sand and brick clay which compose it, nothing more than occasional siliceous pebbles, and fragments of argillaceous grit. This layer is supported by a grand stratum, composed of pebbles, gravel and sand. It is of unknown thickness, wells having been dug to the depth of 80, 90 and 100 feet, without passing through it. The particulars which have been observed respecting its construction are the following: 1. The sand; gravel and pebbles are commonly blended together; but, in some places the sand exists in beds distinctly from the others. These beds are found at considerable depth, and generally exhibit in the position of their particles, a kind of oblique or wave-like stratification; while that of the superincumbent gravel is more horizontal. 2. A large portion of the pebbles of this stratum are opake calcareous carbonate; the rest are semi-transparent, white, blue, brown, and red amorphous quartz; flint; and several varieties of granite, some of which are undergoing decomposition. The calcareous fragments are discoid, the siliceous approach more or less to the globular figure. They are all water-worn, and resemble those found on the beaches of our rivers. 3. In some places these pebbles are cemented by carbonate of lime into breccia. It is somewhat tabular, and horizontally disposed. 4. No fluvial shells, nor exuviae of any kind have been found in this stratum, except a solitary vertebra of the mammoth, which lately was discovered about 20 feet below the surface. It had no doubt been deposited there at the same time with the gravel among which it was found. 5. Veins of loam highly colored, and of fine blue clay, have been occasionally found, more especially along its southern border. 6. In the well of capt. Prince at the depth of 36 feet, that of judge Symmes at 20 feet, and of Jacob Burnet, Esq., at 90 feet, fragments of vegetable matter have been found. In that of the latter gentleman were dug up the stumps or foundations of two trees, one of considerable size, the other smaller. They were represented by the workmen as having grown there, but from the very depressed situation they occupied, and from their resting on sand, it is more probable they were *deposited* there, indicating that to have been once the bottom of a lake or pond, rather than the surface of the dry ground. And this opinion coincides with Mr. Volney's supposition of an ancient lake in this country. 8. The wells of maj. Ruffin, judge Symmes and gen. Lytle, all in a line from the river, have formerly afforded water, considerably impregnated with iron, and probably also with sulphur; both of which might have been supplied by the decomposition of fossil wood.—And where this line intersects the river, the sand and gravel of the beach are cemented into a kind of ferruginous breccia, by oxide of iron.

Such, as far as it has been explored, is the structure of the higher alluvion: that of the lower, or Bottom, differs from it in some respects. The layer of mould is several feet thick, and gradually changes into clay, which terminates about 20 feet from the surface. After this, sand and gravel present themselves, and continue to the calcareous and schistous strata, which underlay the town and adjoining uplands.

Of the geology of the surrounding country but a partial account will be attempted. Its alluvial portions like those already described, consist of mould, loam, clay, sand and gravel, to the depth of several feet. The superior strata of the uplands are mould, from 6 to 24 inches deep, and loam, with loose horizontal limestones and fragments of argillaceous sandstone, to the depth of from 6 to 12 feet. These strata, on this side of the river, are supported by argillaceous schistus (the argilla fissilis of Turton's Linnaeus) alternately and horizontally disposed with calcareous rocks; which construction continues as low as we have yet penetrated. The former substance, in quantity, greatly exceeds the latter, and really gives to this part of our state a *schistous* character. It has a dull blue color, breaks into thick irregular discoid fragments, softens and is diffusible in water, from which it is probably, in certain situations, deposited, forming beds of potter's clay (argilla lithomarga;) it adheres to the tongue, can be scratched with the nail, effervesces with acids, feebly before, but briskly after pulverization, and has 2.55 specific gravity. It contains neither sulphur nor bitumen. The limestone in this region is from 1 to 18 inches thick; is found in oblong or irregular indeterminate angular pieces, of various sizes; has a coarse grain, and is of different densities, with the medium specific gravity of 2.65. The lime obtained from it is said to possess great strength, but adheres slightly, and is not very white, no doubt from the abundance of iron it contains.

An observer, upon examining this calcareo-schistous region is ready to pronounce, that the limestone is nothing but indurated slate; for the change of density and texture, from one to the other, is, in many places, so gradual as to be perfectly imperceptible. This, however, can only be determined by chemical analysis.

Several varieties of marine exuviae, which I am not now prepared to enumerate, are found imbedded in, or impressed on the surface, of these calcareous stones. The slate also, is not without appearances of this kind, though they are not so numerous nor so large.

Along the beach of the Ohio, smooth lumps of sandstone, of different degrees of hardness, and of various colors and sizes, are by no means uncommon. In the lower bank of Licking River, just at its junction with the Ohio, but more especially in some of the river hills, about 15 miles above this town, are huge shapeless masses of breccia or pudding stone. It consists of rolled, calcareous and siliceous pebbles, cemented by carbonate of lime. It is found in hills which appear to have had a secondary formation, and constitutes their nuclei. As we advance into Kentucky, the proportion of argillaceous matter decreases rapidly, until at length dense, thick, almost interminable, calcareous rocks, separated but slightly, form the solid foundation of that state. It has been asserted, that the prevalence of schistous matter ceases at the Ohio river; this may be the case in some places, but it certainly is not everywhere.

Granitical pebbles have been already mentioned, as frequently occurring in the alluvion on which our town is built; but *they* are not *all* the granite this country affords. About fifteen miles north of this place, is a zone or region of larger masses of that compound. It runs from east to west. These masses, some of which are several feet in diameter, are of a reddish color, amorphous, smooth and perpendicularly stratified. It is believed, that similar fragments are bestrewn over most of our state.

The country, between this and lake Erie, is probably erected on a frail calcareo-argillaceous foundation: but little, however, is accurately known concerning any portion of its natural history.

From even this cursory topographical and geological view, we perceive the reason why this state abounds, considerably, in interval lands, and durable springs; while in Kentucky, the springs, though numerous during the rainy season, are transient, and the interval spaces narrow. In that state, the hills are generally steep enough to convey most of the rains rapidly away; and what water does filter through the soil and clay is arrested by the broad impermeable rocks, and conducted to the banks of the numerous creeks, where it bursts out in temporary springs. The surface of this state, in most parts, is so level as to retain a large portion of the water which falls, while the frailer structure of the ground readily permits it to sink below the region of evaporation, where it collects and forms permanent springs. In Kentucky, the rivers and small streams can effect but little, in a lateral direction, against the dense calcareous rocks, which every where abound beneath the surface, and therefore are restricted to narrow limits. In this state, the rivers and creeks are constantly undermining and wearing away their resistless and crumbling banks, and thereby widen their valleys.

About one mile up Licking river, are several copious chalybeate springs, which however are covered with water, except when the river is low. To the east of the town, from under the hill beyond Deer-creek, there bursts out a feeble vein of water, considerably impregnated with sulphate and muriate of soda. But few springs exist about the town. Wells are more common, tho' not very numerous: most of them terminate in sand and gravel. The water they afford is hard, incrusting the vessels in which it is boiled. It contains uncombined carbonic acid, carbonates, muriates, probably nitrates, but no sulphates.

That of the river is softer: In November last, when it was partially examined with chemical tests, it appeared to contain muriate of soda and uncombined soda. At other seasons of the year it no doubt contains other principles.

III. CLIMATE

If a history of the climate of North America, in these middle latitudes, were to be attempted, it might be found both useful and natural, to distribute it into four great divisions: 1. The states lying east of the Alleghenies 2. The countries lying west of the Chippewan, or Stoney mountains 3. The mountains themselves 4. The broad shallow valley, or vast platform comprehended between them; which is drained by the Ohio, Mississippi, and Missouri, with their thousand tributary streams.

Each of these great regions, probably, has a climate somewhat peculiar. That of the first is pretty well known; that of the second has been ascertained to differ from the others; that of the third, from its elevation, no doubt possesses many peculiarities, but our knowledge of them is very limited; that of the fourth is better known, but the aggregate of the *truths,* published respecting it, is by no means equal to a full display of its character. This we may regret, but it cannot surprise us, for accurate, long continued contemporary observations, at different places, must be made, before the climate of any country can be correctly estimated.

The most important results, of those recently made at this point in the great inter-montane region, will now be stated. For the purpose of comparison, they are placed in connection with certain Atlantic observations, the most proper that could be procured. They are not, however, so perfectly comparable as could be wished, but may answer the end of general, though not of critical, information.

I. TEMPERATURE

The mean annual heat of this town, during 4 years, was as follows: In 1806, deg. 54.1—1807, 54.4—1808, 56.4*—1809, 54.4. The average of these, 54.8 deg. may be received, without much hesitation, as the standard temperature of this place: for our deep wells, and copious perennial springs, are constantly about 54 degrees. The mean annual heat of Philadelphia, 50 min. north of this town, from 6 years observation by Dr. Coxe,† was 54.16 deg. That of Springmill, on the Schuylkill, 57 min. north of this, from 2 years observations by Mr. Legeaux,‡ was 55.7 deg. The mean summer heat, for 4 years at this place, was 75 deg. for 6 years at Philadelphia, 75.5 deg. and for three years at Springmill, 73.8 deg. In 1800, the summer heat in Philadelphia, was 75 deg. at Gnadenhutten, § in this state, about 25 min. farther north, it was 72 deg. The afternoon

---

* These three are the results of observations by Jared Mansfield, Esq. Surveyor-general of the United States, which he has kindly permitted me to use. The last year's observations were by myself. Our thermometers (which correspond) were always hung in the shade and in contact with wood. His station was 4 miles north of this place, but in circumstances of local situation and elevation (about 50 feet above the high water mark of the Ohio) nearly the same with mine. Most of the morning observations made by him, however, were at too late a period to indicate the lowest temperature of the 24 hours; but mine were made at or before sunrise, which furnished proper data for correcting his. And if the difference between morning and afternoon heat be the same in different years, the results expressed above may certainly be depended on.

† See Medical Museum.

‡ Rush's Inq. & Obs. and Barton's Med. Phy. Journal.

§ Barton's Journal.

summer heat of this place, in 1805, was 79.7 deg. and in 1808, (a very hot year) 83.4 deg. That of Springmill in the former year, was 89.3. During that summer, at Springmill, the thermometer was 61 times at or above 90 deg. while at this place, during the summer of 1808, it was at or above 90 deg. only 32 times; and for 5 years past, it has been at that point, only 71 times. In the course of the same period, it has been below cypher 10 times at this place; while from 1798 to 1803, at Philadelphia, it was never as low as cypher.* And during the most intense degree of cold experienced in that city, for 20 years previous to 1805, the thermometer was only 5 deg. below cypher;† 6 deg. higher (as will appear presently) than was felt here in 1807, and 2 deg. higher than our last winter afforded, for two successive mornings.

The comparative mean heat of each month and season in 1809, at this place, and 1802 at Philadelphia, (two years which had the same mean temperature) may be seen from the following table:

	Cincinnati.	Philadelphia.		Cincinnati.	Philadelphia.
January,	deg. 25.1	deg. 40.8	July	deg. 73.4	deg. 74.7
February	34.2	34.5	August	73.3	74.5
March	44.0	42.3	September	67.8	67.2
April	57.9	52.9	October	63.3	59.9
May	61.4	59.1	November	44.3	45.6
June	72.7	71.1	December	35.9	33.3
	Cincin.	Philad.		Cincin.	Philad.
Winter	31.73	36.66	Summer	73.13	73.66
Spring	54.42	51.33	Autumn	51.43	54.23

The monthly thermometrical ranges at Cincinnati, in 1809, Philadelphia in 1802, and Springmill in 1805, are exhibited in the following table:

	Cincinnati.		Philadelphia.		Springmill.	
January	from deg. 2	to 47	from 24	to 57	from 2.9	to 63.5
February	13	60	10	50	3.9	65.7
March	16	78	24	66	20.1	84.9
April	27	88	38	70	37.6	92.7
May	38	86	48	71	35.4	88.7
June	48	94	58	86	44.4	95.0
July	56	93	63	88	55.6	100.2
August	54	89	60	88	56.7	100.6
September	43	87	48	84	42.6	98.4
October	33	85	42	76	29.7	88.2
November	5	72	32	57	27.5	74.7
December	11	62	12	64	22.3	66.4

* Med. Museum.
† Rush's Inq. & Obs.

The comparative mean difference, between the morning and afternoon temperatures, for the months, seasons and year, at Cincinnati and Springmill, may be seen from the following table:—

	Cincin. 1809.	Springm. 1805.		Cincin. 1809.	Springm. 1805.
January	deg. 10.8	deg. 16.4	July	deg. 15.3	deg. 23.0
February	11.6	20.5	August	15.0	21.7
March	18.0	23.8	September	21.5	21.1
April	20.6	23.0	October	18.6	19.6
May	24.6	25.4	November	16.0	16.2
June	16.0	22.7	December	11.0	11.1
Winter	11.1	16.0	Summer	15.4	22.5
Spring	20.8	24.0	Autumn	18.7	19.0
			Annual mean difference,	16.5	20.4

The greatest diurnal variations from cold to heat, and from heat to cold, in each month in 1809, at Cincinnati, and of 1805 at Springmill, will appear from the following table:

	Cincinnati.		*Springmill.*	
	From cold to heat.	Heat to cold.	Cold to heat.	Heat to cold.
January	deg. 30	deg. 32	deg. 40	deg. 22
February	31	22	45	29
March	37	36	45	36
April	42	32	40	47
May	36	40	37	37
June	30	25	38	38
July	26	25	35	34
August	21	26	34	30
September	29	30	32	38
October	33	25	39	30
November	38	26	36	36
December	32	28	33	30
Average,	32	28	38	34

Hence it appears, that at both places, the sudden changes from cold to heat are greater than those from heat to cold, which is contrary to popular opinion.

It must be observed that comparisons between the years 1805 and 1809, are not unexceptionable, for the former was much more remarkable for bold and sudden changes than the latter.

The annual thermometrical extremes and ranges, at Cincinnati, during 1806, 7, 8, and 9; at Philadelphia, during 1800, 1, 2, and 3, and at Springmill during 1787 and 1805, were as follow:

	Cincinnati.			Philadelphia.			Springmill.	
Lowest.	highest.	range.	Low.	High.	range.	Lowest.	highest.	range.
9	94	85	10	90	80	5	96.1	91.1
11 below 0	95	106	7	90	83	2.9 below 0,	100.6	103.5
4 below 0	98	102	10	88	78	—	—	—
2 below 0	94	96	14	90	86	—	—	—
—	—	—	—	—	—	—	—	—
2 below 0	95.25	97.25	10.25	89.5	81.7	3.9	98.3	97.6*

* This line shows the mean of the above columns.

These tabular displays of the results of observations, although very limited in their extent, indicate pretty correctly, it is presumed, the annual and monthly standard heat, and the extremes, ranges, and variations of temperature, at this place, and two nearly corresponding situations in the Atlantic states. From which it appears, that the opinion concerning the greater heat of this climate, first expressed by our late illustrious President, afterwards glanced at by Loskiel,* and since supported and extended by Mr. Volney, is not, at least in its full extent, correct. The former published his celebrated NOTES, at a time when but obscure accounts respecting this country had been received; the latter *traveled here* in 1796, and therefore should have possessed more correct information. He however seems to have been sometimes misled by a favorite and ingenious, but not unexceptionable hypothesis.

It is true that these respectable writers have collected several facts, and might even have added more, which apparently tend to support the assertion, "that this country is warmer by three degrees of latitude than the Atlantic States." But as the thermometrical observations made at this place tend considerably to invalidate this opinion, it is necessary, if possible, to refer the phenomena they have observed to other causes. It may be urged, however, that the observations which have been stated, should not be compared with those made at Philadelphia; for the summers of large cities, it is said, are warmer than those of the surrounding country, by several degrees of latitude.† If this were the case, the comparisons that have been made, would indeed be altogether inconclusive. But that it is not, seems highly probable: for in 1805 the summer and annual temperatures were precisely the same at Philadelphia and Springmill, though the latter has 50 feet greater elevation; and the summer heat of 1787 and 1805 at Springmill, and 1793 at Nazareth, bore the same ratio to the annual heat, as the summer and annual heat of Philadelphia ordinarily bear to each other.‡ So that the comparisons which have been made are certainly entitled to considerable confidence.

* History of the Missions.
† Caldwell's Memoirs.
‡ Med. Museum, Med. & Phys. Journ. & Rush's Inq. & Obs.

It has generally been thought that less ice is formed in this country, than in the same latitudes in the Atlantic States. This observation has more especially been applied to the Ohio and Delaware, the former of which is almost every winter frozen over at Philadelphia, while the latter at this place is but seldom blocked up with ice which it floats, and was never known to freeze over. Geography however furnishes an explanation of this fact. The Delaware rises in mountainous lands, and runs nearly a southern course, bringing down with it the temperature of an elevated region of 42 deg. latitude. The Ohio also rises in a mountainous tract, but before reaching this place it meanders for 400 m. in a deep valley, which probably somewhat reverberates the sun's rays, & which in one place is as low as 38 deg. lat. receiving in its course, the Great Kenhawa, from lat. 36 deg. the Big Sandy and other southern streams. In consequence of this, the water is too warm to be reduced extensively to ice, unless the duration of our periods of cold weather (which are frequently intense) were longer. This explanation receives support from the fact, that the great river Mississippi, which rises far north, has had ice sufficiently strong to bear carriages, formed on its surface, in a single night, in 38 deg. lat. This was affirmed to me by the Messrs. Rectors, as occurring at St. Genevieve, in the winter of 1808–9.

The residence, in this country, of the Paroquet, a bird which in the maritime states, does not inhabit a higher latitude than 36 degrees is considered by Mr. Jefferson* a proof of the greater mildness of our climate. But his book furnishes sufficient evidence, that those parts of Virginia which are nearly two degrees farther north than the habitation of the Paroquet, have a milder climate, or at least as mild an one as we know ours to have. So that some other explanation of this fact must be adopted, and that furnished by the learned professor Barton, seems the most plausible. He ascribes this difference of cis and trans-montane latitude to the great length and southern course of our rivers; along which birds in migrating, are fond of traveling.†

This more extensive migration of birds, (animals that are known to contribute very much to the dissemination of seeds) may be one reason why certain vegetables, such as the reed and catalpa, are found farther north in this than the Atlantic countries; but the southern origin of the rivers of western Virginia, Kentucky, and Tennessee, and the fertile, calcareous, alluvial constitutions of our soil, probably are the true causes.

After all, however, it is not denied that this country is possibly warmer than the same parallels in the eastern states, but not by any means in such a degree as has been supposed; and Mr. Volney's tropical summers, during

---

* See his Notes.
† Fragments of the Nat. His. of Penn. as quoted by Dr. Mease.

which the thermometer rises to 90 deg. and upwards, for 60 successive days, have never yet occurred here.*

It is asserted in Loskiel's History of the Missions, and the opinion also prevails here, that as we advance northwardly on this meridian, the increase of cold is in a greater ratio than the increase of latitude. No conclusive corresponding observations have been made, to determine this point, but the most striking of those which have been collected shall be briefly related.

At Lebanon, not more than 25 min. north of this place, according to the information of my friend Dr. J. Canby, and others, the complete evolution of vegetation in the spring, is several days later than at this place. Frost occurred there after the middle of May, 1809, more than a week later than at this place. At Dayton and throughout the champaign country traversed by Madriver, in the latitude of Philadelphia, I am informed by Mr. Joseph Pierce, and other inhabitants of that tract, the snows are much more deep and durable, than at this place; and, Indian corn planted as early as the tardy arrival and establishment of genial weather will permit, is frequently overtaken by autumnal frost. This is strictly true, however, only on the prairies; where the unobstructed and rapid progress of the winds produces great evaporation, and consequent coldness.

At Fort Wayne, which is situated in a prairie, about 41 deg. 10 min. north, it appears from letters from Dr. Abraham Edwards of the army, that in the winter of 1808–9, the surface of the ground was covered with snow from the 1st of December to the 1st of April; whilst during the same period at this place, very little snow, but much rain fell; that on the 2nd of January in the same winter, at sunrise, the thermometer was 17 deg. below cypher, and that at 10 o'clock, A.M. on the 24th of the same month, it was 3 deg. below cypher; whilst at the same period, on the former day at this place, the thermometer was nearly 2 deg. above, and on the latter about 10 degrees. That on the 21st March of the same year, at that place, the snow was 14 inches deep, and the Miami of the Lake was closed up with a heavy body of ice; whilst at this place, no snow covered the ground, and the thermometer rose as high as 40 degrees. The great coldness observed at Michilimackinac by maj. Swan, and at Hudson's sea by Umphraville & Robson, places, however, west of this meridian, is well known.

It would be incorrect from such desultory facts to draw any general conclusion: they however render the opinion mentioned above, probable, and at the same time extend the comparison between the temperatures of the

---

* In *guessing,* this ingenious traveler was not always fortunate, or else he would have lessened the heat of the atmosphere, and increased that of the river, which is not as he stated, 60 or 70 deg. in the summer, but 80 deg. or more.

eastern and western states. They show that the parallel of 41 deg. in our meridians, as in those of Pennsylvania, is the southern limit of steady cold.*

What are the variations of standard temperature, as we advance along a parallel of latitude from the foot of the Alleghenies to the Stony mountains? Future observations in various places can alone determine this. Two or three facts may be mentioned, however, which would induce a belief that it does not at least get warmer, as we advance westwardly. The Wabash and Mississippi freeze over in one night in lat. 38 deg. and lieut. Symmes informed me, that in 1809, the Arkansas froze over in one night, and continued frozen for several days. These facts however, are inconclusive, and we must patiently wait for future cotemporaneous observations.

That our climate has undergone a change, is a popular, and with many, a favorite opinion. The regular observations made here at an early period, are too few and desultory to determine this point accurately; and many of them cannot now be had. The deficiency however, has been supplied in part, by conversation with numerous intelligent persons long resident here, and by an abstract obtained from governor Sargent, through the politeness of maj. John Brownson. The conclusion to be drawn from the whole of which, is, that our summers are about the same, our winters nearly the same, though *possibly* somewhat colder.

The winters between 1785 and '91 are stated to have been uniformly mild. The winters of 1792-3, '95-6, '99-1800, 1805-6, and 1809-10 were also very mild. That of 1791-2 was severe, with deep snow; that which fell in January only, amounting to 24 inches. On the 23d of that month, the thermometer was 7 deg. below 0. The winter of 1796-7, is universally considered the coldest ever experienced here. On the morning of the 8th of January, according to gov. Sargent, the thermometer was 18 deg. below 0, and in the course of the month it was below that point, 4 other mornings. The Ohio that winter was shut up with ice four weeks, and frost occurred as late as the 22d of May. The winters of 1798-9, 1803-4, 1804-5, 1806-7, and 1808-9, were all severe, but not as intense as that of 1796-7. Of the other winters since 1790, nothing certain can be learned: but, it is believed, that they were temperate.

Of the summers since that time, less can be collected than of the winters. The prevalent opinion, however, is, that on an average, they are neither cooler nor warmer than formerly. The summer of 1808 was excessively hot, while that of 1809 was temperate and pleasant.

Respecting spring and autumn, not much early information can be obtained. But it appears, from the manuscript furnished by gov. Sargent, that the

---

* See Rush's Inq. & Obs.

latest vernal and earliest autumnal frosts in 1792, 3, 4, 5, 6, 7, and 8, occurred about the same time that they were observed to occur in 1807, 8, 9 and 10.

From these data, although not so numerous as could be wished, we may conclude, that the temperature of our climate is now nearly the same that it was 15 years ago: for warm and cool summers have lately occurred, and mild and severe winters succeeded each other before the year 1800, as well as since. If, however, these conclusions be rejected, and our winters and summers still be considered more intense than formerly, the cause must be sought in the partial felling of our forests; which admits the N.W. and S.W. winds, the principal sources of cold and heat, to move with increased velocity. And if so, I would say that just as much change has taken place as this cause can produce; which cannot be very considerable, for the face of our country is as it were, only dotted with farms. In future, as Mr. Volney has conjectured, the more unobstructed progress of the S.W. wind will continue to increase the heat of our summers; but the increase of cold will probably cease at no very distant period; for whenever the forests of the territories N.W. of this are cut down, an amelioration of our winters must be the consequence.

2. WINDS

The intimate relation between the temperature of a country and its winds renders it necessary to treat of them as nearly in connection as possible. The subjoined table shows pretty accurately the proportions of our different winds in 1809; but except it, this article will contain but inconsiderable additions to what has been before published by others.

719 observations were made, of which there were

	E.	SE.	SSE.	S.	SSW.	SW.	W.	WNW.	NW.	NNW.	N.	NNE.	NE.	ENE.	Calm.
In Jan.	0	03	00	4	00	17	2	0	23	0	2	0	07	0	00
Feb.	0	08	00	0	03	11	7	0	11	2	1	0	17	0	00
Mar.	1	01	07	3	01	16	11	1	10	1	0	3	05	2	00
Apr.	3	08	00	0	00	20	05	0	08	1	0	1	13	0	00
May	0	07	00	3	00	14	07	2	13	0	3	0	10	0	03
June	0	02	00	1	00	31	01	0	04	2	2	4	09	0	06
July	0	10	00	3	00	14	03	2	05	0	8	0	14	0	06
Aug.	0	04	03	2	00	26	04	0	10	0	0	0	09	0	04
Sept.	4	05	00	6	00	14	01	0	07	0	0	2	11	0	09
Oct.	3	14	03	1	00	32	0	0	01	0	2	0	06	0	00
Nov.	1	10	2	2	00	08	06	0	08	4	0	0	10	0	07
Dec.	6	05	00	1	01	12	10	4	09	0	0	0	00	0	14
	18	77	15	26	5	215	57	9	109	10	18	10	111	2	49

In the following table the proportion which each particular wind bore to all the rest, at this place in 1809, and also the proportion which each wind bears to all the rest in some of the Atlantic states, is exhibited. It obviously, however, cannot be the basis of any very correct general conclusions.*

Winds	Western States	Eastern States	Excess in West. states.	Excess in East. states.
E.	.268	.960	—	.662
S.E.	1.137	1.111	.026	—
S.	.600	.708	—	.108
S.W.	3.200	2.127	1.073	—
W.	.985	.703	.282	—
N.W.	1.616	2.767	—	1.151
N.	.565	.432	.133	—
N.E.	1.632	1.194	.438	—
Southern	4.937	3.946	.447	—
Northern	3.813	4.393	—	.580

With regard to the S.W. wind the following facts may be stated: 1. It prevails more than any other. 2. It commences, generally, sometime after sunrise, and ceases towards evening. If it continue after dark, more especially with any degree of violence, it indicates rain. 3. It seldom follows rain under 12 or 24 hours. 4. It attends, and indeed is the principal cause of our warmest weather. 5. It prevails more here than in the Atlantic states, and is one of the causes of this country being warmer than that, if it be. 6. Most of its phenomena are conformable to the beautiful theory of the very ingenious **Mr. Volney**, but numerous additional observations are still wanting.

Of the N.W. wind, it may be remarked: 1. It is colder than any other, and the longer it continues, the lower is its temperature: 2. At this place it comes from a region altogether south of the great chain of lakes; while in its passage to the middle Atlantic states, it travels over that chain, for several hundred miles. This fact shows the possibility of its being as intense at Cincinnati, as at Washington or Baltimore; although in reaching those places, it has crossed the rampart of the Alleghenies: For that it acquires heat from the lakes, is evident from the warmth of those places to the leeward of Erie: 3. It almost invariably follows rain: 4. It has not diurnal intermissions like the S.W. wind, but

---

*It was conceived that this could be most advantageously done by means of decimals. Ten was assumed as the whole of the observations and each fraction expresses the proportion which the wind, placed opposite to it, bore to all the rest. The observations were reduced to 8 principal points, and the 49 observations when it was calm were rejected. The Eastern observations, all made in the middle states, were supplied by Williams' History of Vermont, and Jefferson's Notes on Virginia. The Williamsburgh northern observations were omitted as the extraordinary prevalence of the N. wind at that place has been supposed to be owing to some local cause.

frequently blows throughout the night: 5. It sometimes brings rain, oftener snow, but in 9 cases out of 10, it is the harbinger, and indeed the cause of clear weather: 6. An inspection of the preceding table shows that if the same parallels be colder, on the eastern and western sides of the mountains, it is as much owing to the undue prevalence of the N.W. wind there, as to the predominance of the S.W. wind here. The surfaces of the Atlantic ocean, and the countries lying between it and the mountains are generally warmer than the mountains themselves, and hence W. and N.W. winds are readily produced: 8. It generally wafts forward our thunder storms, as Dr. Franklin long since remarked. It then, contrary to its usual manner, commences to the windward, and probably may be nothing but a deflected S.W. current.

Concerning the N.E. wind, the following facts may be stated: 1. It may reach this place without passing over any part of the Allegheney mountains, by traveling from the mouth of Davis' Straights, over the river St. Lawrence, Lake Ontario and the eastern end of Lake Erie: 2. It invariably produces rain, or snow, or at least cloudy weather, but not to such a degree as in the Atlantic states: 3. It generally commences in the night, and is commonly nothing more than a gentle but steady breeze; sometimes however (more especially it is said, within 2 or 3 years) it is driving and impetuous. 4. When this wind begins to blow, if the thermometer be high, it sinks; if very low, it rises.

The E. and S.E. winds pass over the Alleghenies in reaching this country. They no doubt deposit much of their moisture on those mountains, but still they almost invariably produce rain or snow. During the night of the 16th of March, 1810, there fell 5 inches of damp snow; the S.E. wind blowing gently during that period, and for most part of the preceding day.

The S. wind may be considered stormy. It is generally impetuous, and as Mr. Volney has remarked, almost certainly produces lightning. The cause of this curious phenomenon has not yet been developed.

Contemporary observations at various places, in the courses of our different winds, would probably show, that the southern generally commences to the windward, and the northern to the leeward; and hence, that one division of them is produced by an a posteriori, and the other by an a priori cause. The northern winds we know commence to the south, and it is inconceivable that a volume of heated tropical air should move northwardly, without a *propelling* cause. Whether the east & west winds have their origin to the leeward or windward, is uncertain. The former is *probably* governed by the same laws with the N.E. for it is generally moist & cool, but the character of moistness belongs also to the S.E. wind, so that it is impossible that the E. has sometimes the same origin with the S.E. The west wind, a very pleasant one, may sometimes be a N.W. at other times a S.W. diverted from its proper direction:

or it may be compounded of N.W. and S.W. currents which have impinged and taken the diagonal of their former courses. The S.E. is one of the principal winds of this country, and it possesses characteristics essentially different from the S. and S.W. in as much as it seldom brings thunder, but precedes and attends moderate continued rains. There can be little doubt, from analogy, and from the facts collected by the learned Dr. Mitchell,* but it is a windward current. Accurate observations, however, at the same time, from Charleston to this place, St. Louis, Detroit, &c., are much wanted, and would be highly interesting.

3. WEATHER

It is difficult to express the proportions of clear and cloudy weather by figures, but the following summary of observations in 1809, is probably worth insertion. In that year, there was an unusual proportion of cloudy weather, and

	clear.	cloudy.	variable.	hazy.	foggy.	rain & mist.	snow, sleet & hail.	thunder & lightning.	quantity of rain & snow in inches.
Jan.	6	15	10	—	—	4	4	2	4.6
Feb.	3	23	—	2	—	8	11	4	5.6
Mar.	9	11	11	—	—	9	9	3	3.1
Apr.	10	7	11	2	2	12	—	6	3.4
May	14	5	9	3	2	4	1	2	4.6
Jun.	15	4	11	—	11	7	—	14	2.8
Jul.	21	7	3	—	10	11	—	4	2.4
Au.	16	3	11	1	12	5	—	4	2.2
Sep.	24	—	6	—	23	2	—	1	.8
Oct.	20	4	7	—	17	2	—	—	.5
Nov.	8	12	8	2	2	2	1	—	1.9
Dec.	11	16	4	—	—	9	3	1	6.
	157	107	91	10	79	75	29	41	37.9

Greatest quantity of rain in 24 hours, December 4th, 3.3 inches. Deepest snow 5 inches.

although an uncommon quantity of rain did not fall, it rained very often. The 6th and 7th columns do not, as probably they should, express only those days on which it rained or snowed somewhat copiously; but they express also several sprinkles of both kinds, which were inconsiderable.

No Pluviametrical observations were made at this place, previous to 1809;

* See Medical Repository.

and I have to regret that those made during that year, and stated in the last column of the above table, are not very accurate. The pluviameter was nothing but an accurately cylindrical tin vessel, of sufficient depth. It was kept in a proper situation, and the water emptied immediately after every rain or snow, kept to the end of the month, and then measured. The column annexed above, if it do not accurately express the quantity of rain during the year, indicated pretty correctly the relative quantities in the different months.

The fogs of the Ohio and its waters are dense, but they rarely continue till 9 o'clock, A.M. They are most constant in June, July, August, September, and October.

The dew in this country is said to be more copious, and the quantity of atmospheric humidity greater, than in the Atlantic states. No comparable hygrometric observations, have, however, been made in the two countries, under circumstances precisely similar; and till that is done, a correct conclusion cannot be drawn. It is highly probable, that on the borders of Lake Erie, and in the depths of those forests which cover level land, the quantity of moisture is greater than in the cultivated portions of the Atlantic States; but a fair comparison cannot be made under circumstances so dissimilar.

The quantity of snow in this country is considerably less than in the Atlantic states in the same latitude; and this is one of the circumstances in which the two countries differ most materially. The E. and N.E. winds, which bring such deep snows east of, & on the Allegheny mountains, in consequence of passing that elevated tract, or from some other cause, produce far less here. We seldom have falls of snow to exceed 6 inches; and generally they are not upwards of four. It is uncertain whether the E. and N.E. or the N.W. produce the deepest snows. The snow which does fall, seldom lies very long, nor are our streams long covered with ice, for our winters are nothing but a succession of mild, and intensive severe weather: So that although their mean temperature be low, the frequent occurrence of transient periods of mild weather counteracts the more powerful operations cold.

We invariably have frost about the termination of the first week of May, and sometimes as late as the end of the third week. Inconsiderable frosts occasionally occur in autumn, as early as the equinox or before; but the more severe ones are not felt till the 15th or 20th of October. There are, however, great differences between different seasons in this respect. On the night of the last of August, 1789, the Indian corn in the northern parts of Kentucky was greatly injured by frost; and on the night of the 9th of the same month, in 1809, frost was observed in the vicinity of this town.

In the evening, during the summer, our horizon is frequently illuminated with broad obtuse flashes of lightning, which are unattended by thunder or rain. In 1809 the most conspicuous of these appearances were noted; which is the reason that the 8th column of the preceding table contains so many figures.

No barometrical observations have yet been made, or at least published, in this country.

4. CALENDARIUM FLORAE

Calendaria Florae furnish interesting indications, respecting the influence of climate, upon vegetables, and the lower orders of animals. They, however, do this imperfectly, unless they be kept for several successive years, and the aspects, elevations, and qualities of the soils in which they grow be noted. They must also, as was long since done, be accompanied by certain meteorological remarks, otherwise they frequently exhibit, upon a comparison with each other, considerable contrariety. For the accession of mild weather, in the latter part of winter, or early in the spring, may bring forward vegetation rapidly, for a time; but the occurrence of colder weather may at length suspend its progress to a very late period. This was the case during the present spring. The later part of February was so mild, that the maple and elm began to flower, the rose and weeping willow to leave, and the buds of numerous other vegetables to swell. March, however, nearly suspended these interesting operations; and it was reserved for the genial month of April to revive and nearly complete them, with a rapidity and luxuriance the most unrivalled.

The following fragment, as far as it extends, may serve to indicate the progress of vegetation, in this part of the valley of the Ohio in 1809. Upon comparing it with similar ones kept in 1807 and 8, the dates appear to hold nearly a middle place between the dates of those two years; being in most stages of Spring, about 10 days earlier than the first, and 10 later than the last. There is a difference between this valley and the adjoining upland country probably of 3 or 4 days, except on southern declivities.

As the meteorological observations made during that year have been already detailed, they will not be stated in the calendar.

CALENDARIUM FLORAE, 1809

Feb. 27 Flower buds of the water maple (acer rub.) swollen.
Do. peach & lombardy poplar beginning to swell.
March 1 Bees out of the hive.
Wild pigeons (col. mig.) geese and ducks (anas can. et bos.) returning northwardly.
4 Flower buds of the water maple beginning to open.
Commons becoming green.

         5 Buds of the weeping willow (s. babylon.) swelling.
           Frogs (ranae) sing.
         9 Buds of the weeping willow unfolding.
       11 Water maple in full flower.
       16 Gooseberry leaf buds beginning to open.
           Grackle (graccula quiscula) arrived.
       18 Flower buds of the elm (ul. amer.) begin to open.
       25 Doves (colum. carol.) mourn.
       28 Elm in full flower.
           Lilac (syringa vulg.) beginning to bud.
       29 Red-headed woodpeckers (pic. erythro.) arrived.
           House flies appear.
       31 Radishes, tongue-grass, peas and lettuce planted.
April   3 Buds of the privet, (ligust. vul.) beginning to open.
           Appletree buds beginning to open.
           Lilac leaves unfolded.
           Red currant buds beginning to open.   [unfold.
        4 Flower buds of the sugartree (acer sac.) beginning to
           Bluebirds (motacilla sialis) building nests.
        5 Purple martin (hirundo purpurea) arrived.
        6 Lombardy poplars in full flower.
        9 Quince leaves unfolded.
           Peach blossoms beginning to open.
           Gooseberry shrubs in full flower.
           Sand swallow (hirun. rip.) arrived.
       10 Leaves of the sweet briar (rosa rubig.) unfolded.
       12 Red currants in flower.
       14 Sugartree leaves opening.
           Barn swallow (hirun. urbica) arrived.
       15 Peachtree in full flower.
           Weeping willow do.
           Catbird (muscicap. carolinensis) arrived.
           Tonguegrass fit for the table.
       16 Dandelion (leonto. tarax.) in flower.
           Peartrees in full bloom.
           Lombardy poplar leaves unfolding.
           Oats (avena) and flax (lin. usita.) sowed since the 9th inst.
       20 Appletree in full flower.
           Chimney swallow (hirun. rust.) arrived.
       23 Quince in full bloom.   [its leaves.
           Black locust (robin. pseud.) beginning to expand
       24 Monthly strawberries beginning to flower.
           Lilac in full flower.
       26 Althaea leaf buds beginning to open.
May   3 Radishes fit for the table.
        9 Dogwood (cornus florida) in full flower.

        12 Racemes of the black locust full grown.
        16 Saw nighthawks (caprim. amer.)
             White clover beginning to flower.
        18 Blackberry (rubus occiden.) beginning to flower.
             Blacklocust in full flower.*
             Indian corn (zea mays) planted.
        20 Woodbine (lonicera caprifol.) beginning to flower.
        21 Peas fit for the table.
        25 Poplar (liriden. tulip.) in full flower.
             Rye (secale) beginning to flower.
        28 Sweet briar beginning to flower.
June   2 Privet beginning to flower.
         4 Red currants beginning to ripen (plentiful.)
        10 Black mulberries (mor. nig.) begin to ripen.
        15 Elder (sambu. canad.) beginning to flower.
             Jamestown weed (dat. stram.) beginning to flower.
        16 Flax beginning to flower.
        20 Raspberries (rub. idaeus) ripe (plentiful.)
        24 Cherries (prun. cer.) ripe (plentiful.)
             Timothy (phl. prat.) harvest begun.
        26 Mullein (verbas. thap.) in flower.
July   4 Rye fit to reap (good crop.)
         7 Poke (phytol. decand.) beginning to flower.
        12 Althaea in flower.
             Blackberries ripe.
             Wheat (triticum) fit to reap (generally good.)
        23 Indian corn beginning to flower.
        24 Oats fit to reap (heavy crop.)
        28 Indian corn in full flower.
August 1 Eupatorium perfoliatum beginning to flower.
         4 Unripe Indian corn in market.
         9 Early peaches ripe.
        14 Aesculus flava beginning to defoliate.
Sept.   4 Wild pigeons beginning to arrive from the north.
        13 Lombardy poplar beginning to defoliate.
        20 Wild pigeons numerous.
        26 Woods variegated.
        29 Wild geese arrive.
Octr. 20 Woods highly variegated.
             But few trees have yet entirely lost their leaves.
        30 Indian corn ripe (great crops.)

   *It is highly probable that the flowering of this beautiful tree, the Robinia Pseudacacia of Linnaeus, indicates the proper time for planting that important vegetable the Indian corn. For several successive years I have observed our farmers generally, to plant the corn during some stage of its flowering. This is from the 10th to the 20th of May.

Nov. 5 Black locust, apple tree, and cherry trees defoliating.
   9 Woods almost leafless.
  22 Weeping willow leaves killed by the frost.

IV. CONDITION OF THE TOWN

The well established fact, that customs, manners and habits exert a decided modifying influence on diseases, renders it necessary, before proceeding any farther, to exhibit a concise statement, the items of which, abstractly considered, are insignificant, but taken aggregately, appear to be of too much moment to be omitted.

CINCINNATI was laid out in 1789. The first emigration was in the preceeding year. About two-thirds of the houses are in the 'Bottom,' the rest on the 'Hill.' It is in squares of 396 feet. The streets, except Broadway (which is an hundred) are 66 feet wide. They intersect each other at right angles, and the meridional vary 17 deg. W. from N. This cannot be considered so favorable to ventilation as an eastern variation, for our prevalent winds are in a line running from S.W. to N.E. None of the streets are paved. Alleys are not numerous. There is no permanent common, except an inconsiderable one between Front-street and the river.—Along some of our side walks trees are planted, but they are not sufficiently numerous. The absurd clamor against the caterpillar of the Lombardy poplar, caused many trees of that species to be cut down; and at present the white flowering locust very justly attracts most attention: it should be cultivated still more generally.

The number of dwelling houses is about 360. They are chiefly built of brick and wood: a few are of stone. Scarcely any are so constructed as to afford habitations for families beneath the surface of the ground; and not many are built with porches.

The town contains two cemeteries. One is for the interment of the deceased of all denominations. It lies between Fourth and Fifth streets, nearly in the centre of the hill population. It has been a common receptacle for the town, for strangers and for the troops in Fort Washington, previous to the erasement of that garrison, since the first settlements here. Its area is something less than half a square. The other place of sepulture is designed for the use of the Methodist society. It was established about five years ago, in the N.E. quarter of the town, on the hill.

There are eight brick yards. They lie in the western part of the Bottom, near the second bank, which is the lowest portion of the site of the town. They abound in pools, the water of which has been drained from almost every part of the town.

The shambles of our butchers are fixed on the bank of Deer creek, to the N. and N.E. of the town. The tanneries are in the same direction.

The population of Cincinnati and its suburbs is 2320 souls. Of which number 1227 are males, 1013 females, and 80 are negroes. The number of children under 16 years is 1051. The number of persons over 45 years is 184. The number who have attained to the Scriptural limit of human life, three score and ten, is not known; but as men who have passed 60 years of age, do not often emigrate to new and distant countries, instances of great longevity are not to be expected here. Indeed from the recent settlement of this place, few or none of its adult inhabitants are its natives. They have emigrated from every state in the union, and from most of the countries in the west of Europe; more especially Ireland, England, Germany and Scotland. The American emigrants have been supplied principally by the states north of Virginia.

A population derived from such distant sources, and so recently brought together, must necessarily exhibit much *physical,* as well as moral diversity. The climate and soil have not yet introduced an uniform constitution of body; nor customs, manners and laws an uniform moral character. The inhabitants are generally laborious. By far the greatest number are mechanics. The rest are chiefly merchants, professional men, and teachers. Wealth is distributed more after the manner of the northern, than southern states; and few or none are so independent, as to live without engaging in some kind of business.

A great portion of the inhabitants are temperate. There are not a few, however, who daily but quietly become intoxicated, and no *very* inconsiderable number have been known to fall victims to that habit. Whiskey is in universal, but not exclusive use, among the intemperate: beer and cider are generally drunk by those of more sobriety. Well water is generally drunk in the summer; and used otherwise by a few, throughout the whole year. But the water of the river drawn up in barrels, is employed for all domestic purposes by far the greatest number, and is drunk throughout half the year by at least half the inhabitants.

The use of tobacco, among the male sex, is much too general. It is not confined to those who might derive benefit or comfort from it, but extends, with the usual number of exceptions, to all ages, from ten years old, upwards.

The diet of the inhabitants is similar to that of the people of the other middle, and eastern states. Green tea and coffee are in general and extensive use. Fresh meats are eaten in great quantities. Beef, more especially in the summer and autumn, is used to the exclusion of most other meats, in a great many families. The market is well supplied with culinary vegetables. Fermented wheat bread is in very general use. It is commonly eaten fresh, but *hot* bread is much seldomer served up here, than in the southern states. Indian corn bread is by no means uncommon. Rye is almost unknown as an article of food. Fish are not a principal article of diet, though the river affords many.

The dress of our inhabitants is similar to that of the other inhabitants of the middle states. The females injure their health by dressing too thin, and both sexes by not accommodating the quantity of clothing to the changes of the weather. The amusments of balls and other evening parties, so destructive to female health in all parts of the United States, are engaged in here, but not to remarkable excess.

No natural or artificial mineral waters are used here in the summer; nor are there any artificial baths. Bathing in the river is practiced by some, but is less regular and general than it ought to be.

V. DISEASES

Having, in the preceeding sections, taken a cursory view of the physical condition of Cincinnati and its vicinity, we are now prepared for a few enquiries respecting the diseases of its inhabitants. These enquiries, however, will be limited to the fulfillment of a promise, incautiously made, upon distributing the previous sections of these memoranda; before the magnitude of such a work as the Medical History of a new region was fully appreciated. Nothing more, therefore, will be attempted, than briefly to indicate the principal endemic diseases, and their supposed sources.

MIASMATA

From the topographical survey in the first section of these Notices, it will be readily seen, that Cincinnati is not *naturally* obnoxious to many sources of MARSH MIASMATA. The river beach opposite the town, is narrow, and, neither it, nor the bank exhibits much decomposable matter. The lower and back part of the bottom, afford some portions of ground, that are yearly overflown by the spring rains; but they might be easily drained, and therefore may be ranked with the artificial causes, which may be always removed. It is to the inundated interval lands about the mouth of Mill-creek, that we are to look for the most prolific source of vegetable miasmata. This miasmata, however, affects the town much less than might be supposed, from the following causes:
1. The drowned lands lie so much to the N.W. that through the summer and autumn the town is but seldom to the leeward of them; the prevailing winds then being from the S.W. During the present autumn, when few or no cases of ague and fever existed in town, a great number of the inhabitants, to the leeward of those grounds, experienced that disease. 2. Of that tract, a large

proportion is covered with trees. It should have been left, as nature prepared it for us, *entirely* covered. Where a tract of wet ground can be rendered *permanently* dry, it should be cleared and cultivated; but when it is subject to annual inundation, the case is different. The more completely the rays of the sun are then intercepted, the lower will be the temperature of the earth's surface, and the less the quantity of noxious gas evolved. 3. Between those intervals and the town, grows a forest of tall trees. There are strong reasons for believing, that the poisonous exhalation from marshes is hydrocarbonate.* Now this substance is readily decomposed by vegetables.† But whether the gas evolved, be a hydrocarbonate, or according to our very ingenious countryman, Professor Mitchell, an oxyd of septon,‡ (which latter substance, however, it has not, I believe, been proven, is decomposed by vegetables) the efficiency of trees in intercepting its progress, and destroying its virulence, is established by numerous authorities.§ This forest should, therefore, be considered in the light of a rampart against a perpetual enemy, and preserved in the most sacred manner.

The artificial sources of miasmata, are not more numerous than the natural, but they are much more operative. The back part of the bottom, throughout its whole length, is a 'hot-bed' of animal and vegetable putrefaction. In some places, it is true, the ground has been raised (not with any regard to health, but to render it cultivable) those parts, however, make much the smallest proportion. The eastern end of this slip of low ground is a broad shallow canal, which conveys the water that falls on the site of the town, saturated with nuisances, to the pits of the brick yards; from whence neither it, nor the putrescent load can escape, except in the form of exhalation or gas. For its escape in this manner the heat of our summer sun, increased by the reflection from the contiguous high bank, is amply sufficient. Upon learning this state of things, observing and reflecting men, who have been accustomed to trace the acknowledged connection between endemic fevers and the spontaneous decomposition of animal and vegetable matter, would not hesitate to pronounce, a priori—that our principal febrile diseases, and more especially the typhous affections that have, as will be stated hereafter, scourged us for a twelve month past, are most probably owing to the exhalations here spoken of. But to proceed cautiously, and avoid all possibility of error in our conclusions, it will be well to take some additional views.

Upon the settlement of this town, fevers of the typhous kind were not uncommon. They arose, as in all newly settled tracts, from the putrefaction

---

* See Chisholm on Fever, vol. 1.
† See Fourcroy's System of Chem. Knowl.
‡ See Medical Repository.
§ See the writings of Rush, Jackson and Barnwell.

which followed the destruction of the forest and exposure to the rays of the sun of a moist fertile surface. As this was a transient cause, the effect was not permanent, and a period succeeded, which was comparatively healthy. But this state of things was not very durable. The flood of emigration to this place, which commenced in 1805, required such a rapid increase of houses, and consequently of bricks, that in less than three years, the number of brick yards, which previous to 1805 did not exceed two or three, was augmented to eight. The accumulation of filth in those pits which were first dug, had been constantly going forward, so that the quantity of exhalation in 1809 and 10 may be estimated at more than ten times as much as it was seven years before. Now it is notorious, that during those years there occurred more malignant cases, of those diseases which are generally, but improperly termed putrid, than had presented themselves for the seven, or even ten preceding years. Further—these typhous affections prevailed most in December, 1809, but during that month not a single case presented itself east of Main-street, which nearly bisects the town. In the course of the ensuing year cases occurred in the other half of the town, more especially in the eastern end, which is to the leeward of a shallow pond, that has been a common receptacle of filth for more than ten years. The western parts, however, have still been more sickly than any others. Again—December, 1809, was a warm moist month, with southerly winds; and there was not only more sickness during that month than any other, but it occurred chiefly to the leeward of the ponds. Towards the close of January it became so cold that the mercury in Fahrenheit's thermometer sunk 7 deg. below 0, and not a single case of typhus occurred in the practice of either Dr. Allison, Dr. Sellman, or myself, for a month afterwards. February was mild, and in the beginning of March, the disease returned. It became more healthy in April and the first half of May, but the latter half of that month was intensely hot, and new cases immediately followed; some of them exhibiting symptoms of great malignity. The rest of the year was temperate, even cool, and cases of the same disease have now and then presented themselves. Thus we see, that those inhabitants contiguous to, and to the leeward of the alledged sources of disease, have been its greatest victims, and that its appearance and disappearance have been considerably influenced by those states of the atmosphere, which were capable of affecting the progress of putrefaction. It is not believed, however, that this is the *sole* cause, that has operated in these cases. Sydenham, more than a century ago, unfolded the existence of sickly states, or constitutions of the atmosphere, during which all the acute diseases that occurred, appeared to partake of certain characteristics in common. Professor Rush, and some other American writers, have, with equal precision and greater science, pointed out the existence of such constitutions in the United States. An atmospheric temperament of this kind appears to have existed in this part of the country for some time past; its tendency seems to have been to favor the production of typhous diseases. This temperament is a predisposing cause. The exciting cause is the miasm or noxious exhalation of which we have been speaking; and wherever such an agent exists, whether in town, or in the adjoining country, these diseases may be produced.

From these tedious but necessary details, it is thought that the opinion of the insalubrity of those ponds is sufficiently corroborated; and it only remains to suggest the means of removing such a potent cause of disease. This is easily done. The gravel, sand and pebbles of the adjoining second bank, form a cheap, convenient and proper material for filling up the pits, except such as are necessary to furnish water for the manufacture of bricks; and it is earnestly hoped that such an important object will no longer be neglected.

Of our cemeteries, it may be remarked, that the one attached to the methodist church, from the limited number that are interred in it, will not very soon evolve much miasmata, and what it may ever produce is too much to the leeward of the town to be a general injury; but the case is different with that of the Presbyterian church. Whenever the population about *it* becomes dense enough to prevent a free circulation of air, and the interments have become double or treble what they now are, its exhalations must inevitably produce disease. No time, therefore, should be lost, in fixing on a new field for sepulture, without the pale of population, whither the contents of the present should be removed.

The shambles of our butchers, and the tanneries, if they be sources of miasmata, are injurious only when the N.E. wind prevails. At present they have no perceptible agency in the production of our endemics.

These appear to comprehend all the sources of koino-miasmata, and it only remains in this part of the subject, to notice two or three cases of the production of idio-miasmata. Typhus fever has been observed, here, as in other places, to be produced by a domestic cause; for the generation of which, want of cleanliness and want of free ventilation, seem necessary. The latter however, probably has most efficiency. I have observed these circumstances to exist in healthy parts of the town without producing typhus, so constantly as in the sickly parts; so that the public and domestic causes seem sometimes to co-operate. In one instance of this kind, where a large family lodged in a close room of an old wooden house, which stood in the western part of the Bottom, one or two cases of typhus mitior, and two cases of malignant and fatal typhus gravior occurred, cotemporaneously, during a warm winter month. Means were employed to effect free ventilation, and no new cases appeared. It is from such instances as these, that the opinion, that typhus is infectious, has arisen. I can assert from observation, that it is not. I have never seen it extend to more than one or two, in a house that was clean, well ventilated, and its inhabitants were lodged in seperate apartments. But it is unnecessary to urge facts against an hypothesis that is already exploded.

If the constant use of fresh beef, and other unsalted meats, in the summer and autumn, be a cause of disease, it must be noticed in this place. But it has not appeared from observation, that they have had much agency in producing the intestinal affections which have prevailed here. Vegetable aliment may produce the exciting cause of that kind of headach which depends on the presence of acetous acid in the stomach, provided that organ be previously debilitated; and if the stomach and intensines be in a state of debility, fresh meats may suffer spontaneous decomposition, the oxyd of septon be generated, and all the varieties of intestinal disease produced. In this way, during the debilitating influence of a koino-miasmata atmosphere, the animal fibre received into the alimentary canal may be chemically decomposed, and produce a disease of the dysenteric kind, which, without such an exciting cause, might have been a fever. In those cases, where a large number of persons have suddenly had dysentery, induced by eating fresh beef, it probably at first acted in a manner similar to that of any other article of diet, to which the stomach and bowels had not been habituated: it excited simple diarrhoea, this debilitated the digestive organs, the production of septous oxyd ensued, and the phenomena of dysentery followed.

VARIATIONS OF ATMOSPHERIC TEMPERATURE

Neither the cold, nor heat, of the climate of this country, appears to produce many diseases. The former is sometimes so great as to freeze the extremities of those who are exposed; but death has seldom or never been produced by it. Goitre and scurvy, if they be dependent upon cold in other latitudes, are certainly not among its effects here. The heat of our summers appears also to produce but few diseases. The coup de soleil, or stroke of the sun is unknown; and death from the inordinate use of well water, so common in Philadelphia, from some cause is scarcely known here. Langour and oppression are, however, frequently experienced to a distressing degree, more especially upon the sudden accession of hot weather in May and June. Rashes, or cutaneous efflorescences of various kinds, appear to depend on the heat of the summer. Children are much more liable to them than adults. They are certainly diseases, but need not to be dreaded, as they are unattended with danger, and their presence *may* protect the system from more formidable complaints.

Febricula, or inward fever, and anorexia, are not uncommon in the hotest weather, but they seldom outlive their cause, and do not often render medical assistance necessary.

But if the extremes of temperature separately, be comparatively harmless, at this place, their sudden alternation is a most fruitful source of disease. By those, however, who skilfully accommodate their dresses and domestic fires to these variations, but little bad effect is ever felt. But among the imprudent, the exposed, and those who are predisposed to the diseases excited by this cause, it produces the worst effects. In the spring and autumn, the diurnal variations, which are greater than in summer, tend to excite intermitting and remitting fever, as has been remarked by Professor Rush. But this is among the most inconsiderable effects of this cause. In the muscles and membranes of the extremities, it produces rheumatism; in the face and throat, toothach, pain of the jaw and decay of the teeth, catarrah, tonsilitis, &c. in the thorax, pneumonia, consumption, croup, &c. It moreover frequently co-operates with marsh miasmata, and produces a disease in which the phenomena, and the indications of cure, are considerably different from any disease produced by those causes separately.

Whether the effects of this sudden alternation be always in proportion to its degree is doubtful. I have observed a great variation sometimes to occur without corresponding bad consequences: other states of the atmosphere may possibly modify its effects.

Changes from heat to cold, appear to be more prejudicial than those of the opposite kind. One reason of which, probably, is, that the system relieves itself from the effects of a sudden application of heat, by perspiration, but possesses no such resource in the other case.

The natural tendency of this cause seems to be to produce diseases that are *purely* inflammatory; but the winter of 1809–10 furnished opportunities of observing, what had been remarked before by others, that there is in epidemic constitutions, a kind of omnipotency, as it respects other causes of disease. The pulmonary affections, of that sickly season were few, and bore the lancet indifferently.

MISCELLANEOUS ARTICLES

Fogs are by many considered an active cause of disease. Dr. Jackson, in his Treatise on the Fevers of Jamaica, seems to have put this opinion in its proper light. A fog may be the vehicle of marsh miasmata, but is not of itself deleterious. It is nothing but elevated water, and can produce no effects beyond those of simple moisture. This a priori decision accords with fact, for in this town, those who are most exposed to the fogs, certainly are not more sickly than others. Both fog and dew, however, may be sometimes the *exciting* causes of fever. By conducting off the heat, and lessening *directly* the excitement of the system, they increase the excitability, and thereby augment the efficiency of miasmata. The internal use of river water has by some people been deemed unhealthy. Its degree of saline and aerial impregnation, is certainly much less than that of well water, or even that of spring water; but there does not appear to be any just foundation for the opinion of its insalubrity. It produces, so far as observation can determine, no disease, excepting diarrhoea in those unaccustomed to it, which is nothing more than spring and well water produce on those who have been habituated to the use of river water. The occurrence of that disease, is no proof therefore, of the unhealthiness of any water. In some diseases, however, although the river water be not positively unhealthy, the greater benefit resulting from the use of well water, makes it seem so. These are cases of dyspepsia. In this disease the carbonic acid, the carbonates, and other salts of the well water produce very salutary effects. A lady in this town has repeatedly had all the symptoms of dyspepsia aggravated and palliated, by the alternate use of river and well water.

Before concluding the consideration of the causes of disease, it may not be amiss to observe, that some progress has been made, in the discovery of the cause of the endemic disease, announced in the appendix, to the sections of these Notices, which were printed last spring. The people who live where it prevails, are of opinion, that the milk of the cow is poisoned by some unknown deleterious plant on which the animal feeds. It has not yet been discovered; but the experiments which have been made, and the facts which have been collected, seem almost sufficient to command our full assent.

MIASMATIC DISEASES

In specifying the diseases of this place and its vicinity, it will be proper to commence with those endemics which are ostensibly excited by miasmata. They are the following:—Ague and Fever, Periodical headach, Intermitting and Remitting bilious fever, Typhus mitior and gravior, Cholera morbus, Cholera infantum, Diarrhoea, Dysentery, Jaundice and Ophthalmia.

AGUE AND FEVER. A tertian ague has been considered the simplest form of fever; and if unity of cause, greater regularity in the trains of diseased action, and more uniformity in the disorders consequent upon those trains, entitle any febrile affection to a character of greater simplicity than the rest, it certainly belongs to this disease. Its legitimate cause appears to be generated by the decomposition of vegetable matter alone; its empire in the system is more limited than that of most fevers; and the same consequence, dropsical effusion, more constantly results from it, when protracted, than almost any consequence from any other disease. In a series of notices, therefore, respecting our endemics, this disease constitutes the most proper commencement.

In the adjoining state, Kentucky, the thirsty calcareous ridges and dry narrow valleys are unfavorable to the production of ague and fever, and it is but seldom felt, except in the vicinity of some of the larger streams. But in this state, especially in the central, northern and western parts, a leveler surface, with a diminished quantity of calcareous and an increased proportion of argillaceous matter, admits of a more frequent production of this disease. Even here, however, it is rarely fatal; and except in a few situations, its prevalence or malignity has never rendered it a serious evil, nor retarded in any perceptible degree, the current of emigration.

Concerning its symptoms, but little need be said. It generally assumes the quotidian type; sometimes the tertian, and more rarely the double tertian, or quartan. When left to itself, it commonly produces hepatic affections of a mild kind, with ascitic or anasarcous effusion; but under the ordinary treatment, it seldom proves obstinate, except where its remote cause continues to act. In such cases, when the removal of the patient has not been attended to, it has sometimes resisted the combined action of the most powerful remedies, and proved fatal. Emetics, cathartics, and the bark, with opiates and gentle diaphoretics, are generally found sufficient. In a case of protracted quotidian, the cold fit of which was so intense as to threaten life, my respectable friend and preceptor, Dr. Goforth, administered 4 oz. of the bark in substance, during a single apyrexia. The patient recovered.

With arsenic, exhibited according to the formula of Professor Barton, I have sometimes succeeded; and during the present autumn (1810) a gentle salivation, as suggested by Professor Rush, effected a cure in two cases, which had obstinately resisted many other remedies. The great tendency in this disease to produce hepatic affections, would seem to point out mercury as a principal medicine, in long continued cases.

PERIODICAL HEAD-ACH. As it is deemed correct to range with the ague almost any disease that has diurnal paroxysms, the "sun-pain" or periodical head-ach may be introduced here. In its most regular form, it consists of a pain in the lower part of the os frontis on one side, near the orbit of the eye, commencing early in the morning, and continuing through a part or the whole of the day. But these symptoms are not constant. There does not appear to be any inflamation in the pained part, and the arterial action is generally defective. The stomach and bowels are commonly overloaded with bilious matter. From observations at this place, it prevails more in winter than summer. It is generally sporadic; but in the winter of 1803-4 so many were affected with it, as to entitle it to the appellation of an epidemic. Antispasmodics are absolutely inadequate to the cure, as are also sinapisms and blisters. The latter, however, are a good auxiliary. Evacuations from the stomach and bowels with the subsequent use of the bark, as in ague and fever, are the most certain, and generally the only remedies necessary. I have never known it prove fatal. It does not appear to affect those of any age, sex, or condition, exclusively.

INTERMITTING AND REMITTING FEVERS. To make room for the anomalous affection of which we have just spoken, the higher grades of bilious fever have been arbitrarily separated from the ague, of which they are merely extended and more intense degrees—augmented effects of the same cause. The assertion, which was first made by that illustrious pathologist, Professor Rush, is amply supported by the phenomena which these diseases have exhibited at this place. From the simplest 'shaking ague,' with a febrile paroxysm of two or three hours, to an intense bilious fever with a remission scarcely perceptible, I have observed symptoms of the same kind. In the ague the cold fit is considerable; in what is popularly called the dumb ague, and here denominated intermitting fever, the chilliness is less regular and violent, and in the more ardent remittent, the cold stage is feeble or wholly absent. The danger therefore is generally in an inverse proportion to the intensity of the cold stage. A diminution of both chill and fever is favorable, of the chill alone, unfavorable.

As it is only designed in these Notices to announce some of the principal phenomena of our diseases, a detailed account of the symptoms of these fevers will not be attempted, and the following limited remarks may suffice.

They are invariably attended with an undue excretion of bilious matter. In the present state of pathological science, this excretion is not regarded as the cause of the disease, but it certainly produces some of the secondary symptoms, and aggravates the whole. It also tends to prevent the action of sudorifics, sialagogues and tonics, and I am convinced from experience, notwithstanding the plausible reasonings of that eminent chemist, Professor Mitchell, that it ought to be expelled from the system as early as possible. It has not been proven that the vitiated secretion of the liver contains soda; and if it do, in these cases it will probably be better to alkalize the alimentary canal by some more unexceptionable agent.

The state of the pulse in these affections at this place, has not appeared to vary *very* much. It is commonly full, frequent and tense, but seldom hard or depressed. It has frequently tempted to the use of the lancet, but not always with the anticipated benefit. Indeed our bilious fevers in most cases, although apparently of an inflammatory character, do not admit of copius venesection.

For some time past at least, the tendency to typhus has been so great, that the lancet has been almost wholly laid aside. Every autumn is not however alike in this respect, and these diseases have occasionally been presented in a form that unequivocally indicated, and really required extensive bloodletting.

But venesection in these complaints is the only evacuant that is not uniformly beneficial. Emetics and cathartics, diaphoretics and sudorifics, diuretics and sialagogues are all of great consequence. The two first are indispensable. Emetics however cannot be safely employed where the degree of inflammatory action is great; but it appears to me that in the reformed practice of medicine in the United States they are by many physicians, too much neglected. I have repeatedly observed cathartics to fail evacuating the stomach, and in bilious fevers of the milder kind, one or two emetics would probably always be beneficial. The employment of this medicine, however, will not remove the necessity for cathartics; and in all cases they should be administered, and generally repeated till the discharges exhibit a healthier aspect. This is the method which has usually been pursued here, and with satisfactory advantage. It has not, however, always been possible to procure good discharges, even where medicines, to supercede morbid action, have been employed at the same time. Such cases have usually proved fatal. The choice of cathartics has not been deemed a matter of great moment, provided calomel be not omitted. From the disordered state of the biliary system in these diseases, that medicine seems to be peculiarly required. It is also required as a sialagogue; and when it can be made to produce a good salivation at the same time that it evacuates the bowels, it does all that can be expected from medicine—it invariably cures the patient.

Sudorifics and diaphoretics have been employed in these fevers, after due evacuation from the stomach and bowels, and from the blood vessels in some cases, with manifest advantage. And diuretics have been frequently found serviceable. It appears to me that sal nitre, which in the quantity of a scruple or half a drachm every hour, is no contemptible remedy in the milder bilious fevers, produces its good effects chiefly by operating as a diuretic. After sufficient evacuation and reduction of the tone of the system, the combination of opium with this salt forms a valuable sudorific and anodyne.

Blisters have been employed in these affections with the usual benefit.

Cases of bilious fever have occasionally presented themselves, in which the bark could not be taken even during convalescence; but in most instances, after due evacuation, that medicine has been found beneficial. In general, the probability of its being serviceable, is in proportion to the violence of the cold stage. In some cases, where neither the chill nor fever was considerable, I have seen cream of tartar and the bark combined, given throughout the whole twenty-four hours with evident advantage. But these cases should be properly referred to the ague and fever.

Having found but little good effect from nitric acid in other diseases, I have never tried it in bilious fever.

TYPHUS MITIOR AND GRAVIOR. These diseases seem to bear the same relation to each other that is observable in intermitting and remitting fever. They are also in this country closely connected with those affections, and furnish a good proof of the correctness of that pathological idea, which questions the doctrine of diagnostics. The difference between a case of inflammatory remittent, and one of typhus gravior, is indeed very manifest, but these are to be regarded as the extremes: many of the milder cases are so complicated, that the pathognomonic symptoms of neither disease appear to predominate.

The more characterized cases of these typhus affections frequently exhibit nearly the same derangement of the biliary system with the fevers already noticed. But they are attended with many phenomena not common in those simpler affections, such as inactivity of the functions of the brain, oppression of the thorax, and the exhaustion of the muscular energy. They are also generally accompanied with diarrhoea, in which the discharges are constantly vitiated; and almost invariably with complete anorexia. In the more violent cases the pulse is small, intermitting and frequent, and the pains and anxiety of the thorax and abdomen are very great. In milder cases the pulse is fuller, but always frequent, and the restlessness gives way to profound stupor. The tongue is generally dry, and sometimes covered with a dark colored hard crust that appears cracked into fissures. In two cases there occurred an eruption of pimples, which in a few hours became filled with pus. They both proved fatal. Concerning their other phenomena, the limits of this work will not admit of any detail.

As these diseases consist in a more extended series of morbid actions, than those we have before considered, they are of much more difficult management, and have not unfrequently proven fatal at this place, during the last two years. Before that time they occurred more seldom. Their cure has been attempted nearly in the same way with that of the bilious fever, except the early administration of tonics and stimulants, and the total omission of venesection.

These medicines, with mercury and cold water, would probably in most cases effect a cure, could they be retained in the system; but the tendency to diarrhoea has generally been so great as to preclude the copious exhibition of sudorifics or of mercury, and require the constant use of astringents, demulcents, and alkalies. Of the former class of medicines, saccharum saturni and geranium root (*geranium maculatum*) have been employed with most advantage. Of the latter, the alkaline earth, magnesia alba, has been commonly preferred. When mercury has not been employed, these complaints have generally had their full course, the typhus gravior a shorter, the typhus mitior a longer one. The ordinary remedies in many cases appear to have saved life, but not cured the disease. Mercury, however, has done both. In the few instances in which a genuine salivating effect has resulted, the disease has yielded and the patient recovered, some cases have occurred in which mercury ulcerated the mouth without producing ptyalism, and then it did but little service.

Blisters have been a constant remedy in these typhus affections. Much advantage has frequently resulted from them; but it has been considerably diminished, by the strong tendency to gangrene which the blistered places, in a great number of instances, have shown. Among the effects of blistering in a case of typhus mitior, may be mentioned the total suspension of a copious ptyalism attended with sore mouth, for two days, and its return upon the cessation of inflammation in the blister.

A *local* application of cold water has been frequently made, with obvious advantage. But a general affusion, as recommended and practised by several ingenious physicians of the present day, has never been resorted to here. Many cases of our mixed fevers, appear to be very analagous to those in which Dr. Jackson found the cold affusion so beneficial; but at this place medical intrepidity has heretofore yielded to the invincible prejudices of the people. In May, 1810, I had a case of typhus mitior, in which the patient was exposed, covered with a single sheet, to a constant and copious current of fresh air, except a few hours of the latter part of the night: His recovery, which was unusually rapid, appeared to depend much more on that than on the medicines employed.

CHOLERA MORBUS & CHOLERA INFANTUM. These affections having essential symptoms in common, and probably depending on a similar mode of action of the same cause, may be considered together. Their phenomena, however, are not perfectly identical. The first is generally a disease of adults. When it has appeared at this place, it was attended with inconsiderable fever, but with copious bilious discharges, and ultimately with spasms and cramps. It has usually terminated in health in 24 or 48 hours. The cholera infantum is commonly attended with fever, which is sometimes intense; the discharges are not uniformly bilious; stupor and insanity are apt to supervene; it sometimes terminates in health, or in death, in two or three days; but generally has a protracted course, producing, with great debility, a peculiar, sunken and languid state of the eyes. These two varieties of cholera agree, however, in being apparently excited by an irritating material, exerting a strong impression on the stomach and duodenum.

The former of these diseases is much rarer at this place than the latter. It appears sporadically during the warm part of the year, but has never yet been epidemic. Of its treatment I have nothing to observe, except that in one protracted case, in which the discharges were very bilious, a salivation, induced, principally, by mercurial frictions, suddenly removed all the symptoms.

The Cholera Infantum prevails every summer, in this town and its vicinity, and may be regarded as the principal disease to which our children are liable. As in other parts of the United States, it precedes the other summer and fall endemics, generally beginning in June, and sometimes much earlier. In this disease I have seen calomel in small doses, with, or without opium, according to the state of the pulse, as recommended by Dr. Miller, of more service than any thing else. Cold applications to the abdomen, and head, have also proved very advantageous. I have never tried the cold immersion, as practiced by some physicians. At the same time, that the refrigorating applications are made to the head, sinapisms to the feet have been useful. When the evacuations have been very copious, and the child's strength is very much reduced, calomel and opium, with a milk decoction of the geranium root are invaluable. This complaint, however, has frequently resisted the powers of these and other medicines, and either proved fatal in two or three days; or assumed a protracted form, and yielded to nothing but the frosts of the succeeding autumn.

DIARRHOEA & DYSENTERY. An epidemic diarrhoea, has never been known here. This disease, however, occasionally presents itself throughout the whole summer; and appears like the other endemial affections of the warm season, to depend on miasmata. Its cure has generally been attempted with rhubarb, and other cathartics, followed by alkalies, farinaceous preparations, geranium root, and other astringents; aided, in obstinate cases, by the cold bath, flannel next the skin, and exercise on horse-back.

The Dysentery is a more formidable disease. Every summer and autumn furnish sporadic cases of it, and in 1808 it was epidemic. In the month of July, of that year, it was more prevalent, than any disease has ever been at this place, except the influenza. Fevers, during its predominance, were not observed to occur, and the simple diarrhoea and cholera infantum, which appeared cotemporary with it, did not long preserve their pathognomonic characters. Notwithstanding this power of banishing, or assimilating to itself other diseases, this epidemic was mild, and proved fatal in but few instances. It was not often attended with fever, and the appetite, generally, was unimpaired. The morbid cause appeared to exert a very limited power on the system, mucous and sanguinious discharges, with gripings, constituting the principal symptoms. Large portions of Ol. ricin. alternated with opium, or opium and ipecac, were chiefly relied upon; and when aided by amylaceous and glutinous preparations, were generally sufficient. When astringents were required, the geranium root was employed with success. The carbonates of potash and magnesia were exhibited in several cases, but not with very marked advantage. In the dysenteries of some parts of this country, however, they have been found more efficaceous. My friend, Dr. Canby, has employed them along with the usual remedies, with a success, as honorable to himself and his profession, as to the respectable Professor, who first pointed out their modus operandi in this disease, and insisted on their exhibition. The dysentery of 1808 was so mild, that calomel was scarcely resorted to. It had been epidemic previous to that year, but has not been since.

JAUNDICE. This is one of our endemics, but is seldom very prevalent. Throughout the whole of the year 1808, cases of it presented themselves more frequently, than before, or since. It was generally attended with a dull pain in the pit of the stomach. I heard of its proving fatal in one case, in the vicinity of this town. In one instance it was connected with a slight eruption, and violent itching in the skin, attended with a synocha pulse, and required bloodletting. Generally the pulse was weak, and the whole system appeared to partake of the inaction of the alimentary canal. It affected adults more than children.

As a remedy for this disease, the puccoon root (*sanguinaria canadensis*) has been recommended by Dr. Schoepf.* The people in this part of the country employ a tincture of it, for the same purpose, and from experience I can declare it almost a specific. But I prefer giving it in substance.

OPHTHALMIA. On the arrangement of this affection among the miasmatic diseases, it is by no means intended to insist. The following are the reasons for which it was referred to that head, and physicians can estimate them, as they deserve. 1, The ophthalmia is an endemial disease of this country, which like our other endemics, appears sporadically every summer, and occasionally becomes epidemic, affecting great numbers, especially children. 2, It occurs as much, if not more, along our water courses, and in the depths of forests, as on open plains or uplands; and therefore neither dust, not reflected light, has any agency in its production. 3, When epidemic, it appears and declines about the same time, with our other summer miasmatic diseases. 4, It has been prevalent before the annual burning of the woods, which invariably takes place in some parts of this country, and therefore is not occasioned by smoke. 5, In the summer of 1807, I was assured of two cases, in which this disease alternated

---

* See Barton's Collections, part 1.

with cholera infantum; the ophthalmia prevailing at night and the cholera infantum in the day. Similar cases have been mentioned to me by Dr. J. Canby. 6, This disease has diurnal exascerbations. It is generally worst at night, even where the eyes have not been exposed to the light. In one case, the subject of which (a man of veracity and observation) communicated the account to Dr. Este, of Hamilton, it assumed a tertian type. During the paroxism, which had about the length of a common fit of fever and ague, light and every exertion of the eye were intolerable; but during the intermission, he was entirely free from those morbid sensibilities. The same physician has also lately met with a case in his practice at that place, in which ophthalmia had true tertain paroxisms. 7, Topical applications are seldom adequate to the cure, and means calculated to operate on the general system, must be resorted to, in all violent cases. 8, It is somewhat difficult to conceive, how, either directly, or through the medium of the general system, the action of miasmata can be concentrated in the eye; but there does not appear to be in it, any physical impossibility.

Of the local remedies in this disease, I have generally seen the stimulating, the most beneficial. Cold water seldom gives permanent, and frequently not momentary relief. I had lain it aside before reading the experiments of Dr. Wilson, which prove inflammation to consist in defective, instead of excessive action. Of the general remedies, blood letting and purging are frequently necessary. They reduce the action of the system, at large, when necessary, and prepare it for the exhibition of opium and sudorifics. The former is necessary in all obstinate cases. I have seen from two to six grains given during a single night, with obvious and permanent advantage. In protracted cases, a salivation would probably be of great service. In a case of several months standing, in which the eyes were covered with films, to such a degree, as to produce total blindness in one, and very impaired vision in the other, after various collyria, blistering, repeated cupping, sternutatories, cathartics, opium and tonics (the pulse being weak) were employed for several weeks, with inconsiderable advantage; a salivation suddenly removed most of the inflammation, and promoted the absorption of the films, so far as to restore one eye entirely, and render colors perfectly distinguishable by the other.

DISEASES CONNECTED WITH VARIATIONS OF ATMOSPHERIC TEMPERATURE

The diseases comprehended under this head, are not *exclusively* produced by changes in the temperature of the atmosphere; but this cause so frequently excites them, that they may with propriety be referred to it. The principal ones which have been observed to occur here, are Catarrh, Consumption, Pleurisy, Peripneumony, Rheumatism, and Tooth-ach.

CATARRH. This is the most ordinary and simple effect of the above cause. It does not appear to be more frequent or obstinate here than in other parts of the United States. The schneiderian membrane appears to be first affected in most cases of this disease. From thence the morbid action extends to the pharynx and larynx, and the pulmonary affection follows. In children, this disease is sometimes attended with such symptoms, that it can scarcely be distinguished from the genuine croup, except by the facility with which it yields to medicine. The common catarrh, upon a reference to its cause, appears to be essentially different from the influenza, and should probably be always regarded, in the language of Dr. Sydenham, as an intercurrent disease; yet it sometimes becomes almost as prevalent as the epidemic just spoken of; and there is some foundation for believing that in the causes of the two diseases there is an intimate connexion. But as one of them results even from a trifling variation in the state of the circumambient caloric, and the other traverses whole continents, uninfluenced by any changes of that kind, it is difficult to perceive in what the connexion consists.

The catarrah is frequently a harrassing and protracted disorder, but is formidable, chiefly, as an exciting cause of consumption, of which we will now proceed to speak.

PULMONARY CONSUMPTION. From Dr. Spalding's bills of mortality, it appears that in Portsmouth, New Hampshire, a fifth of the deaths are from this disease. In Philadelphia it carries off between a fifth and a sixth. In this town, from several years observation, I am confident that a tenth or twelfth of our deaths from consumption, is a liberal estimation. So that if we make due allowance for the skilfuler treatment of this deplorable malady by our more enlightened fathers and brethren of the maritime cities, we may conclude that consumption occurs nearly three times as often in those places as in this town. It has, however, been a more frequent and fatal disease since the influenza of 1807, than before. Its subjects are generally women, between the ages of 15 and 30 years.

I have not had the satisfaction of seeing this disease cured by a salivation. In several cases mercury, in conjunction with the usual auxiliaries, has been exhibited to such an extent as to produce ptyalism for several weeks: It has appeared to mitigate, but in no instance, whatever, to remove the disease. From digitalis no greater benefit has been derived. In cases of legitimate phthisis, its exhibition has been continued unceasingly for several months; and it has sometimes moderated the pulse, but never superseded the cough or hectic fever. In two instances the vegetable alkali was given for many weeks in large quantities; but no advantage resulted. Of the efficacy of those nearly obsolete remedies, carbonated hydrogen gas and azotic gas, mixed with atmospheric air, I can say but little from experience. But in the vicinity of this town, nearly a whole family has been swept off by consumption, while living in a situation, the atmosphere of which, must have abounded, at least, with the former of these gases; and in 1808, a phthisical patient was put under my care from an aguish part of the country, whose hectic fever was preceded, every other day, or every third day, by a chill and shake, so violent, that her friends supposed her to have the ague. The two diseases, indeed, appeared to be combined.

PLEURISY. This disease was more prevalent here previous to the visitation of the influenza, than since. It has seldom presented itself in such a shape as prohibited the use of the lancet. It is almost invariably attended with a preternatural excretion of bile, and not unfrequently with a very obvious degree of hepatic affection. Bleeding, blistering, and the common antiphlogistic regimen are inadequate to the cure in such compound cases, and a liberal use of mercury must be resorted to. It has been given so as to evacuate the bowels freely, and also to excite a ptyalism as early as possible. Upon the accession of that effect, that symptoms have almost invariably yielded. Mercury, indeed, is wholly indispensable in these bilious pleurisies, and when combined with the ordinary antiphlogistic treatment, is seldom unsuccessful.

PERIPNEUMONY. That singular epidemic the influenza, whilst it diminished, at this place, as has just been stated, the number of cases of pneumonia pleuritis, seems to have invited a more frequent occurrence of the pneumonia peripneumonia; for since the autumn of 1807, the latter disease has been much more common, than previous to that period. Its most conspicuous phenomena are, a frequent elastic pulse, cough, obtuse pain in some part of the thorax, or the total absence of all pain in that region; frequent and difficult, but not painful respiration, and inability to lie with the head and shoulders level with the body. In one case which terminated in vomica, not the slightest pain was at any time felt above the diaphragm; but there was a constant pain in the lower part of the left hypochondrium, attended with vitiated alvine discharges. In this complaint, there is not, as in the pleurisy, any crisis on the 5th, 7th, or 9th day, but it continues until, probably from congestion or disorganization of the lungs, it terminates in death, at no specific period; in vomica; or in health, from the successful exhibition of medicine. It is, like the pleurisy, occasionally attended with derangement of the biliary system.

In the treatment of this complaint, blood-letting and the ordinary antiphlogistics, are indispensable; but it is seldom possible to reduce the morbid force and frequency of the pulse by them alone. From the progress and termination of several cases, it is rendered probable, that mercury and digitalis are the most efficient medicines that can be superadded to the common debilitating means. The first of those active substances should be given so as to produce a ptyalism, which in part effects the reduction of the pulse, and appears to prepare the system for the reception of the second. In the administration of the digitalis, a constant regard should be had to its effect on the pulse. If it do not produce a slow, pausing pulse, it is of but little advantage. During the convalescence from this disease, I have felt the pulse of a young adult, at 52 and 54 or 56 strokes in a minute, with very remarkable intermissions: when it was in that state she felt active and comfortable; when a relaxation in the exhibition of the medicine permitted the pulse to rise to 70 or 80, dyspnoea, & oppression at the breast rendered it difficult for her to lie down, or to make any considerable exertion. A salivation preceeded its use in this instance, and indeed in almost every case of peripneumony, in which it has appeared to be serviceable. The following case will in part confirm this, and may be somewhat interesting in other respects.

W. W. aged 26 years, with a flat chest, and distant shoulders, was seized in July with a severe cough, and inability to lie with his head and shoulders low. After trying the use of some popular remedies for several days, with no good effect, he applied to me. Finding his skin cool, his pulse slow and weak, his thorax entirely free from pain and stricture, and that he had no thirst, and could walk about, I did not at first suspect the existence of inflammation. An emetic and cathartic, with the subsequent use of anodynes, and a plaster of Burgundy pitch, were employed without any advantage whatever. In three or four days he was unable to lie down at all. His exemption from pain, and weak pulse continued; but it was determined to bleed him.—About eight ounces were taken, which exhibited some slight traces of buff. A blister was then applied to his side. His pulse did not rise, from bleeding, but as he felt rather better, the next day, it was repeated to the quantity of twelve ounces. The blood drawn this day was more sizy; and after the operation, his pulse rose a little. On the succeeding day he was bled again. The blood exhibited much inflammatory crust, and after the operation his pulse became full, tense and frequent. His cough continuing, the administration of calomel, with squills and nitre, was now commenced. Venesection, to the quantity of fourteen or sixteen ounces, was continued every day, or every other day, from this time for a week, the pulse beating 120 strokes in a minute, with a great degree of energy. The blood was remarkably cupped and sizy. By the expiration of that time, a salivation came on. No considerable reduction of the pulse followed, but he was able to lie with his head and shoulders lower. The use of digitalis was then begun. It was given in substance. In three or four days the expected intermissions in the pulse occurred, and it was soon at 60 and 54 in a minute, having sustained an equal reduction in its force and fullness. The cough soon became more moderate, expectoration increased, and his amendment was unequivocal. The digitalis has been continued ever since (a period of six months) in such quantities as generally to keep his pulse in a state of defective action; he has taken exercise on horse-back, and at this time has as good a prospect of complete restoration, as is consistent with a malformed thorax.

Was the pulse depressed in this case? Is it not more probable that the disease was at first local, and that the arterial system did not sympathize for some time. In the fanciful manner of Dr. Darwin, it might be said, that depletion increased the sensorial power of association, and brought the general system into excessive action, much sooner than it otherwise would have come.

Judging, which, however, is improper, from the event of a few cases, I am not disposed to ascribe much efficacy, in this disease, to the carbonic acid and carbonated hydrogen gases, as recommended by Dr. Withering and Dr. Beddoes.

CROUP. The cynanche trachealis, or hives, is here, as well as in the middle and northern maritime states, one of the principal diseases of children. It prevails more in autumn, winter, and spring, than in summer, and more in some years than others; but it has never assumed that malignant and epidemic character which, according to Dr. Dick, it exhibited at Alexandria in 1799. It is almost invariably attended with fever, and as constantly with a disordered state of the bowels, the alvine excretions being green or blackish.

In one case, only, have I employed blood-letting to any considerable extent. The infant had labored under the disease 16 or 18 hours, but still had a vigorous pulse. The quantity taken was so considerable, as to produce partial deliquim. Many other of the usual remedies were employed, but the patient died. I have seen the violent operation of strong emetics at the commencement of the disease, as recommended by Dr. Rush and others, of great service. In one case that was fully formed, more than a dozen motions were procured by an emetic, in less than an hour; and the little patient began to recover immediately. After the operation of a strong emetic and cathartic, I have found the exhibition of a decoction of Seneca root, as recommended by Dr. Archer, of more benefit than anything else. Unusual quantities of emetic medicines are necessary to produce vomiting in these diseases; and the same observation may be made respecting the Seneca decoction. It should be very strong, and in most cases given in larger quantities than are recommended by Dr. Archer. In one case that was about to terminate fatally, such a free exhibition of this decoction was made, as to dislodge from the glottis great quantities of thick phlegm, tinged with blood. The irritation throughout the whole system was so great, for a few minutes, as almost to produce convulsions, but the urgent croup symptoms were mitigated, the threatened dissolution averted, and the child recovered. Would not the roots of the Sanguinaria Canadensis (which indeed have been employed) the Lobelia siphilitica, and the Jeffersonia binata, produce the same effect? The warm bath and blistering are excellent auxiliaries in this disease; but the first should never be employed until the intensity of the fever is abated by evacuants.

RHEUMATISM. This disease frequently presents itself in this country, but not often in a formidable shape. It appears to result from exposure to vicissitudes of the weather. Now and then it assumes the form of lumbago. In one instance it terminated in white swellings of various parts of the body. Among many other remedies, a protracted salivation, with a subsequent course of the volatile tincture of gum guiac, was employed in this case without any good effect.

In the treatment of the milder cases of rheumatism, the people use the Seneca oil, a bituminous substance brought down the Allegheny river. Concerning the remarkable efficacy of this liquid, in removing the numerous cases of rheumatism and stiffness in joints, in a detachment of troops, here is a note by B. Lincoln, Esq. in the first volume of the American Museum.

The poke (*phytolacca decandra*) and the prickly ash (*zanthoxilum fraxinifolium*) are popular remedies of considerable estimation; but the *actea racemosa*, or squaw root, will probably supersede them. This powerful medicine has received too little attention from physicians. In two instances in this town, in which it was taken to excess, it produced the most violent and alarming effects. One of them I had an opportunity of witnessing. In about an hour after the tincture was taken, by a person able to go about, and of an inflammatory diathesis, violent pain in the epigastric region came on, with vomiting, intense head-ach and delirium. The face was flushed, and the pulse full, frequent and tense. The loss of fourteen or sixteen ounces of blood, followed by a portion of paregoric, and the subsequent use of a cathartic, carried off these disagreeable symptoms. The people no doubt frequently err, by using this medicine when too much inflammation exists.

TOOTH-ACH. Pain of the jaw, decay of the teeth, and tooth-ach, are common here, but by no means so frequent as in some of the states. According to Dr. Hazletine, these diseases constitute an eighth of the morbid affections incident to the inhabitants of the province of Maine.

Dr. Foot, in an ingenious paper, inserted in the Medical Repository, has rendered it highly probable that the undue prevalence of these maladies in the United States, is referrible to the sudden vicissitudes in our climate. The action of septic acid, generated in the mouth, may account for the destruction of the teeth in some particular instances; but it is difficult to believe that the teeth of the inhabitants of a whole country can suffer from that cause. I have seen a fine set of teeth apparently decomposed and very much injured in the course of a year, during which time the person labored under a high degree of dyspepsia, and frequently ejected a very sour liquid. If it be possible for oxalic acid to be generated in a human stomach, it probably was in this case, and in its passage through the mouth effected the decomposition of the teeth.

It frequently happens, that those who have decayed teeth, are seized with pains in the jaws, or some other parts of the face. These pains are often extremely severe. They are not fixed, but attack almost every part of the jaw and sometimes all the teeth in succession; but the most remarkable circumstance is, that the decayed teeth are quite as much and in some cases more exempt from pain than the rest. The immediate exciting cause of this kind of pain, is exposure to cold; but the agency of the decayed teeth appears manifest, from the impossibility, in many cases, of removing the pain without extracting them. In one instance, the pain, after attacking most parts of the face, at length affected the whole anterior part of the head. The use of snuff, however, soon transferred it to the face again, where it obstinately resisted the application of galvanism and many other stimuli: upon extracting two decayed teeth, in which scarcely the slightest pain had been ever felt, the whole disease instantaneously vanished. In another case, the pain of the face was attended with many of the phenomena of hysteria; and likewise appeared to have a periodical type, recurring many times, in the forenoon. The pulse was weak during the paroxism. The bark and volatiles afforded considerable relief. A blister was drawn, on the neck: it moderated the pain of the face, but became affected itself with a most insupportable sensation, which was likened to the action of needles or of animalculae in the flesh, and at the same time it became very much inflamed.

There can be no doubt but that, as Dr. Darwin has asserted, this pain of the membranes which invest the jaws and alveoli, contributes to the destruction of the sound teeth; and therefore those teeth which are already decayed should be extracted as soon as any pain in the face is felt.

EPIDEMIC DISEASES

Concerning these diseases but little will be said. They are, Measles, Mumps, Hooping-Cough, Angina maligna, Scarlatina anginosa, and Influenza.

It is not pretended that these affections can, properly, be grouped together; and by most physicians, the following superficial reasons for this arrangement will be deemed wholly insufficient:—They are seldom or never sporadic, but when they occur, it is almost invariably in an epidemic form; and as they do not appear to have the same origin with our miasmatic endemics, which are either local or general, according to the extent of their causes, they cannot be ranked with that tribe. Thus they have some agreement in cause. In their symptoms, although it have not been generally remarked, there is also some loose analogy. They all affect the throat and lungs chiefly. But in the type of the fever, which accompanies them, the similitude fails, for some are generally attended with synocha, and others as continually with typhus. How they are arranged, however, in this series of memoranda, is a matter of little moment.

MEASLES. This disease has occasionally prevailed here, but not having seen many cases of it, I am unable to say whether its symptoms have exhibited any peculiarities. It does not appear to have proven fatal in any instance at this place.

MUMPS has also prevailed in Cincinnati and the adjoining country several times. In the year 1807, it was more general than at any other period. A few cases of its retrocession then occurred. In one instance of that kind, blood-letting, cathartics, and saturnine applications to the sympathetic tumour, which was large, were of very essential service.

HOOPING COUGH has affected the children of this part of the country more than either of the preceding diseases. For several years, indeed, it appears never to have been extinct in every section of this part of the state. In its symptoms nothing very peculiar has been observed, except that during the state of atmosphere which produces croup, the hooping-cough has sometimes assumed or counterfeited the phenomena of that disease. When medical aid has been applied for, emetics and cathartics, with the subsequent use of expectorants, tonics, and the cold bath, have generally been found sufficient. In one case, blood-letting appeared of decided advantage. A sweetened decoction of colts-foot or wild ginger (*tussillago*) is a popular remedy here, as in other parts of the United States. The precise value of this medicine appears to remain yet to be ascertained.

ANGINA MALIGNA & SCARLATINA ANGINOSA. The former of these diseases had an extensive and fatal prevalence in this country 18 or 20 years ago. It does not appear to have been epidemic since. The latter probably never has prevailed generally here. Within two years, however, there has been at this place a tendency to both these diseases. Within that period, a few bad cases, and a considerable number of very mild ones, more especially of the former disease, have occurred. It would seem as though their cause or causes were or had been among us, but in a state too diluted or unformed, to excite an extensive or fatal epidemic.

In the course of this constitution, cases of sore mouth have been common. Small blisters, ulcers and redness of the tongue, gums and inside of the lips and cheeks, with a burning sensation in those parts, were the usual symptoms. They were probably produced by such a gentle action of the cause of scarlatina, as was conjectured to produce the submaxillary abscesses mentioned by Dr. Rush.

In the treatment of these affections, Dr. Allison, Dr. Sellman and myself pursued nearly the same course. Emetics, calomel, and the bark, with blisters, mercurial frictions, and emollients externally to the throat; and astringent, alkaline, saline and pungent gargarisms, were the remedies generally employed. They were attended with considerable, but not invariable success—a few cases of the angina maligna proving fatal. In the smarting and blistered mouth, Dr. Sellman found magnesia alba a serviceable remedy. He conceived that the contents of the stomach were in a state of morbid acidity, and that the affection of the mouth was thereby aggravated. In a few cases, the stimulating gargle recommended by Dr. Farquhar, was employed, but not with the advantage experienced by him in the West Indies.

For six months past, few or no cases of these diseases have presented themselves.

INFLUENZA. The few remarks which follow, relate entirely to this disease, as it appeared in 1807. Cases of it occurred in town the two or three last days of September; but it was not general before the 5th or 6th of October. On the 29th and 30th of September, a great proportion of a regiment of militia, which was encamped in the open air, about seven miles from town, became affected. The dust and smoke and night air to which they were exposed, probably acted as exciting causes. This disease affected adults chiefly, but not exclusively. In a great number of its subjects, it was so mild as not to require any medicine. Intermittents which were prevalent at the time of its appearance, immediately declined, and it was soon left almost the only disease. During its reign, a sudden diminution of atmospheric temperature effected the supervention of pnemonic inflammation in several persons. These cases as well as many others, were attended with a very redundant secretion and excretion of bilious matter. The pulse was synocha, and they required copious blood-letting, with all the remedies generally employed in bilious pleurisies. In one of these cases, an ague which had left the patient just before the attack of influenza, returned for three successive days: but unequivocal signs of pulmonary inflammation existing, copious blood-letting, blistering and the antiphlogistic regimen were employed. To this treatment the combination yielded, and abortion was prevented, though the period of gestation had more than half elapsed. But these compound cases were not the only ones in which bloodletting was advantageously employed. Whenever the pulse was excessive, the loss of blood was found to afford great relief. Emetics, cathartics, and febrifuges were likewise employed with considerable advantage. After the excess of morbid action had subsided, and in those who experienced no excess, gentle anodynes and stimulants were found of great service. In some persons a troublesome cough continued for many weeks after the other symptoms of the disease had subsided, and did not appear to be much affected by any of the common remedies for that complaint. It proved fatal to few or none. About the first of November, the disease began to occur more seldom, and was entirely gone by the middle of that month. After its disappearance, the town was very healthy.

The influence of this epidemic on the pulmonary diseases which have preceded it, has been already mentioned.

The equinoctial storms of the ensuing spring were unusually violent, and the temperature was low. Immediately after the equinox, a catarrhal affection, but little inferior to the influenza, in its violence and the numbers that were affected with it, appeared in this town and the adjoining country.

MISCELLANEOUS REMARKS

The ITCH (*Psora*) or cutaneous affections nearly resembling it, are remarkably common in many parts of this country. They are ascribed by some to the water, but the cause of their general prevalence does not yet appear to be ascertained. They are treated by the people, with the different mercurial preparations, and in most cases with success; but sometimes they prove very obstinate. Occasionally these affections are in all probability attended with a scorbutic diathesis.

The Lepra Grecorum, now and then presents itself, generating, in its usual manner, immense quantities of large branny scales. I have observed it not to affect, at least in any considerable degree, those parts of the body that are uncovered. Venesection, cathartics, and low diet, with the subsequent and long continued use of mercury or arsenic, will generally effect a cure.

An affection, called by the people, catarrh, or guittar, now and then presents itself. It is a deep seated farunculus of the joint that connects some one of the fingers with its metacarpel bone. It appears oftenest to attack the ring finger. Stimulating poultices and a free vent for the pus, produce a cure very readily.

A few cases of mortified gums, in children, have been observed. The first that presented itself was attended with a depraved state of the primae viae, and a moderate fever, inclining to the typhoid character. A great variety of astringents and stimulants was applied to the sphacelated parts, without arresting the progress of the disease. At length I discovered that the silver spoon, with which the mouth was examined, was tarnished by the action of an acid, which must necessarily have been the septic. I directed a solution of carbonate of potash, which in a short time produced a cure. The same application has been since made, in similar cases, with corresponding success. Since the occurrence of that case, I have read Dr. Harrison's paper on the diseases of Chillicothe, and find that cases of the same disease have occurred at that place, and were treated by him in the same way, with the best effect.

The Goitre is an endemic of some of the N.E. portions of this state, but it is not known here. The Scrophula, Rickets, and Scurvy, are very rare diseases. Canine madness has not been epidemic for many years, and Hydrophobia has not been observed to occur as a symptom of any of our other diseases. Insanity seldom presents itself; but the protean disease, Hysteria, is frequently met with. Hydrocephalus internus but seldom occurs. Tetanus is rarer still. But one case has occurred here for many years. It was produced by a wound in the hand. It proved fatal in less than three days after the spasms came on; probably from their extending to the muscles of the glottis. Cancers occasionally occur with their usual fatality. Calculus, Arthritis and Apoplexy, are rare diseases. Dropsies are more common, but generally appear only as one of the consequences of intermitting fever.

No bill of mortality has yet been kept in this place.

APPENDIX

1. "COLUMBO ROOT"

In different parts of the western country, a bitter root, said to be the officinal columba, has excited considerable attention. In 1805, upon presenting that great botanist, professor Barton, with a specimen of it, he informed me that it was the root of the Frassera Carolinensis of Walter; the Frassera Walteri of Michaux. I have not been so fortunate as to obtain the works of those botanists, but the authority of the professor is unquestionable. The plant therefore is not a nondescript, as has been said. It however does not appear to be recognized in Turton's edition of the Systema Naturae.

It is sometimes found pentandrous, at other times tetrandrous, always monogynous. The stamina are erect, the antherae are deeply cut longitudinally, on both sides, and fixed obliquely. The style is simple, erect, and longer than the corolla, the stigma is somewhat bilobed. The calyx consists, when the plant is tetrandrous of four, when it is pentandrous of five, ovato-lanceolate, permanent leaves. The corolla has 4 or 5 lanceolate petals; with a nectariferous? radiated spot of bristles in the centre of the superior surface of each.* The stem (caulis) is from 2 to 7 feet high, smooth and straight. The cauline leaves are verticilate, glabrous, lanceolate, and from 4 to 8 in number. The radical ones of the same form. The branches are axillary, as numerous as the leaves of the whorl from whence they rise: they incline upwards, and send out opposite, axillary, one-flowered peduncles. The root of the young plant is fusiform and branching, of the old, tuberous and irregular. It is *said* to be triennial: it certainly is not annual. It is chiefly found in thin soils, among oak timber, or in prairies surrounded by that timber. I have never seen it in *fertile woodlands*. It flowers from the 25th of May to the 20th of June.

Every part of the plant is bitter, but the root is generally chosen for medicinal purposes. Dr. J. Canby, and others, inform me, that they have found it, more especially in its recent state, to possess considerable laxative power.

It gives out its bitterness both to aqueous and alcoholic menstrua, but more fully to the latter; the reverse of which is the case with the colomba. Its spiritous tincture also suffers decomposition, upon the addition of water, indicating that it contains resin, which the colomba does not. And the addition of decoction, or alcohol of galls, to its watery and spiritous infusion, caused no precipitate of cinchonin, the principal constituent of colomba.

Hence it appears to be essentially distinct from the substance to whose name it probably owes a portion of its reputation. It is, however, a medicine unquestionably entitled to attention,† and will no doubt be found equal, and possibly superior to most of our other indigenous bitters.

II. NEW DISEASE

In the spring of 1809, Dr. Barbee, of Virginia, on returning from a visit to the Madriver country, in this state, gave me some information concerning a new and formidable disease which had appeared among the settlers of that tract. Since that time, I have been able to collect several additional facts respecting it, from different persons, more especially Mr. William Snodgrass, and Mr. John M'Kag, two intelligent and respectable inhabitants of that country, who have several times, experienced the disease in their persons and families. A summary of the whole, is here given, that physicians may determine how far it deserves the appellation of a new disease.

It almost invariably commences with general weakness and lassitude, which increase in the most gradual manner. About the same time, or soon

---

* Whether this be a specific or generic character, I do not know.
† See Barton's Collections and Journal.

after, a dull pain, or rather soreness, begins to affect the calves of the legs, occasionally extending up to the thighs. The appetite becomes rather impaired, and in some cases nearly suspended; sensations of a disagreeable kind affecting the stomach: upon taking a little food, however, a greater disposition for it is generated, and more agreeable feelings are introduced throughout the whole system. Intestinal constipation in this, as in all the subsequent periods of the disease, exists in a very high degree. A strong propensity to sleep occurs, and according to Dr. Barbee, the pulse is "full, frequent, round, and *somewhat* tense, but regular." During this stage, exercise of any kind is highly detrimental, and if persisted in, soon induces loathing and nausea at the stomach. If the patient repose, upon first experiencing these symptoms, they generally cease, and he is allowed a longer exemption from the *vomiting* that awaits him. Sooner or later, however, that symptom almost invariably succeeds the predisposition we have described, and either proves fatal in 1, 2, 3 or more days, or leaves the patient in a most exhausted state, from which he recovers only to sustain, at no distant period, a repetition of the same attack.

The matter ejected is sometimes bilious, but much oftener sour, and so acrid, that its action on the throat, in one case, (which proved fatal) was likened to that of boiling water. Towards the close of mortal cases, it is occasionally very dark colored so that it has been compared to that very convenient and fashionable object of similitude—coffee-grounds. At this time the intestinal constipation is very great: Mr. Snodgrass knew one patient in whom it continued for 9 days, throughout which he took no food whatever, and vomited during six of them. After such an attack, the propensity for sleep is destroyed, and an uncommon degree of watchfulness is produced. The patient remains languid, and his face and person generally become rather tumid. His skin is cool, palish, and frequently affected with clamminess. He has a disagreeable burning sensation in his stomach, and hot eructations are very troublesome. The thirst is considerable. The breath is peculiarly disgusting, even loathsome. The appetite is generally poor; and the inclination to costiveness remains. These symptoms often continue for several months, during which the patient experiences frequent returns of the vomiting. But at length, more especially upon the approach of winter, they gradually wear away, leaving the patient considerably worse than they found him, and liable to a fresh attack the ensuing summer.

Nothing like *regular* periodical exascerbations is observable in this disease; no chilliness occurs; the color of the skin and eyes does not deviate widely from that of health, and gives no striking indication of bile; there is no pain in the region of the liver, nor in the shoulder; it does not terminate in dropsy; nor are there any symptoms which bespeak it a disguised or anomalous intermittent. It however prevails (though not exclusively) in aguish situations, and intermitting diseases are thought to have declined since its appearance.

It affects all ages, conditions, and both sexes, indiscriminately; except probably very young children. They however are not wholly exempt from it. Emigrants are not peculiarly liable to it. It was first observed in the summer of 1806, and is thought annually to extend its geographical range, and to become more intense. It sometimes commences in July or before, but oftener in August, and continues till the approach of winter, when it generally, but not always subsides.

The cure of this disease seems hitherto to have been left chiefly to the people, who have not yet discovered any certain method. Purging was a remedy that naturally suggested itself; and by some it has been thought very serviceable, more especially when effected by aloes; but others assert that they have frequently known a cathartic to increase the vomiting, and therefore rely more on enemata. All agree however, that the intestinal obstructions are to be overcome; and that the less the means made use of, affect the stomach, the better. Vomits evidently do harm. Blisters to the gastric region are considered the most efficient remedy. Tonics have been used, but no great benefit appeared to arise from them. Wine and salted meats, however, have appeared to do good, and are relished beyond any thing else. Indeed, eating a little frequently, whether an inclination exist or not, has been found a good palliative: It relieves the stomach from the knawing which so perpetually exists. Alkaline lye has been used in one case: it gave some temporary relief, but not more than almost any other substance which might be received into the stomach. Bleeding has occasionally been resorted to, but with doubtful advantage. Ardent spirit appears to render the disease worse: It is not however, much sought after, all inclination for it, generally being destroyed. Tea and coffee, also, with several other articles of diet, which were agreeable before the disease, are in many cases disliked for a long time after.

This disease is unequivocally observed to affect four domestic animals: the horse, the cow, the sheep, and the dog. It is often fatal to the two former; but not so fatal to the latter. It as frequently attacks horses in the winter as summer, and sometimes kills them in 24 hours.

It prevails chiefly in the neighborhood of Staunton on the Great Miami, and in the country south of Madriver, between Dayton and Springfield. In those tracts, ponds and marshes occasionally occur, more especially in the former. The soil and water are calcareous. The timber generally oak.

Anniversary Address

to the School of Literature and the Arts*

(1814)

As early as August 1810, prior to the unfriendly reviews accorded his *Notices concerning Cincinnati* by western editors who found the work pompous and pointless, Drake took note of the difficulties which attended the pursuit of science in the West. "The taste of our citizens at large is not for physical disquisition," he said. "Any work that is purely physical, however preeminent its merits may be, will have in this country a very limited number of readers; and it is only by connecting it with theology, ethics, politics, or belles lettres, that its general celebrity can be insured. This connection is sometimes natural and convenient; but in a country so new, so interesting, and intrinsically so little known as ours, inquiries into the productions, the laws and the operations of nature are of the first importance, and should have popular sanction, without the aid of a connection with popular and fashionable topics."[1] The organization of the School of Literature and the Arts by Drake and others in 1813 was an attempt to meet these difficulties by combining the activities of a scientific society with those of a lyceum or debating club. The school, because of the emphasis placed by its members on the investigation of scientific subjects, and despite its name and the poetical recitation which was a regular part of its meetings, stands as the first scientific society in Cincinnati, and perhaps in the West, and may be credited with laying the groundwork for the establishment of more pointedly scientific institutions in later years.

* *Anniversary Address, Delivered to the School of Literature and the Arts at Cincinnati, November 23, 1814. Published by Order.* Cincinnati: Printed by Looker and Wallace. 1814.
[1] "Strictures on Volneys 'View of the Soil and Climate of the United States,'" *The Port-Folio* (Philadelphia), n.s. IV (December 1818), 587.

YOU HAVE EQUAL reason with myself, to regret the absence of that distinguished member,* to whom was assigned the delivery of an oration on this evening. Having, so recently, been appointed his successor, I should not venture to exhibit my crude and desultory performance, which has been hastily executed in the midst of pre-existing engagements, but for a conviction, that his removal has imposed on every member an obligation to augmented industry. Under this impression, and in the hope of your indulgence, I shall proceed with confidence in the execution of my task.

Our first year's labors were closed, by the interesting discourse, which has just been read. During that period, we have assembled, for literary exercise, more than twenty times; and our President has delivered, on Astronomy and Natural Philosophy, a variety of Lectures, equally eloquent and perspicuous. He has deduced from them sentiments both amiable and exalted, such as a philosophical survey of the works of GOD invariably excite; and has interspersed them with many impressive recommendations of the pleasure conferred by the acquisitions of knowledge. Thus, from his labors have resulted both instruction of the understanding, and improvement of the heart. The objects and character of our infant association have been defined and established; an impetus has been given it, and regular exertion only is wanting, to raise it into notice and respectability.

The essays of the members, if less learned and profound, have equalled all *reasonable* expectation. Some of them consist chiefly of original matter, while others manifest a degree of research, which is honorable to their authors, and auspicious to the School.

It would be amusing to review their contents, but being restricted to limits too narrow for that undertaking, I will substitute a catalogue of their titles, that, by a single glance we may see the number and diversity of the subjects to which our attention has been directed. I shall enumerate them in the order of their delivery:

1 An Essay on Education—2 on the Earthquakes of 1811, 1812 and 1813—3 on Light—4 on Carbon—5 on Air—6 on the Mind—7 on Agriculture—8 on Caloric—9 on Gravitation—10 on Instinct—11 Notices of the Auroræ Boreales of the 17th of April and 11th of September, 1814—12 an Essay on Water, considered chemically and hydrostatically—13 on Common Sense—14 on Heat

---

* Josiah Meigs, Esquire.

—15 on the Mechanical Powers—16 on the Theory of Earthquakes—17 on Enthusiasm—18 on the Geology of Cincinnati and its vicinity, illustrated with mineral specimens and a vertical map—19 on the Internal Commerce of the United States—20 on Hydrogen—21 on Rural Economy—22 on the Geology of some parts of New-York—23 on General Commerce.

The third and subordinate portion of our exercises, poetical recitation, has been strictly performed; and our ALBUM of poetry already exhibits specimens indicative of a cultivated taste. The proposition to connect with the pieces recited, such critical remarks as they may suggest, has received some attention, and promises to give to this branch of our performances an interest and dignity which were not originally anticipated.

Such, briefly, is the character of our introductory labors. Their retrospection cannot fail to excite a portion of complacency and hope in ourselves, tho from our fellow citizens they may extort neither the meed of approbation, nor the humbler reward of occasional attendance. These, however, are certainly attainable. Our lot, gentlemen, is cast in a region abundant in but few things, except the products of a rich and unexhausted soil. Learning, philosophy and taste, are yet in early infancy, and the standard of excellence in literature and science is proportionably low. Hence, acquirements, which in older and more enlightened countries would scarcely raise an individual to mediocrity, will here place him in a commanding station. Those who attain to superiority in the community of which they are members, are relatively great. Literary excellence in Paris, London or Edinburgh is *incomparable* with the same thing in Philadelphia, New-York or Boston: while each of these, in turn, has a standard of merit, which may be contrasted, but cannot be compared, with that of Lexington or Cincinnati. Still, comparative superiority in Europe, the Atlantic states, or Back-woods, is equally gratifying; and gives to him who possesses it, the same influence over the community to which he belongs.

But it will, perhaps, be asserted, that in a state so young as this, *no* literary distinction is attainable, that would outvalue its cost; that academies and colleges are as yet scarcely instituted; that libraries, philosophical apparatus and scientific teachers are equally rare and imperfect; that associations for improvement, animated and impelled by a persevering spirit, can find no habitation in these rude and chequered settlements; and, lastly, that our countrymen are accustomed to look with frigid indifference on every species of literary effort. This is, indeed, pouring cold water on the flame of literary ambition: but that noble passion is not to be thus extinguished; and if a single spark remain, it will enable us to perceive, through the Gothic darkness which envelopes our literature and science, the certain tho narrow paths to a brighter region.

New countries, it is true, cannot afford the elegancies and refinements of learning; but they are not so unpropitious to the growth of intellect, as we generally suppose. The facilities of improvement which they furnish, differ from those of an old country, more in kind, than degree. In new countries, the empire of prejudice is comparatively insignificant; and the mind, not depressed by the dogmas of licensed authority, nor fettered by the chains of inexorable custom, is left free to expand, according to its original constitution. But the sources of information are fewer, than in old countries; and in balancing between the exemptions of one and the advantages of the other, it must be acknowledged that the latter has a great ascendancy. New countries, however, possess some positive and *peculiar* aids to the developement of understanding. Of these, the principal are to be found in the composition of their society. St. Pierre, in the preamble to his Arcadia, has made a beautiful allusion to this, when depicting an imaginary community of immigrants, assembled on the river of the Amazons.

"They abjured (says he) the national prejudices which had rendered them, from infancy, the enemies of other men; and especially that which is the source of all the animosities of the human race, and which Europe instils with the mother's milk into each of her sons—the desire of being foremost. They adopted, under the immediate protection of the Author of Nature, the principles of universal toleration; and by that act of general justice, they fell back without interruption into the unconstrained exercise of their particular character. The Dutchman there pursued agriculture and commerce into the very bosom of the morasses; the Swiss, up to the very summit of the rocks; and the Russian, dexterous in managing the hatchet, into the centre of the thickest forests. The Englishman there addicted himself to navigation, and to the useful arts, which constitute the strength of states; the Italian, to the liberal arts, which raise them to a flourishing condition; the Prussian, to military exercises; the Poles, to those of horsemanship; the reserved Spaniard, to the talents which require firmness; the Frenchman, to those which render life agreeable, and to the social instinct which qualifies him to be the bond of union among all nations. All these men, of opinions so very different, enjoyed, through the medium of toleration and intercommunication, every thing that was best in their several characters, and tempered the defects of one by the redundancies of another. Thence resulted from education, from laws, and from habits, a combination of arts, of talents, of virtues, and of religious principles, which formed of the whole but one single people, disposed to exist internally in the most perfect harmony, to resist every external invader, and to amalgamate with all the rest of the human race."

A state of society analogous to this is actually presented by Ohio. In no country of the same age and numbers, do the immigrants exhibit more diversity. The sister states, from Georgia to Maine—the Canadas and West-Indies—the united kingdom of Great Britain and Ireland, and the empires of Europe, from the shores of the Baltic to the Mediterranean, have contributed to increase and variegate our population. With the currents of emigration from those countries, have flown into this, many peculiar customs, manners and sentiments, which furnish the elements of a new national character, and display, if not the works of each, at least the principles on which they were designed and executed. A society, thus compounded, has *within itself* no indifferent substitute for travelling; and exhibits, in the lapse of time, what belongs naturally to change of place.

It requires but little reflection on the comparative influence of these causes, and those, operating among a people regulated by a confirmed system of laws and customs, where the national character is uniform and the authority of precedent indisputable, to perceive, that the developements of mind which they effect, are on very different principles; and that the former have many advantages over the latter. To illustrate their relative effects by a metaphor, it may be said, that the operations of intellect, in an old country, are like the waters of a deep canal, which, flowing between artificial banks, pursue an equable and uniform course; while in a new country, they resemble the stream which cuts its own channel in the wilderness; rolls successively in every direction; has a current, alternately swift and slow; is frequently shallow; but always free, diversified and natural. The former is eminently useful for a *single* purpose—the latter can be made subservient to *many*.

For these differences, gentlemen, an additional reason may be assigned. Old states are abundant in the means of imparting elementary knowledge; new ones, in occasions for applying it practically. This is preeminently the case in the natural sciences, and to these chiefly I propose to advert. The larger and more common objects of a country are soon described; but these are not always the most wonderful or numerous. Productions exhibiting great complexity of structure, and connected with each other and with man by very surprising and complicated relations, are frequently minute, or neglected. In Great Britain, the strata, which had for centuries sustained the footsteps of successive tribes of geologists, have recently yielded, to the observations of Mr. William Smith, results equally new and interesting. In France, the vicinity of Paris afforded to Jussieu, more than a hundred new plants, after having been explored for forty years by the penetrating eye of Tournafort. More recently, the environs of the same city, for a thousand years the metropolis of the empire of physical science, have rewarded the genius and industry of Cuvier, with a rich museum of quadruped remains, the living archetypes of which the world cannot at this time furnish. If, in Europe, such enviable conquests were reserved for the naturalists of the present day, what vast acquisitions may not be made in the region of the Ohio, where the germs of civilization have not been planted more than half a century!

On this subject, gentlemen, our enthusiasm can scarcely rise to excess. We are surrounded by a boundless region, redundant in objects the most novel and inviting—where the strong may exhaust their mightiest energies; and the weak may find, in the luxuriance of the harvest, a substitute for strength—where gleaning is neither necessary nor practicable, and the time elsewhere employed in search of fruitful fields, is devoted to selection in the midst of universal plenitude. But let us descend to particulars.—The climate of this country exhibits many singular phenomena: To note and compare them with those of other climates, and thereby to ascertain the laws peculiar to each and common to all, are objects of great interest to the Meteorologist, and remain to be accomplished. To observe the symptoms peculiar to our diseases, investigate their causes, and assign their remedies; to mark the succession of epidemics, and point out the means of preventing endemics, are duties of the first consequence, which the physicians of this country have yet to perform. To analyze and compare the varieties in our soil, and assign to each its appropriate species of culture; to ascertain the extent and diversities of the great calcareous strata which support this region, classify their marine exuviæ, and investigate their marbles, their saline deposits and metallurgic precipitates; to explore the tracts of sandstone which are occasionally found, and bring to light their beds of coal; to survey and disintegrate our extensive alluvions, determine the process of their formation, their richness in iron ore, in copperas, alum, clays and ochres, their antiquity, and their vegetable and animal remains; to collect and arrange specimens of the granite, mica-slate, gneiss and other primitive stones; which, detached from their kindred strata in the depths of the earth, are here scattered over the surface in profusion; to discover the region from which they were derived, and assign the species of convulsion which transported them hither; to examine and point out, to the infirm and disordered, those mineral springs which possess a healing power; and lastly, to analyze the waters of our salines, and increase their utility, by extracting from them, the sulphates of soda and magnesia, with other valuable medicines, are objects which offer to the Geologist and Chemist, and to society generally, a recompence of the highest order. To discover, examine and describe the plants peculiar to this region; to compare its general botany with that of other countries between the same parallels; to determine the latitudes of certain plants common to this and other states; to investigate and bring into notice such of our indigenous vegetables as would be useful in medicine and the arts;

to search for species, which, by proper cultivation, would become articles of nourishment; and lastly, to enrich our pastures with some of the numerous grasses which adorn our fertile prairies, would immortalize the names of a greater number of Botanists than the United States can at present boast. Finally, our Zoology, in the classes of quadrupeds and birds, would not furnish much novelty; but in the departments containing the more imperfect animals, many curiosities might unquestionably be found. Our fishes, reptiles, insects and vermes remain to be examined; and promise to those who undertake it, the reward of a distinguished reputation.

These, gentlemen, are some of the desiderata in our Natural History. Their number, variety and magnitude are scarcely surpassed by those of any country of the earth. They are at once the objects and stimuli of industry, the springs of ambition, and the fuel of enthusiasm.

To the Naturalist, they furnish the means not only of applying, but of extending, correcting and improving his elementary knowledge.

To the Philosopher of expanded views, they offer a theme for the sublimest contemplation. Directing his eye to the strata on which he treads, their marine origin is obvious, and he is instantly carried back in astonishment, to the era, when this great region, now overshadowed by lofty forests or embellished with farms and villages, presented nothing but a deep and interminable waste of waters. While eagerly attentive to the process, by which the habitations of its animals were converted into stone, at the bottom of this ocean, he is interrupted and suddenly called to speculate on the causes which produced its entire dispersion. The surface of the new made earth being exposed to view, his curiosity is excited by the formation of rivers. Where the grand and stately Missouri, Ohio and Mississippi now roll their currents, he sees nothing but depressions, abounding in ponds and morasses. Swelled by copious rains, he perceives these waters surmount their barriers, and accumulate in the south, until by their irresistable weight, all obstacles are borne down, and the impetuous torrent mingles with the ocean. Descending from this period, he is occupied in contemplating the plants and animals of the new continent. Where submarine groves of red coral but lately grew, he sees forests of majestic oak arise. Where the sponge attached its slender forms, he perceives the luxuriant *maize** shoot forth, to nourish future nations. The tracts which were once encumbered with myriads of shell-fish, he now sees verdant with shrubs and herbage, infested with the rattle-snake and wolf, enlivened with deer and elk, or pressed by the gigantic forms of the elephant and mammoth, long since extinct. Lastly, in the lapse of time he is brought to contemplate the arrival of

---

* The Indian name for corn.

man. Observes, his progress from the north-west, his temporary locations by the way, and more lasting settlements on the banks of the Ohio. His enclosures of earth for permanent residence—mounds for the erection of temples—embankments for defence; and his manufactures of clay and shells, of stone and copper, constituting the achme of his perfection in the arts. Finally, to the unwilling view is presented the gloomy spectacle of exterminating wars, and decline in civilization; with his ultimate degradation into the present savage, his exile to some distant country, or entire annihilation.

Such, gentlemen, is the series of amazing and inexplicable events, which this country presents for examination; and which cannot fail to attract much of our attention. There are, however, many other subjects, which, if less brilliant, are of equal or greater utility. Whatever relates to the improvement of our agriculture, manufactures and commerce; to the perfection of our political and social institutions; to the economics, statistics and history of our infant state, is of the greatest consequence. But I have only time to expatiate on the last a single moment. Were the most intelligent young men of Ohio interrogated concerning the Indian war which closed in 1795, they could scarcely do more than relate, that Harmar was repulsed; that St. Clair was disastrous; and that Wayne conquered. With those details in which all *true* knowledge consists, they have no acquaintance, nor can they at present obtain it. With the schemes and stratagems, the inroads, murders and plundering of the enemy; with the cruel and insidious co-operation of Great Britain; the extensive combinations among the tribes, and their ferocious perseverance; with the captivities and suffering of our mothers and sisters; the watchfulness and intrepidity of our fathers and brethren; the hardships and courage, the defeats and victories of our troops; the character of our commanders, and the expenditures of our government, during that predatory and barbarous war, most of us are less acquainted, than with the campaigns of Bonaparte or Alexander.

To collect from the surviving actors in those tragical scenes, and from other authentic sources, the materials necessary for a true and minute history of that period, would be an undertaking worthy of an older institution than ours. Those who accomplish it, will appease the manes of many neglected heroes slain in battle; they must receive the gratitude of society, for supplying a great desideratum, and in future times will be honored, as the fathers of our history.

Gentlemen of the town—The members of the School of Literature and the Arts solicited the honor of your company this evening, to exhibit before you a specimen of their labors; and to make you acquainted with the plan and objects of their humble association. They have done this, fearless of the imputation of vanity, for the sole purpose of engaging your good will towards an institution, which, if continued, may, under abler guidance, be made of public utility.

That you will not withhold the cheering reward of commendation, when their labors deserve it, your conduct this evening, with your general character, is an ample pledge. With that encouragement, they will diligently fan the spark that has been kindled, until it shall rise into a more bright and durable flame.

Natural and Statistical View,
or Picture of Cincinnati and the Miami Country*
(1815)

A reworking of his *Notices concerning Cincinnati* of 1810, Drake's *Natural and Statistical View* sought to combine information of interest to scientists concerning the climate, geology, and natural history of a new and unexplored country with the more practical information desired by potential emigrants concerning the quality of life in the greater Cincinnati region. This volume of 1815 as a result is of interest both because of its extensive description of Cincinnati on the verge of the period of its most rapid development, and because of the evidence it provides of Drake's maturing conception of the relationship of cities and their hinterlands, and of the influence of physical environment on the distribution of plants and animals and on social development.

Although Drake felt it necessary to comment apologetically on the extensive attention devoted to natural history in the volume, when compared with contemporary accounts describing the advantages of life in one or another of the cities of the East, this particular combination of kinds of information was in fact in the great tradition of the emigrant guides of the late eighteenth century. Then the leading scientists of Europe and America debated the habitability of North America, ostensibly on theoretical grounds but always with an eye to the encouragement or discouragement of European emigration to the new United States. They raised in the process many of the questions which Drake sought to answer in the *Natural and Statistical View:* was not the climate more extreme than in older settled regions, and hence less hospitable to agriculture and the establishment of sophisticated social organization; were not the plants and animals native to the area more "primitive" in their taxonomic characteristics and smaller in size; were not the aboriginal inhabitants few in number because the environment made them less fertile than Europeans, and primitive in their society and culture because the environment did not permit greater

---

* *Natural and Statistical View, or Picture of Cincinnati and the Miami Country. Illustrated by Maps. With an Appendix Containing Observations on the Late Earthquakes, The Aurora Borealis, and Southwest Wind. By Daniel Drake.* Cincinnati: Printed by Looker & Wallace. 1815. Pages v-ix, 26-33, 50-57, 129-61, 166-97, 219-32.

development?[1] In all cases Drake answered in the negative. The climate was comparable to that of the Atlantic Coast, and hence to that of Europe. The plants and animals were of the same size and complexity, although the larger mammals had fled before the westward movement of settlement to the wilderness at the headwaters of the Missouri and the Arkansas, and many of the native plants had been displaced by food crops or eaten by cattle until they had virtually disappeared. The number of Indians had been reduced by warfare and disease, but far from being an inherently primitive people, they were related racially to the nations of Central and South America which had constructed the high civilizations of the Aztec and the Inca before the white man came to America. The responsibilities of the scientist, to participate in the persistent quest for knowledge, were thus met simultaneously with the responsibilities of the citizen, to assist in the erection of a more satisfactory and more viable social order.

[1] Cf. Gilbert Chinard, "Eighteenth Century Theories of America as a Human Habitat," *American Philosophical Society Transactions* 91 (February 1947): 27–57.

PREFACE

IN THE YEAR 1810, the Author of the following work, composed a pamphlet on the Topography, Climate and Diseases of Cincinnati; a few copies of which were printed and distributed, chiefly, among his medical and scientific friends, for whom only it was designed. The perusal of it, however, was not confined to them; and several applications were made to obtain copies for the use of travellers in quest of information concerning this country. It was these applications which, two years ago, suggested the advantage that would result from a more extended, and less professional work, of a similar kind; and a prospectus was accordingly sent abroad. For more than a year, it remained doubtful whether sufficient patronage would be afforded to warrant the risk of publication; and as this was an indispensable prerequisite, the preparation of the manuscript was consequently suffered to languish. The causes which have since deferred its completion are many and imperious; but as they no longer interest the public, it would be useless to detail them.

With respect to the subjects which compose the book, it may be observed, that an account of a village in the woods, necessarily differs from that of a populous city, as widely as their landscapes vary from each other. The former dwells on natural objects and advantages; while the latter exhibits the progress of improvement, and expatiates on the works of art. They are, moreover, read for different purposes:—We desire to know what there is in a new country, that can recommend an emigration thither: in a city, we seek for that which is worthy of imitation or adoption. Thus the PICTURE OF CINCINNATI will be found to contain a larger proportion of natural history, than any of the works which have lately appeared east of *The mountains,* under similar titles. The author does not apprehend that this will diminish the value of the book, however unusual it may be considered; but he deeply regrets his not being able to assert that in this portion of the work there is that accuracy, fulness and perspicuity which the interests of science require. To those who are experimentally acquainted with the difficulties attending the acquisition of elementary knowledge in chemistry, geology, botany, and the other physical sciences —without apparatus, with but few books, and no arranged collections;—or even to those who have felt the minor embarrassments attending the practical study of these branches, without practical works, he need make no apology. In the other chapters, it is hoped, that not many errors or exaggerations exist; as it has been the author's constant aim to write a history, and not a panegyric. Still, as accidental associations and local attachments, are liable to give an *undue* degree of meanness or excellence to many of the objects among which we have spent the greater part of our lives; he is unwilling to flatter himself that he has not made some statements which may be pronounced partial, or even erroneous. To point out these, is peculiarly the province of that domestic criticism, which he chooses to invite, rather than deprecate.

There may be readers who will consider the work as extending further, on some points, than can be justified by its title; but this impression would be erroneous. No subject is introduced that has not a connexion with the town, and were any such to be omitted, the plan would obviously be defective. A book of this kind should contain whatever it is desirable to know, concerning the spot of which it professes to treat. The *relations* of a town with the surrounding country, are an essential part of its history, and cannot be understood without studying both. The Author is by no means so confident, that he has adopted the best mode of exhibiting this information; and in giving it a formal distribution under the heads which have been employed for geographical delineations of greater extent, he does not expect to escape the charge of a precise and finical devotion to method; but with the hope, that the opportunity it afforded of disposing the materials in that state of arrangement which will facilitate a reference to any particular subject, he felt no disposition to pursue a different plan, merely to avoid so harmless a criticism. A more ample field of animadversion, will perhaps be found in the examination of his style.

'Tis true, the merit of a topographical work, composed chiefly of facts and observations, does not depend altogether on the choice and collocation of the words in which it is expressed; but still it is the sacred duty of every writer to improve, rather than corrupt his language. The Author performs, therefore, merely an act of justice to himself, when he declares that the imperfections in his style have arisen neither from indolence, nor contempt of public opinion, but from causes which lie beyond the sphere of his control; and at the same time, it is equally due to the reputation of his fellow townsmen, that he should protest against the reception of this performance as a fair specimen of their literature.

The map of the Miami country, which includes also the adjoining parts of Kentucky, so as to exhibit an entire view of the tracts dependent on Cincinnati as their emporium, has been compiled with much care by Mr. Thomas Danby, from the following materials, furnished him by the Author: 1. The correct and beautiful map of Ohio, published in 1807, by the late captain J. F. Mansfield, from the official returns in the office of the Surveyor General of the United States; 2. Transcripts from the plats in the office just mentioned, of such parts of this district as lie west of the state of Ohio; 3. A manuscript map of the counties watered by the eastern branch of the Little Miami, procured from the Auditor of the State; 4. A map of Campbell county, in Kentucky, furnished by General Taylor; and 5. Personal observation and research through most of the district, with oral and manuscript information from various persons. The plan of the town has been executed by the same gentleman, on a scale of 800 feet to an inch, from materials obtained chiefly at the office where the surveys of the town plat are recorded; and has therefore all the accuracy which can be conveniently given to such a work. The plate representing geometrically the comparative temperatures of each month in the Atlantic and Western States, which was promised in the prospectus; has been omitted for want of the proper eastern observations.

The two first papers of the Appendix were read before the School of Literature & the Arts, in 1814; and have been extracted, by permission, from the register of that society. The third has already appeared in The Port Folio; but as only a small proportion of the inhabitants of the Miami country have an opportunity of reading that valuable Magazine; and as the Author has collected some additional facts, he considered the reprinting of it not improper.

It only remains for the Author to make a public acknowledgement of obligation, to those gentlemen who have aided him by their communications, in the difficult and tedious business of collecting small facts; and to those young friends, whose assistance, in the correction of the work, has brought it

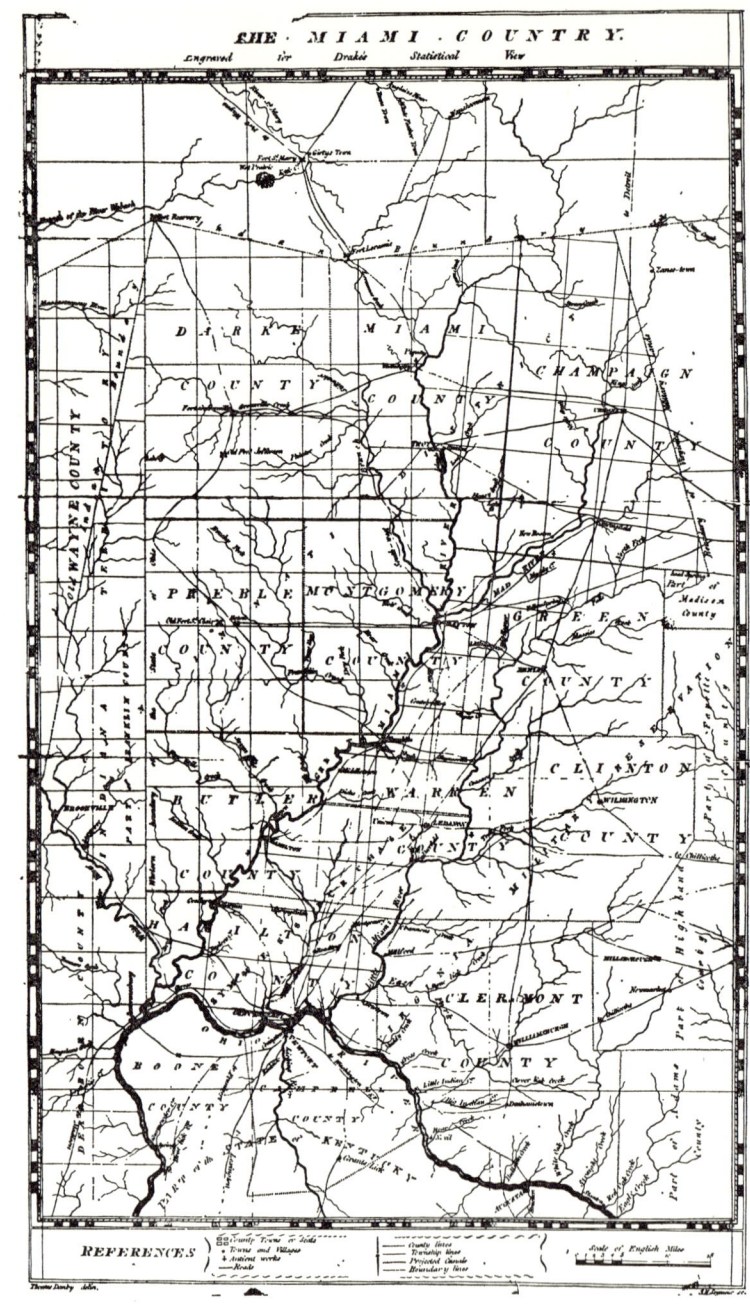

before the public in a more perfect condition that it would otherwise have attained.

Cincinnati, Ohio, September, 1815.

FROM CHAPTER I, GEOGRAPHICAL AND HISTORICAL
INTRODUCTION
POPULATION

The principal inducements for immigration to this state are, the fertility of its soil; the low prices of lands, and entire security of titles; the high price of labor, and the exclusion of slavery. For several years the Indian war opposed the operation of these inducements, but the Greenville treaty of 1795, brought them into full effect. Fortunately, they happened to attract most attention, in the Northern and Middle states, which are at all times able to furnish the greatest number of emigrants. In the Northern, especially, where the means of subsistence bear the smallest proportion to the population, these advantages have been fully appreciated, as appears from the prevalence of the manners and customs of New-England, over most of this state. The extraordinary emigration from that quarter cannot be wholly contributed to these inducements, but has arisen in part from a portion of the north of this state being owned by Connecticut. In the same way the retention, by Virginia, of her right to the soil between the Little Miami and Scioto rivers, has been an additional motive with the people of that state for migrations to this. The prohibition of slavery has contributed greatly to the population of this state. The operation of this cause has not been confined to those states in which the practice of slavery is abolished, but has extended throughout the south, and is likely for many years to continue in full operation. It has even turned the current of European emigration from Kentucky and Tennessee, and spread it widely over Ohio.

The progress of increase, in this state, has been equally rapid with that of immigration. From the abundance of subsistence, the preventive checks to population do not operate, and marriages are both early and productive. Males frequently marry before twenty-one, and females before seventeen. The positive checks are neither numerous nor powerful. The diseases peculiar to new countries, and incidental to those who change their climate, have an effect, not susceptible of estimation, but which is unquestionably considerable. This however, is the only cause to which much should be ascribed. From 1794 to 1812, there was no Indian war. The loss of lives, in the campaigns of 1812 and

1813 was great, but is not perceptible to observation; nor will it produce a sensible diminution in the ratio of increase. Those who perished were chiefly young men, a portion of our population that is always in excess, and which when reduced is soon restored by immigration.

It may not be uninteresting to compare the progress of population in the new transmontane states—Tennessee, Kentucky and Ohio. Lying nearly in the same meridian, and almost equally remote from the parent states, their settlement may be supposed to have proceeded on similar principles, although not commenced at the same period. Within the limits of Tennessee, formerly attached to North Carolina, there were 2000 inhabitants as early as 1775. In the same year, the settlement of Kentucky, then a county of Virginia, commenced, and twelve years later, in 1787, the settlement of Ohio was begun.

The following table exhibits the population of these states, as ascertained by the three successive enumerations of the general government:

	In 1791	*In 1800*	*In 1810*
Tennessee	35,691	105,602	261,727
	In 1790		
Kentucky	73,677	220,960	406,511
Ohio (by estimation)	3,000	42,156	230,760

From tables, founded on these data, and constructed on the principles of a regular geometrical ratio of increase, it appears that the population of Tennessee increased, from 1791 to 1800, at the rate of twelve and three-fourths per cent. and doubled in six years; from 1800 to 1810, at the rate of nine and a half per cent. and doubled in eight years. Since that time, if the rate of increase has diminished regularly, it amounts to about six and three-tenths per cent. and will cause the population to double in little more than eleven years.

From tables of a similar kind for Kentucky, it appears that the population from 1790 to 1800 increased at the rate of about eleven and six-tenths per cent. and was doubled in less than seven years; from 1800 to 1810, at the rate of six and three-tenths per cent. and doubled itself in something more than eleven years. Since 1810 it probably increases at the rate of three and one-third per cent. and will require, for the period of doubling, about twenty-three years.

In Ohio, the population was augmented at the rate of thirty and one-fourth per cent. and doubled in less than three years between 1790 and 1800: from the latter period to 1810, it advanced at the rate of eighteen and a half per cent. and nearly doubled every four years. Since 1810, it probably increases at the rate of seven and eight-tenths per cent. and will double itself in less than ten years.

From these rates of increase, the population of the present year (1814) in round numbers must be nearly as follows: Kentucky 420,000, Tennessee 334,000, Ohio 312,000. In 1820, it will probably approach to the following: Kentucky 453,000, Tennessee 481,000, Ohio 492,000.*

These statements exhibit the greater proportional advancement of population in Ohio, than either Kentucky or Tennessee; and disclose to us the interesting fact, that at no very distant time, we shall outnumber either of our southern sisters. Deducting black population, we are, indeed, at this time, more numerous than Tennessee, and approach nearer to Kentucky than is generally supposed.

An enquiry into the causes of increase in the future population of these states, is neither within the power of the author, nor the plan of this work; but it may not be improper to devote a moment to the consideration of the leading causes that will secure a rapid augmentation of population in Ohio.

The cheapness of land and the high price of labor will continue to promote immigration to this state until the lands owned by the United States are principally sold and settled. The effect of these causes will then be lessened; but the general fertility of our soil, the security of land titles, and the prohibition of slavery, as already enumerated, are inducements equally strong and durable.

The extinction of the Indian title to the region watered by the Sandusky and Maumee rivers, will have a very beneficial operation on the progress of our population. The uniform richness of soil in that tract; the facility of removing to it over Lake Erie; its numerous creeks, bays and harbors on the north, and proximity to the future capital of the state on the south, must inevitably effect a rapid settlement. The formation of commercial and trading establishments on its northern border, and the construction of good roads, if not canals, between the navigable streams of the Lake and Ohio river, will also contribute greatly to a dense population in the interior.

Lastly, the erection of manufacturing establishments will co-operate in the future augmentation of our numbers. To convert into manufacturers the hands engaged in clearing and improving a new country, would be a mistaken policy; and if adopted, must soon correct itself. In the case in which a new country is *contiguous* to an older, of dense population, which can exchange manufactures for subsistence, it may even be advisable to defer manufacturing

---

* As this prediction will certainly not be considered probable, it may be well to observe, that having ascertained from a comparison of the population of 1790, 1800 and 1810, the rates of increase in each state, for two periods, the rates for the third are assumed as bearing the same proportion to the second, that the second did to the first. But it must be acknowledged, that before 1820, these ratios may vary so materially as to give very different results. The maximum of population in the different states will indeed depend *mainly* on their agricultural produce, and in this respect Kentucky will doubtless have the advantage of Tennessee, though not of Ohio.

in the former to a late period. But where a new country must transport its surplus agricultural products to a great distance, and import the necessary manufactures from shops equally remote, it may be advisable to commence manufacturing much earlier. It must not, however, attempt to convert its farmers into tradesmen. They should be imported instead of their manufactures. The ranks of agriculture would then remain entire, the simple process of barter at home be substituted for expensive and hazardous commercial operations, and the immigrant manufacturers with their increase become an addition to the population. The situation of Ohio seems to recommend this policy, and it is already adopted. Manufactures have been commenced in various places, and are principally conducted by foreigners, or persons from the Atlantic states.

It appears by the census of 1810, that the proportion of males to females in Ohio, is,

Under 10 years of age, as	100 to	94.7
Over 10 and under 16	100	93.1
16 26	100	99
26 45	100	85.3
45	100	61.2
The average of which is	100	86.7

The following is a comparative view of this mean proportion with that of some other states:

Ohio	100 males to	86.7 females
Kentucky	100	90.9
Tennessee	100	93
Rhode-Island	100	104.8
Northern States	100	100.7
Middle States	100	95
Southern States	100	97
Mexico (*according to Humboldt*)	100	95
France (*by the same*)	100	103

These numbers would seem to indicate, that in this state an extraordinary number of hands, in proportion to the whole population, are employed in masculine avocations, which may be the case in comparison with some of the countries named, but not with all. The black population of the Southern states, of Tennessee and Kentucky, which is chiefly employed in agriculture and the coarser mechanical occupations, is excluded from these estimates. In Tennessee, at the last census, this population amounted to 45,852, bearing to the white population the proportion of twenty-one and a quarter to one hundred, and making of the whole, about one-sixth. In Kentucky, the negroes amounted to 82,274, bearing to the whites the proportion of twenty-five and one-third to a hundred, and making of the whole mass, nearly one-fifth. While in Ohio the blacks were only nineteen hundred, being to the whites as eighty-three to one thousand, and making of the whole population less than a hundred and twentieth.

The proportion of inhabitants above forty-five years of age, to the total white population, was, by the last census, in Ohio, as nine to one hundred; in Kentucky, as nine and five-tenths to one hundred; in Tennessee, as eight and eight-tenths to one hundred; in Connecticut, as sixteen and six-tenths to one hundred; and in the United States, as twelve and four-tenths to one hundred. Had there been no migration to or from any section of the Union, these proportions would prove the Western States less favorable to longevity than the others. As it is, no such conclusion is deducible. The difference is produced by the continual emigration of young persons from the latter to the former, increasing the proportion of the aged in the east, and diminishing it in the west. Nothing, indeed, is more difficult, than to derive from such comparisons between an old state and a new one, any correct information on this point, as will be manifest from a reference to Connecticut and Ohio. Few persons above 60 ever emigrate to this country—let us *suppose,* then, that none exceeding that age have arrived here, since the year 1800; in this case, it is evident that all who are now 74, or older, must have been 60, or upwards, in 1800, when our population was but 42,156. At this time it is greater than that of Connecticut, but the number which have attained to 74 years is much less, for they have grown out of the 42,156, while those of the same age in Connecticut are the residue of 251,000, the population of that state in 1800. These numbers are to each other as sixteen and seven-tenths to one hundred, so that the amount of aged population in Connecticut *should* be at present nearly six times greater than that of Ohio; and if this be not the case, we are warranted in considering the former as not more propitious to old age than the latter.

* * * * *

LAND TITLES

These are all derived from the government of the United States; but in the manner of their transfer to the occupiers of the soil, there are some varieties which deserve notice.

1. VIRGINIA MILITARY RESERVATION. It has been already stated, that in

ceding to the United States her portion of the Northwestern Territory, Virginia reserved the lands between the Little Miami and Scioto rivers, for the payment of her line of troops, serving on continental establishment in the revolutionary war. The following is the course pursued in locating and patenting these lands: The Secretary at War, according to a law of Congress, made to the Executive of Virginia, a return of the names of such officers and soldiers, as were, by the laws of that state, entitled to these bounties, and the Governor issued warrants to the same. When these warrants are located, a return of the surveys is made to the Secretary of State of the United States, and the patents of the President obtained. When it is found that a survey includes land previously located, the holder of the warrant is permitted to locate it elsewhere. Interfering claims, therefore, but seldom produce litigation. A large number of warrants, it is expected, remain to be located; and it is equally uncertain when they will be completed, and whether the tract reserved by the state of Virginia will be of sufficient extent: should this not prove to be the case, the General Government will undoubtedly furnish other lands.

2. SYMMES' PATENT. In the year 1787, John Cleves Symmes, of the state of New-Jersey, made a successful application to the General Government, for the purchase of a tract of land immediately north of the Ohio, between the Miami rivers. A bargain was made with the Commissioners of the Board of Treasury for a tract, which it was expected would contain a million of acres, but which was found to embrace less than 600,000. Of this, the purchaser made payment for no more than 248,582 acres. In 1794, he received the patent of the President for 311,682 acres, 63,100 acres being reserved in pursuance of sundry acts of Congress. These reservations were, 15 acres around Fort Washington, in the town of Cincinnati, which were sold in 1808—a complete township, to be located as near the centre of the tract as possible, for the benefit of an academy; which, however, was sold by the patentee, and replaced by the Government with a township west of the Great Miami—section 16, in each township, for the use of schools—section 29, for religious purposes—and sections 8, 11 and 26, for the future disposal of Congress; and which were, in 1808, by law directed to be sold.

For the lands contained in this patent, the deeds of the patentee are indisputable; but prior to the year 1794, he sold several tracts lying north of his patent, tho within the limits of his original purchase. These sales the Government refused to sanction, but granted pre-emptions to the purchasers, and compelled them to make payment to the Receiver of public monies at Cincinnati, and take out patents in the usual way.

3. UNITED STATES' LANDS. The other lands of the Miami country, south of the Indian boundary, have, by the Surveyor General, acting under the

direction of the Secretary of the Treasury, been divided into townships, sections and quarter sections, by lines according with the cardinal points. These have been executed with great accuracy, and constitute, with the other surveys of the Government, a more regular and beautiful system than any other country perhaps can boast. For the sale of Miami lands, excepting the 16th section in each township, reserved for the support of schools, a law was passed in 1800, creating the Cincinnati District, and establishing the offices of Register and Receiver. Payment for a tract being completed in the latter office, a final certificate is forwarded from the former to the Commissioner of the General Land-office, who returns the President's patent.*

PRICES OF LAND. These have been constantly, tho not regularly, increasing, ever since the first settlement here. In 1787, John C. Symmes paid to the United States two-thirds of a dollar per acre. Their uniform price, since that time, has been two dollars, except at public auctions, when from competition, the prices are frequently raised much higher; and except reserved sections, which were at one time fixed at eight, but afterwards reduced to four dollars.

Within 3 miles of Cincinnati, at this time, the prices of good unimproved land, are between fifty and one hundred and fifty dollars per acre, varying according to the distance. From this limit to the extent of 12 miles, they decrease from thirty to ten. Near the principal villages of the Miami country, it commands from twenty to forty dollars; in remoter situations, it is from four to eight dollars—improvements in all cases advancing the price from 25 to 100 per cent. An average for the settled portions of the Miami country, still supposing the land fertile and uncultivated, may be stated at eight dollars; if cultivated, at twelve.

Of tracts that have the same local advantages, those alluvial or bottom lands, which have been *recently* formed, command the best price. The dry and fertile prairies are esteemed of equal value. Next to these, are the uplands, supporting *hackberry, papaw, honeylocust, sugartree* and the different species of *hickory, walnut, ash, buckeye* and *elm.* Immediately below these, in the scale of value, is the land clothed in *beech* timber; while that producing *white* and *black oak* chiefly, commands the lowest price of all.

These were not the prices in 1812; the war, by promoting immigration, having advanced the nominal value of land from 25 to 50 per cent.

AGRICULTURAL PRODUCE

GRAIN. The principal kinds are Indian corn, wheat, rye, oats and barley. The first is found on every plantation, but flourishes best in a fertile, calcareous

---

* See Land laws of the United States.

soil; where, with good culture, it will yield from 60 to 100 bushels per acre; but an average crop, for the whole region, cannot be higher than 45. Wheat is raised almost as generally as Indian corn, and is perhaps better adapted to the soil of most parts of the Miami country. Twenty-two bushels may be stated as the average produce per acre, tho it sometimes amounts to 40. Its medium weight is 60 lbs. the bushel. The bearded wheat, with reddish chaff, seems latterly to be preferred, as least liable to injury from the hessian fly and weavel. The cultivation of rye is much more limited, as it is only employed in the distillation of whiskey, and as provender for horses. For the former purpose, it is mixed with Indian corn. Its average crop may be estimated at 25 bushels per acre. The common crop of oats is about 35 bushels, and that of barley 30. The latter was not extensively cultivated till since the erection of two large breweries in Cincinnati.

FRUITS. An extensive variety of excellent *apples* have been introduced, and succeed well, in the Miami country. As in other parts of the United States, they are occasionally injured by vernal frosts. In the valley of the Ohio this is less frequently the case, than on the uplands. Cider, of a good quality, is annually made in large quantities. *Peaches* attain to great perfection, and are found on every farm. *Pears, cherries* and *plumbs,* of different kinds, are common: some finer varieties of the two latter, however, as well as the *apricot* and *nectarine,* have not yet been successfully cultivated. The *vine* has not been planted, for the purpose of making wine; nor has its cultivation in gardens been continued long enough to ascertain whether the soil and climate of this quarter be adapted to its growth.

FLAX AND HEMP. The first is raised on every farm. It is said not to be so good as that of the Atlantic states. The seed, especially, is inferior, yielding much less oil than the flaxseed of those states. Hemp, a few years since, was cultivated to some extent, and found to succeed well in bottom lands, but from a depression in the price, it is now neglected.

MEADOWS. These are generally luxuriant. Timothy, red and white clover, and spear-grass, are principally cultivated, and yield a good crop. Two tons per acre, are considered the medium produce of the two first. *They* are not found, except when sown; but the latter spring up spontaneously on every farm, after the cultivation of a few years, and afford excellent pasture.

Before the settlement of this country, the woods abounded in grass and herbage proper for the subsistence of cattle, but these have long since disappeared, except in remote situations. In the prairies, however, where the whole energy of the soil is employed in producing grasses and herbaceous plants, instead of trees, the pasture is still luxuriant, and the business of grazing extremely profitable. It is chiefly of Champaign and Green counties, that this remark is true. In the former, one hundred thousand dollars, it is estimated, are annually received for fat cattle. The prairies are likewise found to support *hogs;* which grow and fatten on the numerous fleshy roots, with which those tracts abound. *Sheep,* both domestic and foreign, are already diffused extensively through the Miami country. They are in general healthy, and rather prone to excessive fatness. Their flesh is said to be superior in flavor to that of the sheep of the Atlantic states.

The agriculture of this, as of other new countries, is not of the best kind. Too much reliance is placed on the extent and fertility of their fields, by the farmers, who in general consider these, a substitute for good tillage. They frequently plant double the quantity they can properly cultivate, and thus impoverish their lands, and suffer them to become infested with briars and noxious weeds. The preservation of the forests of a country should be an object of attention, in every stage of its settlement; and it would be good policy, to clear and plant no more land in a new country, than can be well cultivated.

* * * * *

CHAPTER III, CIVIL TOPOGRAPHY
PROPRIETORS

Cincinnati is built upon one entire and two fractional sections; numbered 18, 17 and 12, in the fourth township and first fractional range, as surveyed by the patentee, John Cleves Symmes. The two first of these, viz. the entire section No. 18, and the fraction No. 17, lying between it and the river, were sold by the patentee to Matthias Denman, of New-Jersey, whilst they were still a wood. Not long after this purchase, Denman transferred to Robert Patterson and John Filson, of Kentucky, an undivided third part each, making them joint proprietors with himself; but Filson being killed by the Indians, before complying with the terms of this bargain, his interest reverted to Denman, who sold it to Israel Ludlow, of the same state with himself. A plan for the intended town was then designed, and in January 1789, Mr. Ludlow executed a survey of that part which extends from Broadway to Western Row. The proprietors then proceeded to sell the lots, and in conformity to a previous arrangement, the purchasers received their deeds directly from J. C. Symmes. In the ensuing year the patentee laid out several blocks of lots on the fraction No. 12, lying east of the first town plat. In the year 1808, the reservation around Fort Washington was divided into lots by the Surveyor General, acting under the direction of the Secretary of the Treasury, and sold at public auction by the Register and Receiver, on the 2d of March. In addition to these original owners, several persons have since divided out-lots, or tracts adjoining to the first town plat, and are therefore to be considered as proprietors.

PLAN

Philadelphia seems to have been the model after which that portion of this town first laid out, was planned. Between Broadway and Western Row there are six streets, each 66 feet wide, running from the river north 16° west, and lying 396 feet asunder. These are intersected at right angles by others of the same width, and at the same distance from each other; except Water and Front streets, and Second and Third streets, the former of which are nearer, and the latter, on account of the brow of the *Hill,* more distant. Not a single alley, court, or diagonal street, and but one common, was laid out. The blocks or squares were each divided into eight lots, 99 by 198 feet, except those lying between Second and Third streets, which made ten lots each; and those between Front and Water streets, the size of which may be seen by a reference to the frontispiece. The out-lots, 81 in number, contain four acres each, and lie chiefly in the north of the town. This plan was not deposited in the public archives for record until the 29th of April, 1802. The streets in that part of the town laid out by John C. Symmes, are but 60 feet wide. Those intersecting the river run north 44 degrees west, and lie at the same distance from each other as the streets in the original town; but the cross streets are nearer, and hence the lots of this quarter are shorter. The plan of this survey was not recorded by the proprietor till the 12th of September, 1811. The reservation of the General Government was surveyed so as to connect the plats just described. The different subdivisions will be best understood by a reference to the engraved plan.

The DONATIONS by the original proprietors are, a tract between Front-street and the river, extending from Broadway to Main-street, for a public common; and a square west of Main-street, between Fourth and Fifth streets. The south half of this was conveyed to the First Presbyterian Congregation; and the other to the Commissioners of the county; an amount in each case, nearly equal to the value of the ground, being paid.

PRICES OF LOTS

For several years after the settlement of this place, the lots along the principal streets were sold for less than $100 each. They gradually increased in price until the year 1805, when, from a sudden influx of population, they rose for a short time with rapidity. Their advancement was then slower, till 1811; since

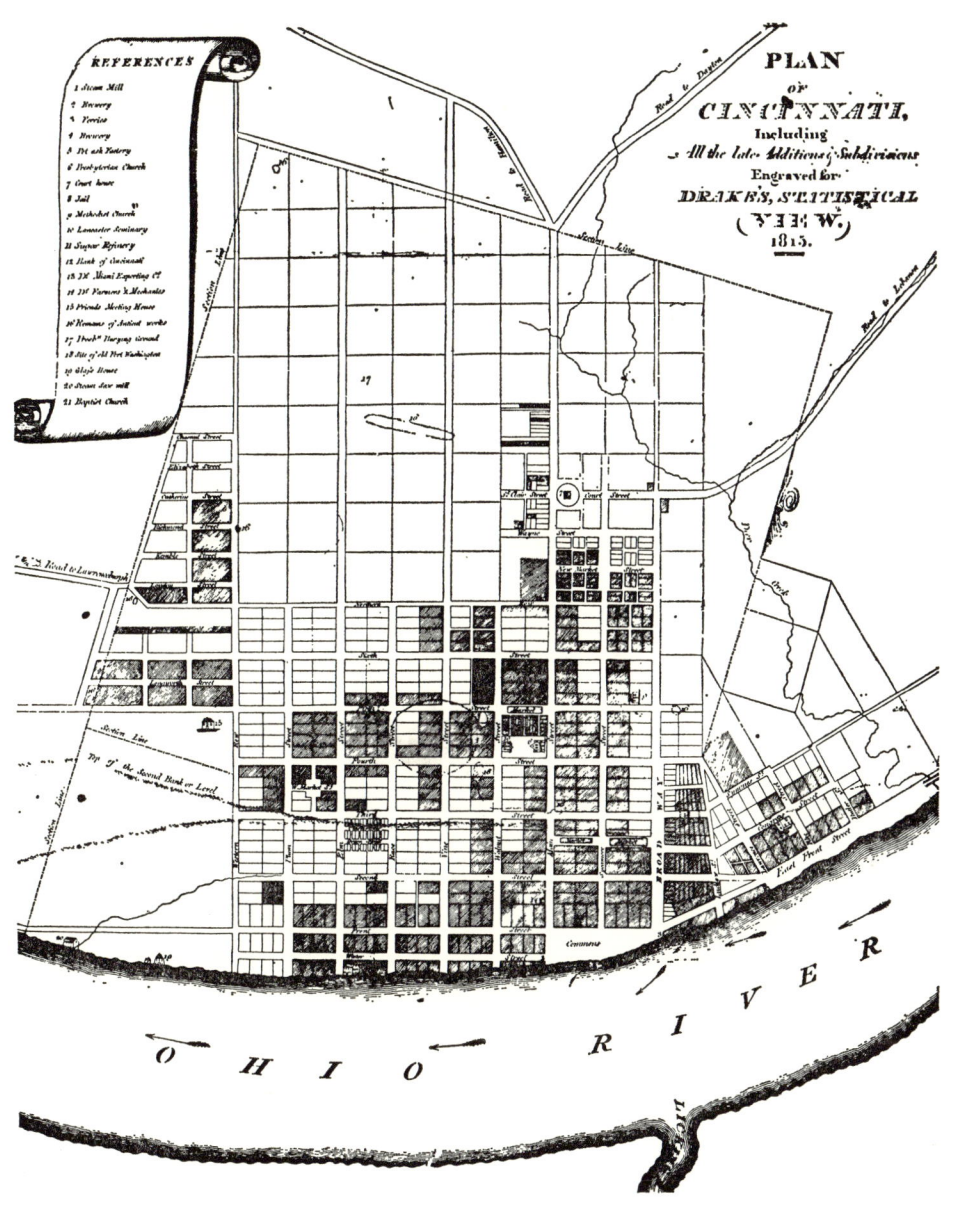

which the rate of increase has been so high, that for a year past the lots in Main, from Front to Third streets, have sold at $200 per foot, measuring on the front line; from thence to Sixth-street, at $100; in Broadway, Front and Market streets, from 80 to 120; and on the others, from 50 to 100, according to local advantages. Out-lots, and land adjoining to the town plat, bring from 500 to 1000 dollars per acre.

GRADUATION, AND DRAINING OF THE STREETS

One part of the town being elevated from 40 to 60 feet above the other, it has long been an interesting question, whether the streets running from the river should be graduated to a steep or gentle ascent. The latter method has at length been adopted, and Main-street rises by degrees from Second to Fifth street. The earth and gravel at the intersection of Third-street on the brow of the *Hill,* and beyond it, as far as Fifth-street, being hauled and washed down to raise the surface below. The angle of ascent varies, by estimation, from 5 to 10 degrees. Broadway, Sycamore & Walnut streets, are partly completed on the same plan. To the constant change of level which the streets have undergone for many years, from the descent of gravel into the *Bottom,* is to be ascribed the want of pavements and sidewalks, which the town so strikingly exhibits. Preparations are making for the pavement of Main-street, from the river to Fourth-street, the ensuing year; which will no doubt be followed by a general improvement of the town in this respect.

Concerning the points at which the water falling on the town plat should be discharged into the river, there are two opinions. The first and most natural is, that it should be conducted down Second-street, and emptied into the river below the town, through the same ravine which formerly carried it off. The other opinion is, that each street running to the river should be so graduated as to convey its own water. But the obvious injury which the banks, the beach and the water would sustain, from the discharge of these sluices of filth immediately opposite the town, together with the enormous expense attending it, seem to be procuring for the other method a general preference; and it is probable that all the gutters west of Broadway will be discharged into a common sewer in Second-street, along which in an open canal the water now indeed runs.

It has been already stated, that the north-west part of the Bottom is occasionally inundated by great floods of the Ohio. To prevent this, it has been proposed to throw up a *levee* along the western border of the town plat. The cost of this could not be very great, as it would not have an average height of more than six feet, nor exceed two hundred yards in length; and having no current to stem, it need not be very strong. No measures, however, have yet been taken to effect this important object.

MATERIALS FOR BUILDING

Cincinnati is eligibly situated for obtaining these. The beds of Licking and the Ohio afford excellent limestone, which, however, can only be quarried when those rivers are low. Marble of a fine quality can be brought by water from the cliffs of Kentucky river; and freestone of a grey color and good texture is already freighted, for a small sum, from near the intersection of the Big Sandy and Scioto with the Ohio, where inexhaustible quarries exist. The clay of the lower part of the town makes excellent brick, about five millions of which are annually used in this place. The lime afforded by the common limestone is dark colored, but the silicious limestone pebbles, which are abundant in the alluvial grounds, make lime of a fine quality and pure white. Oak, ash, poplar, walnut and other native timber trees, squared or sawed into boards, plank and scantling, are brought to market in waggons, boats or rafts, and delivered on moderate terms. But the Allegheny mountains furnish the most valuable, and must long continue to afford the most abundant supplies of timber. From those mountains, the white pine, either in the form of logs, boards or shingles, is annually floated down in immense quantities, and sold in all the towns on the Ohio, at a lower price than domestic timber.

The different kinds of masonry, carpentry, painting, papering, and Venetian blinds, are executed in a firm and handsome style.

BUILDINGS

On the plat of Cincinnati, there is at this time (July 1815) nearly 1100 houses, exclusive of kitchens, smokehouses and stables. Of these, more than 20 are of stone, 250 of brick, and about 800 of wood. Six hundred and sixty contain families; the remainder are public buildings, shops, warehouses and offices. The great proportion of frame houses seems to be owing to the vast immigration within a few years—a wooden house can be erected in a shorter time than a brick, and at seasons when brick work cannot be done. The dwelling houses are generally two stories high, and built in a neat and simple style, with sloping shingled roofs, and Tuscan or Corinthian cornices. Several have lately been erected with an additional story, and exhibit, for a new town, some magnificence. A handsome frontispiece or balustrade occasionally affords an evidence of opening taste; but the higher architectural ornaments—elegant summer houses, porticos and colonnades, are entirely wanting. Very few of the frame houses are painted, which is the more remarkable, as the timber of which they are built is so perishable as to require seclusion from the weather.

PUBLIC AND MANUFACTURING EDIFICES

The first COURT HOUSE in this place, stood on the eastern end of the public ground. It was erected in the year 1802, and burned down early in 1814, while a company of soldiers were using it as a barrack. It was built of limestone, on a plan furnished by Judge Turner, in the form of a parallelogram, 42 feet in front by 55 in depth; the height of the walls, including a parapet, being 42 feet. It had a wooden cupola with four projecting faces, arched and balustraded, 20 feet high, terminated by a dome, and resting on a basement 20 feet square. From the ground to the top of the cupola was 84 feet. A couple of two story wings, to be made fire proof, for the purpose of public offices, and connected with the body by corridors, formed a part of the *design* which remained to be executed.

Since the conflagration of this edifice, the Commissioners of the county have sold out, on perpetual leases, the whole of the public ground; and accepted of a lot near the intersection of Main and Court streets; in the centre of which they are now engaged in the erection of a second court house, 56 by 62 feet; with fire proof apartments for the different offices of the county.

The new PRESBYTERIAN CHURCH is a very spacious brick edifice, measuring 68 by 85 feet. Its eastern and narrower front looks towards Main-street, and is cornered with square turrets crowned with cupolas. From the rear is an octagonal projection, for a vestry. The roof is of a common form. The height from the ground to the eaves is only 40 feet, to the top of the cupola 80, which is less than either side including the towers, and hence the aspect of the building is low and heavy. The stair cases are in the basements of the turrets, and are entered without passing into the house. The inside will be divided into one hundred and twelve pews, and five capacious aisles.

The BAPTIST CHURCH, in Sixth-street, is a handsome and commodious brick edifice, 40 by 55 feet, well furnished with doors and windows, ornamented with a balustrade, and finished inside with taste.

The METHODIST CHURCH, in Fifth-street, is a capacious stone building, one story high.

The FRIENDS MEETING HOUSE, near the western end of the same street, is a temporary wooden building.

The CINCINNATI LANCASTER-SEMINARY, on Fourth-street, in the rear of the Presbyterian church, is an extensive two story brick edifice, built, with some alterations, on a plan furnished by Isaac Stagg. It consists of two oblong wings, extending from Walnut-street, 88 feet deep. Near the front, they are connected by an apartment, for stair cases, 18 by 30 feet, out of which arises a dome capped peristyle, designed for an observatory. The front of this intermediate apartment is to be decorated with a colonnade, forming a handsome portico, 12 feet deep and 30 feet long. The front and each side are ornamented with a pediment and Corinthian cornice. The aspect of the building is light, airy, and might be considered elegant, were the doors wider, the pediments longer, and divested of the chimnies, which at present disfigure them. One wing of this edifice is designed for male, the other for female children; and between them there is no connecting passage, except through the portico. The lower stories are finished entire, and calculated for the reception of 900 children. Each upper story is to be divided into three apartments, two in the ends 30 feet square, and one in the centre of 25, with a sky light, and the appurtenances of a philosophical hall. When completed, the whole building can receive about 1100 scholars.

Cincinnati has three MARKET HOUSES—the two older are supported by a double, the newer one by a triple, row of brick pillars. The latter extends nearly the whole distance from Broadway to Sycamore-street being upwards of 300 feet in length. The others are both shorter and narrower.

The BUILDINGS of the Cincinnati Manufacturing Company, on the bank above Deer-creek, are numerous and extensive; the main edifice is 150 feet long, from 20 to 37 feet wide, and from two to four stories high.

The most capacious, elevated and permanent building in this place, is the STEAM MILL, erected in the years 1812, '13 and '14, under the direction of William Green, an ingenious mason and stone cutter, on a plan furnished by George Evans, one of the proprietors. It is built on the river beach, upon a bed of horizontal limestone rocks, and in high floods is for its whole length exposed to the current. The foundation is 62 by 87 feet, and about 10 feet thick. Its height is 110 feet, and the number of stories nine, including two above the eaves. To the height of 40 feet, the wall is *buttered,* or drawn in; above, it is perpendicular. The cornice is of brick, and the roof of wood, in the common style. It has 24 doors and 90 windows. The limestone with which it was built were quarried at various places in the bed of the river, and measure in the wall 6,620 perches. Besides this, it swallowed up 90,000 bricks, 14,800 bushels of lime, and 81,200 cubic feet of timber. Its weight is estimated at 15,655 tons. Through the building there is a wall dividing each story into two unequal apartments—the one designed for manufacturing flour; the other for receiving wool and cotton machinery, a flax seed oil mill, fulling mill, and several other machines.

It is equally creditable to the prudence of the superintendent and the temperance of the laborers, that during the erection of this house, not one serious accident occurred.

PRESERVATION FROM FIRE

The means of accomplishing this, are few and inefficient. They are not therefore introduced on this occasion for imitation, but admonition. In the year 1808, the Select Council purchased a fire engine, and an association called the *Union Fire Company,* comprising nearly all the men in town, was formed. The engine proved indifferent, and the organization of the company still worse. For two years it has not had a single meeting. A second fire company was lately organized, which it is reported, intends to do some good. In 1813 a tax was assessed for the purchase of another engine, but it has not yet been obtained. The ordinances of the corporation require each house to be furnished with a fire bucket, but this requisition is disregarded by the majority. They also require every male citizen, between the ages of 15 and 50 years, to attend on the cry of fire; a provision finely calculated, if enforced, to augment the rabble which infest such places. A more important requisition, considering the absence of hose companies, is that each dray man shall furnish at every fire at least two barrels of water. Bonfires, and all other conflagrations on the streets or in-lots, are expressly but not successfully forbidden.

WATER

The borders of the town plat have a few indifferent *springs,* and on the surrounding hills there are others; but none afford water sufficient for distribution. The *wells* are of various depths — those east of Broadway are from 30 to 50 feet, in the Bottom from 40 to 60, and in some of the north-western parts of the Hill, from 20 to 40. Between Third and Sixth streets, and west of Broadway, they are from 70 to 100. The water afforded by some is slightly impregnated with iron, and the whole contain the several salts which abound in the wells of all countries. *Cisterns* are common, and from the general absence of coal in our fires, afford good water. But a large proportion of all that is used, is drawn up in barrels from the river. This is often impure, and requires time to settle; but for most domestic purposes, it is preferred to well water. The proprietors of the steam mill contemplate distributing water from the river over the whole town; a plan so interesting, that its execution will constitute an important era in our public improvements.

FUEL

Wood is the chief article of fuel at this place. Beech, ash, hickory, sugar tree, oak, red maple, honey locust and buckeye, are most in use. The first, from its excellence and profusion, will long continue to be burnt in larger quantities than any of the others. Many teams are constantly employed in hauling wood into town from the surrounding hills; but the principal part is rafted and boated down the Ohio and Licking rivers—the channels through which this important article will be mainly received in future.

As no coal has been discovered near to Cincinnati, but little of it is yet consumed here, except by manufacturers. It is brought from Pittsburgh, and sold on the river shore at 10 or 15 cents per bushel. The English chaldron seems to be unknown in the measure of this article on the Ohio.

MARKETS

Cincinnati has four market days in each week; two mornings at the small market house between Main and Sycamore streets, the oldest in town; and two afternoons at the market house in Fifth-street. That between Broadway and Sycamore-street, is not yet attended.—The Town Council have enacted a long and complicated ordinance regulating the markets, and keep an officer termed a clerk to carry it into effect; but violations are constantly suffered to pass unnoticed. At one or the other of these market houses, fresh meats can be had, except in the midst of winter, on every day in the week but the Sabbath. On the regular market days, however, the shambles are much more abundantly stored, and exhibit beef, veal, pork and mutton. The last is of superior excellence; the first, tho' generally good, is said to be inferior in flavor to that of the maritime states, which if true, is no doubt to be ascribed to a difference in the mode of fattening. The poultry is fine. The supply of fish is not great, tho' in the Ohio they are abundant. Perch, pike, eel, yellow-cat and swordfish are most esteemed—to these may be added the soft-shelled turtle, which is considered a great delicacy. Venison is brought from the woods, during the proper season; and bear meat is now and then offered. The quantity of butter and cheese is in general not equal to the demand, and much of both is of an inferior quality, which arises from the want of better dairies and a greater number of good cows than have yet been introduced into the fertile pasture grounds of the Miamics. Of vegetables, our markets afford an abundance. Among these, are a great variety of fruits, both native and cultivated. Of the former, blackberries, crab apples, pawpaws, fall, winter and fox grapes, mulberries, plumbs, wild cherries, cranberries, and the nuts of the walnut, hickory and chesnut are the principal. Of the latter may be enumerated many fine varieties of apple, peaches of a delicious flavor, pears, cherries, plumbs, quinces, raspberries, currants, gooseberries, strawberries, grapes,* and various kinds of fine melons. All the culinary roots, herbs and pulse of the middle states, with the sweet potatoe of the south, are plentiful and delicious.

Within four years, the prices of many articles in our markets have advanced; which indicates a rate of increase in the population of the town greater than that of the surrounding country. The effect of this will be, an increase in the number of grazing farms, the erection of larger dairies, and the cultivation of more extensive gardens, for the whole of which the vicinity of this place is most eligibly situated.

MANUFACTURES

As this town is *older* than the surrounding country, it has at no time had a surplus of laboring population or of capital. The former have been required to assist in clearing and improving the wilderness; the latter has been invested in lands, which from their low price and certain rise, have held out to capitalists a powerful inducement. The conditions which are said to constitute the basis of manufacturing establishments, have not, therefore, existed in the same degree as if the town had been *younger* than the adjoining country. Notwithstanding this, some progress has been made, as will appear from the following sketch, which embraces the manufactures most worthy of notice.

Cincinnati has no iron foundery; but is well supplied with blacksmiths, who fabricate in a neat and substantial manner, every article which those tradesmen usually make, and many others which belong to the whitesmith. Several shops are devoted to the manufacture of cut and wrought nails, which are made in sufficient quantities for the town and adjacent settlements. Stills, tea kettles and other vessels of copper, with a great variety of tin ware, are made in abundance. Rifles, fowling pieces, pistols, dirks and gun locks of every kind are manufactured. It is six years since a manufactory of cotton and woollen machinery was established, in which time 23 cotton spinning mules and throstles, carrying 3,300 spindles; 71 roving and drawing heads; 14 cotton and 91 wool carding machines; besides wool spinning machinery to the

---

*These are either brought from the vineyard of gen. Taylor, in Newport, Ky. or the more extensive plantations at Vevay, in the Indiana Territory. This place is chiefly inhabited by a body of immigrant Swiss, who employ themselves in little else than the cultivation of the vine, the manufacture of wine, and the distillation of spirits. Their wine has made its way into all the principal towns of the western country; but from some defect in its preparation, keeping, or the quality of the grape from which it is made, most likely the former, it is apt to become sour.

amount of 130 spindles; twisting machines and cotton gins, have been made. Plated saddlery and carriage mounting of all kinds, many different articles of jewelry, and silver ware of every sort—after the most fashionable models and handsomely *enchased,* are manufactured. Swords, dirks, &c. are mounted in any form, and either plated or gilt. Clocks of every kind are made, and watches repaired.

Sills, chimney pieces, monuments, and in short all the varieties of stone cutting, are executed with neatness and taste. Common pottery, of a good quality, is made in sufficient quantity for home consumption. A manufactory of green window glass and hollow ware, is about to go into operation; and will be followed by another of white flint glass the ensuing summer. Clean sand, of a beautiful white color, has been found in abundance near the mouth of the Scioto; but no clay proper for crucibles has been discovered as yet on the Ohio, and that article has to be brought from the state of Delaware.

The principal manufactures in wood are the following: sideboards, secretaries, bureaus, and other articles of cabinet furniture; all of which may be had of a superior quality, made either of our beautiful cherry & walnut, or of mahogany freighted up the Mississippi. Fancy chairs and settees, elegantly gilt and varnished. Waggons, carts and drays; coaches, phætons, gigs and other pleasure carriages, trimmed and ornamented. Plane stocks, weaver's reeds, and the different productions of the lathe, comprehending wheels, chairs, screws, &c. The various kinds of cooper's work, for the execution of which a machine has been erected and is now in full operation. The author of this invention is William Baily, of Kentucky, who in 1811 obtained a patent. The power is given by one or two horses, which with a man and a boy can dress and joint, in a superior manner, the staves necessary for one hundred barrels, hogsheads or pipes, in twelve hours. It can also be employed in shaving and jointing shingles, with equal advantage. The proprietors of the establishment in this place are making arrangements for the exportation of dressed staves to New-Orleans.

To the productions in wood may be added, the STEAM SAW MILL, erected on the river bank, below but adjoining the town. The principal building is a strong frame, 70 by 56 feet, and three stories high. The engine drives four saws in separate *gates,* acting at the rate of 80 times in a minute, making the product of each saw about 200 feet of boards an hour. The carriages run upon *cast racks,* are propelled by the improved *short hand,* and *gigged* backwards by *bevel wheels,* in the manner of the best mills. The logs to be sawed are chiefly brought in rafts to the beach, and drawn up the bank and into the mill by power from the engine. Other branches of business will be carried on in this establishment. The engine is estimated at 20 horse power, and of Evans' patent, except the condenser, which the proprietors have abandoned, as being attended with a degree of trouble and expense altogether disproportionate to its advantage. In place of this, they pour on the waste steam a current of cold water, which becoming instantly heated, is employed to replenish the boilers. The Steam Mill Company, and Cincinnati Manufacturing Company, have adopted the same alteration, with great success.

There are four cotton spinning establishments, most of them small. The whole contain upwards of 1200 spindles, which are moved by horses. Wool carding is performed in several places; and an extensive woolen manufactory, designed and calculated to yield 60 yards of broad cloth per day, will be in operation the ensuing winter. It is owned by the Cincinnati Manufacturing Company. The machinery is driven by an engine of 20 horse power. The products of the loom at this place have not been great; but several handsome pieces of carpeting, diaper, plaid, denim and other cotton fabrics, deserve to be mentioned. Cables, the various kinds of small cordage and *spun yarn,* are made in two extensive ropewalks. The latter has for some years been an article of exportation. Wool hats are not manufactured here; but fur hats, of a good quality, are made in such quantities, as to give a surplus for exportation to the Mississippi, where they are exchanged for peltry. The tanning and currying of leather is carried on at six tan yards in this place and its vicinity; and the manufacture of shoes, boots and saddlery, is extensive. Skin-dressing in alum is executed with neatness. Trunks covered with deer skin and oil cloth, leather gloves, and a great variety of brushes, are made, of a good quality. Blank books, and all kinds of common and extra binding, are executed with neatness.

The Cincinnati Manufacturing Company have embraced in their plan, manufactories of white and red lead, of such extent as will yield six or seven tons per week. The latter is not yet completed; but the former, which is the third that has been erected between the Mississippi and the mountains, is in operation, and produces white lead of an excellent quality. It must indeed be superior to that brought from the Atlantic states, as it has no mixture of *whiting,* with which the imported white lead is always alloyed.* Arrangements for a sugar refinery were made early in the present year; the buildings have been commenced, and the establishment will be in operation in a few weeks. Tobacco and snuff are manufactured in four different shops. Pot and pearl ash, soap of various kinds, and candles, are made in such quantities as to give a large surplus for exportation.

The rectification of spirit and distillation of cordials, are prosecuted to such a degree as to give an ample supply of the latter for domestic use. But these

---

* See Cooper's Emporium of Arts and Sciences.

establishments, both in extent and utility, are eclipsed by our breweries. The first was erected on the river bank, in the lower part of the town, four years ago, and uses the river water; the other was established since, on a smaller scale, and derives its water from wells and cisterns. The two are calculated to consume annually 30,000 bushels of barley. Their products are beer, porter and ale, of a quality at least equal to that of the Atlantic states. Large quantities have been exported to the Mississippi, even as far as New-Orleans, the climate of which they are found to bear very well.

The manufacture of flour, at the steam mill, will be carried on to a great extent. The machinery is all on the plan of Oliver Evans, and driven by an engine of 70 horse power. Four pair of six feet burr stones will be run. Two pair have been in motion for several months, and produce about 60 barrels of flour per day; the whole when in operation will, it is expected, afford 700 barrels a week. The flour is generally of a superior quality.

In the year 1814 a mustard manufactory was erected above the town; but has not yet got into such extensive operation as to supersede the importation of that article.

In the fine arts we have not any thing to boast; but it is worthy of being mentioned, that all kinds of labeling, sign and ornamental painting, together with the engraving on copper of official and other seals, cards of address, and vignettes, is executed with taste and elegance.

COMMERCE

VESSELS. Flat bottomed boats, keel boats and barges, are the vessels in which the commerce of this place has hitherto been carried on. The first will long continue to be employed in transporting heavy articles down the Ohio; but the latter, it is probable, will be in a great degree superseded by steam boats; of which two kinds are coming into use on the western waters. From these inventions the people on this river anticipate many substantial advantages; more, perhaps, than will be realized; but all must admit, that no country on earth, equally fertile with this, can be more benefited by such boats. The reduction of the voyage from New-Orleans to Cincinnati from a hundred, to thirty days, is equivalent to an approximation of the two places, or to the annihilation of two-thirds of the distance; and superadds to the security and abundance of a temperate interior region, the productions of the south, and of all foreign lands.

EXPORTS. Of these, flour is the chief article, and several thousand barrels are annually exported from the Miami country to New-Orleans. After this follow pork, bacon and lard; whiskey, peach brandy, beer and porter; pot and pearl ash, cheese, soap and candles; hemp and spun yarn; walnut, cherry and blue ash boards; cabinet furniture and chairs; to which might be advantageously added, kiln dried Indian meal, for the West Indies.

IMPORTS. The different kinds of East Indian, European and New-England goods, with several manufactures of the middle states, are received from Philadelphia and Baltimore, but chiefly from the former. It is not difficult to foresee, however, that at no distant time the ingress of foreign merchandise will be through other channels. A portage of three hundred miles, over high and rugged mountains, must at all times be more expensive than ascending a navigable river five times the distance. Whenever the General Government shall complete the road from the navigable waters of the Potomac to those of the Ohio, the expense of transportation by land will be so far reduced, that factories and other mercantile houses, will perhaps at no distant period be established on the former of these rivers. Should New-York execute the canal which it has projected, the metropolis of that flourishing state will probably become one of our inlets for foreign goods. But the great emporium of the western country in future must be New-Orleans. To effect this change in the current of importation, but three things are necessary—more extensive and wealthy mercantile houses in that city; an improvement in the navigation of the Ohio at the Falls; and an increased number of steam boats. Even under existing circumstances, many articles are brought from thence at a lower price than from the eastern cities; of which coffee, salt fish, claret and some other wines, copperas, queensware, paints, mahogany and logwood, may be cited as examples. In addition to these, we obtain from the state of Louisiana, of its productions, sugar and melasses, cotton, rice, salted hides, and some other articles.

Our imports from the Missouri Territory are lead, peltry and skins—from Tennessee and Kentucky, cotton, tobacco, salt petre and marble—from Pennsylvania and Virginia, bar, rolled and cast iron, with several of the manufactures of that metal; millstones, coal, salt, glassware, pine timber and plank. Castings of an excellent quality are brought from Zanesville and Brushcreek, in this state. And furs are obtained from the waters of the Great Miami, Wabash and Maumee.

The goods brought for consumption in this quarter are kept in more than seventy shops. Of these about sixty contain dry goods, hard, glass and queens wares, liquors and groceries. The others are stores for iron, shoes and drugs.

Cincinnati was made a port of entry in 1808, but the business of building ships having been discontinued on the Ohio, no vessel has yet *cleared* from this place.

BANKS

MIAMI EXPORTING COMPANY. This is the oldest banking institution in the Miami country, being incorporated in 1803 for forty years. The original object of the company was the exportation of agricultural produce to New-Orleans; but the charter permitting issues of bank paper, an office for that purpose was opened in this place, and on the 1st of March 1807, the bank went into full operation, all commercial projects having been previously relinquished. The capital is divided into shares of $100, and $450,000 have been paid in by one hundred and ninety persons, the present number of shareholders. The affairs of the company are managed by eleven Directors, chosen annually, one of whom is elected President. The reputation and notoriety of this institution are equal to that of any bank in the western country; and its dividends correspond, having for several years fluctuated between 10 and 15 per cent. Oliver M. Spencer and Samuel C. Vance are the President and Cashier.

FARMERS' & MECHANICS' BANK. This was established in the year 1812, and incorporated in 1813 for five years; at the expiration of which time the charters of all the banks in the state, except the Miami Exporting Company, will expire. The shares are $50 each, and the amount of capital as prescribed in the law $200,000. The number of Directors is thirteen; one-third of whom must be practical farmers, and the same proportion practical mechanics. The President is elected out of their own body. The paper of this institution has acquired an extensive circulation, and its dividends have varied from 8 to 14 per cent. The officers are William Irwin, President, and Samuel W. Davies, Cashier.

BANK OF CINCINNATI. This was founded in 1814, and made its first issues of paper in the month of June of that year. Its shares are $50 each. Eight thousand eight hundred have been sold, to three hundred and forty-five persons. $140,000 have been paid in. It has not yet been chartered, and is governed by twelve Directors, chosen annually, one of whom is declared President. Its notes are in excellent credit; and the dividends, for a new institution, very good—having advanced during the first year from six to eight per cent. The President is Ethan Stone; Cashier, Lot Pugh.

NEWSPAPERS

The first newspaper printed north of the Ohio river, and the third or fourth west of the mountains, was issued at this place November the 9th, 1793, by William Maxwell. It was on a half sheet royal of quarto size. Its name was THE CENTINEL OF THE NORTH-WESTERN TERRITORY—its motto, *Open to all parties, but influenced by none.* In the summer of 1796, Edmund Freeman purchased the establishment, and changed the name of the paper to the FREEMAN'S JOURNAL, under which he continued it till the beginning of 1800, when he removed to Chillicothe.

On the 28th of May 1799, Joseph Carpenter issued the first number of a paper entitled THE WESTERN SPY & HAMILTON GAZETTE, which was continued by various editors for ten years. The name was then changed by Messrs. Carney & Morgan to THE WHIG; 58 numbers of which were published—when it passed into other hands, and had its title altered to THE ADVERTISER, under which it was continued till November 1811, when it expired.

In September 1810, Mr. Carpenter re-commenced THE WESTERN SPY, which has been regularly published ever since. At present it is of a super royal size, has about 1200 subscribers, and is edited by Messrs. Morgan & Williams.

A paper called LIBERTY HALL AND CINCINNATI MERCURY, was established in 1804 by John W. Browne. The first number came out on the 4th day of December. Its present editors are Messrs. Looker & Wallace, who print it of a super royal size, and have upwards of 1400 subscribers.

In the month of July 1814, a paper entitled the SPIRIT OF THE WEST was commenced; but continued only for 41 numbers. On the 15th of July 1815, the first number of the CINCINNATI GAZETTE was issued by Thomas Palmer & Co.

None but weekly papers have yet been published here. The offices of Liberty Hall and The Western Spy have each an extra press, for book printing, which is executed with accuracy and neatness. Ten years ago, there had not been printed in this place a single volume; but since the year 1811, twelve different *books,* besides many pamphlets, have been executed. These works, it is true, were of moderate size; but they were *bound,* and averaged more than 200 pages each. The paper used in these offices was formerly brought from Pennsylvania, afterwards from Kentucky, but at present from the new and valuable paper mills on the Little Miami.

POST OFFICE

As furnishing data for estimating the state of political curiosity and information, this office is worthy of notice. The number of mails that arrive every week is nine; by which are brought for distribution in the town about seventy different papers, making three hundred & fifty sheets. Besides these, a great number of papers and documents, franked by members of Congress, and most of the eastern periodical works, are received through the same channel.

The office was established in 1793. Abner Dunn was the first post master;

and his successors have been William Maxwell, Daniel Mayo, William Ruffin, and William Burke, who at present holds the office.

EDUCATION

One thirty-sixth part of the state of Ohio has been granted by the General Government for the support of schools; besides two or three townships for college education. Of these donations our legislature is the guardian, and has enacted several laws respecting them. In most parts of the state, the 16th, which is one of the four central sections in each township as originally surveyed, is the one assigned for this purpose. This is the case in the Miami country. In each township there have been, or should be elected, three trustees and a treasurer, who possess corporate powers as it respects the school section; which it is their duty to lease out to different persons, for periods of 15 years, and to divide the rents among the schools of the township, according to their relative number of scholars. What advantage the people of the adjoining country have derived from these donations, I am not prepared to state. To the inhabitants of this place they could be of no benefit, as the township of Cincinnati is fractional, and does not include the section numbered 16. The proprietors of the town must have known this, but they made no donation for the support of education, not even a site for a school house. The business of tuition was therefore generally conducted by strangers, and transient teachers, in rented rooms, till the year 1811; when ten or twelve individuals purchased a small lot, erected a couple of school houses, and employed two or three teachers; but notwithstanding their laudable exertions, this academy has not flourished, and is likely soon to be superseded by an institution, of which I will now proceed to give some account.

CINCINNATI LANCASTER-SEMINARY. A project for establishing in Cincinnati a school on the plan of Joseph Lancaster, of Great Britain, was agitated more than three years ago, by the reverend Joshua L. Wilson, to whom a teacher residing in the Atlantic states had written on this subject. Nothing, however, was done at that time; and early in the year 1814, Edmund Harrison, of the state of Tennessee, who had been instructed by one of the pupils of Lancaster, came to this place and proposed to the Methodist Episcopal Church, of which he is a member, to undertake a school on the Lancasterian plan. His proposition was readily accepted by that public spirited body, and the reverend Oliver M. Spencer drew up a body of articles for the government of the association, under which the school was to be organised. In these articles, no provision was made for instruction in the higher branches of literature; and a majority of the trustees, it was declared, should be at all times members of that church. Exception being taken by some persons both to this provision and defect, a modification was proposed. This, after some negotiation, failed; and a rival institution was formed, under the name of the *Cincinnati Lancaster-Seminary*. By the mediation of the teacher, the two schools, not long after, were united under the same articles, and in the ensuing winter a law of incorporation was obtained. The monies subscribed in 1814 for the benefit of the seminary, amounted to nearly $9000, payable in shares of $25 each. Since the commencement of the present year, about $3000 more have been contributed; and the Banks of the town, with a laudable desire for the promotion of learning, have agreed to loan to the institution, on an extended credit, the sums necessary for the completion of its edifice. A suitable site for the building became a desideratum, but this was soon supplied by the Presbyterian congregation, which in the true spirit of christian benevolence, executed to the Directors of the seminary a lease for 99 years, of the ground on which the building is erected, without any other compensation than the privilege of selecting annually for instruction, 28 poor children, to be considered as charity scholars. By the charter and by-laws, the seminary consists of a Junior and a Senior department, each subdivided into a male and a female school. The Junior department to be organised on the plan of Joseph Lancaster; and the Senior according to such plans, and under such teachers and professors as the Board of Directors may choose. The surplus revenue from the Junior department, after defraying its expenses, and deducting the tuition of those who may be considered as objects of charity, is to be applied to the purchase of books and philosophical apparatus for the Senior department. The price of schooling in the former is reduced to eight dollars a year; and in addition to the charities already mentioned, it is provided that if a shareholder upon dying, shall leave his children without the means of purchasing an education, they shall be entitled to a regular course of instruction in the lower department. The institution is governed by seven Directors, elected annually by the shareholders. These Directors may be of any, or of no religious society; they elect a President from their own body, and have the exclusive management of all the pecuniary and literary concerns. Jacob Burnet has been President from the commencement.

On the 17th of April 1815, one of the lower rooms being completed, a school composed of children of both sexes, was opened, and in less than a fortnight 420 were admitted; when the apartment being sufficiently filled, many subsequent applicants were rejected. By the indefatigable exertions of the teacher, order and method were at length introduced, and the proficiency of the scholars has equalled all reasonable expectation. A second school, on the same plan, for females only, has just been commenced, and promises to be well filled.

The Board of Directors, by a late resolution, have decreed the establishment of a school for children of color, in a separate house; but no teacher has yet been procured.

CINCINNATI UNIVERSITY. In the year 1806, a school association was formed in this place, and in 1807 it was incorporated. Its endowments were not exactly correspondent to its elevated title, consisting only of moderate contributions; and an application was made to the legislature for permission to raise money by a lottery, which was granted. A scheme was formed, and great part of the tickets sold: they have, however, not been drawn, and but little of the money which they brought refunded. On Sunday the 28th of May 1809, the school house erected by the corporation was blown down; since which it has become extinct.

MIAMI UNIVERSITY. In the year 1809 the legislature of this state, which by an ordinance of the General Government holds the school and college lands in trust, enacted a law creating and incorporating the *Miami University*. By this act, the Governor was authorised to appoint three commissioners to fix on the site of the institution; and Lebanon was selected. The succeeding legislature revoked this decision, and by a liberal, if not an unwarrantable construction of its powers, removed the site of the edifice to the land with which the college is endowed, lying, as we have seen, west of the Great Miami, and beyond the limits of Symmes' purchase. By the same disastrous law, the Trustees were directed to lay off a town, which was of course named Oxford, and to sell out on leases as much of the township as they should consider expedient; all of which was so amply performed, that in less than two years nearly one-third of this valuable endowment was disposed of, on terms which will not yield a revenue adequate to the support of a grammar school! It was however to be reduced still lower; and a succeeding legislature passed a law exempting for a number of years those purchasers who might become actual settlers before the year 1816, from the payment of a large proportion of their annual rents.

Previous to this, as if sagaciously anticipating a defalcation, and with the laudable ambition of erecting on the ruins of the wigwam an edifice devoted to literature and the sciences, the Trustees appointed John W. Browne, minister of the gospel, to solicit alms. The reverend missionary set out, and after two years of devious and thorough travelling in the East, returned richly freighted with more than four hundred dollars in cash, besides a ponderous cargo of venerable volumes; great part of which, as being obsolete, worn out, or otherwise unworthy of preservation for a college library, the present directors have wisely ordered to be sold at auction.*

In the year 1814, the Trustees authorised the purchase of a quantity of building materials, and contracts were accordingly made for brick and timber; but their successors finding the treasury almost empty, and calculating that the annual revenues for many years would not be adequate to the completion of a building, have suspended further purchases.

Such are the progress and present state of this institution, in the history of which I have been the more explicit, on account of the erroneous information which has gone abroad respecting it. That it will attain to the rank of a second rate college, in the course of the present century, where it is now fixed, no well informed person has the courage to predict. The general opinion is, that both the interests of the seminary, and common justice to the people for whose benefit it was expressly designed, require its restoration to Symmes' purchase; where the funds necessary to the erection of suitable edifices could be promptly raised by subscription; and a college organised in time to benefit the rising generation. Whether this will be done, depends on the wisdom of future legislatures.

LIBRARY

It was not until the year 1809 that any efforts were made towards the establishment of a public library in this place. A petition was then forwarded to the legislature of the state for a law of incorporation, but it proved unsuccessful. In the summer of 1811, a paper was circulated by Judge Turner, who obtained subscriptions for several hundred dollars. A meeting of the shareholders was held, and a constitution adopted, which they ordered to be sent on to the next legislature as the basis of a charter for the society. This was not done till a subsequent session, when a law incorporating the association under the name of the CIRCULATING LIBRARY SOCIETY OF CINCINNATI, was enacted. Owing to various causes, however, the library was not opened until April 1814. Since that time, a second and more perfect charter has been obtained, and the institution is at present in a flourishing state. It has about 800 volumes, which are arranged under the following heads: *Arts and Sciences, Belles Lettres and Rhetoric, Biography, Botany, Chemistry and Medicine, Drama, Education,*

---

* In justice to several members of the Board, it should be observed, that they were opposed to inflicting on their infant seminary the stigma of mendicity. They moreover believed, that the people of the United States had, through their government, made such donations in land for the support of education in this state, as should exempt them from applications for charity.

Geography, History, Law, Metaphysics and Moral Philosophy, Natural History, Natural Philosophy, Novels, Philology, Poetry, Politics, Theology, Veterinary Art, Voyages and Travels, Miscellanies, and continued Periodical Works.

Among the more valuable scientific books, are *Rees' Cyclopædia* and *Wilson's Ornithology*. About 60 volumes have been received as donations. The affairs of the society are managed by seven Directors, elected annually, one of whom is designated as President. The library is kept open one day in each week.

SCHOOL OF LITERATURE AND THE ARTS

This is an association for literary and scientific improvement; composed chiefly of young men, who formed themselves into a society in 1813, and elected Josiah Meigs, an accomplished scholar, their first President.* Their constitution provides for frequent meetings, at which the exercises are of three kinds: a lecture from the President—an essay from one of the members—and a poetical recitation from another. On the 23d of November 1814, the school held its first anniversary meeting, at which an oration was delivered by appointment. From this discourse, it appears that many interesting lectures and essays have been delivered, and that the infant institution is probably the germ of a permanent and respectable society.

* * * *

STATE OF SOCIETY

This cannot, of course, be pourtrayed with the same facility and exactness as in older communities. The people of the Miami country, may in part be characterised, as industrious, frugal, temperate, patriotic and religious; with as much intelligence, and more enterprise, than the families from which they were detached.

In Cincinnati the population is more compounded, and the constant addition of emigrants from numerous countries, in varying proportions, must for many years render nugatory all attempts at a faithful portraiture. There is no state in the Union which has not enriched our town with some of its more

---

* This gentleman was at that time Surveyor General of the United States; but is now Commissioner of the General Land Office, at Washington City.

enterprising or restless citizens; nor a kingdom of the west of Europe whose adventurous or desperate exiles are not commingled with us. To Kentucky, and the states north of Virginia—to England, Ireland, Germany, Scotland, France and Holland, we are most indebted.

Among such a variety, but few points of coincidence are to be expected. Those which at present can be perceived, are industry, temperance, morality, and love of gain. With a population governed by such habits and principles, the town must necessarily advance in improvements at a rapid rate. This, in turn, excites emulation, and precludes the idleness which generates prodigality and vice. Wealth is moreover pretty equally distributed, and the prohibition of slavery diffuses labor—while the disproportionate immigration of young men, with the facility of obtaining sustenance, leads to frequent and hasty marriages, and places many females in the situation of matrons, who would of necessity be servants in older countries. The rich being thus compelled to labor, find but little time for indulgence in luxury and extravagance; their ostentation is restricted, and industry is made to become a characteristic virtue.

It need scarcely be added, that we have as yet no epidemic amusements among us. Cards were fashionable in town for several years after the Indian war that succeeded its settlement; but it seems they have been since banished from the genteeler circles, and are harbored only in the vulgar *grog-shop* or the nocturnal *gaming-room*. Dancing is not infrequent among the wealthier classes; but is never carried to excess. Theatrical exhibitions, both by *amateurs* and *itinerants,* have occurred at intervals for a dozen years; and a society of young townsmen have lately erected a temporary wooden playhouse, in which they have themselves performed. But as the tendency of their institution to encourage *strollers* and engross time, has been deprecated by the more religious portion of our citizens; and as the members have failed to realise their anticipations, with regard to the accumulation of a fund for the relief of indigence, they will be likely soon to relinquish the pursuit, and leave their stage and its trappings to some future votaries of Thespis. During the winter, select parties are frequently assembled; at which the current amusements are social converse, singing and recitation—the latter of which has been lately predominant. Juvenile plays and diversions are sometimes resorted to; which are generally such as promote a rational exercise of the mental faculties. Sleigh riding and skaiting are rarely enjoyed, on account of the lightness and instability of the snow and ice. Sailing for pleasure on the Ohio is but seldom practised; and riding out of town for recreation, on horseback or in carriages, is rather uncommon, for want of better roads. Evening walks are more habitual, in which the river bank and adjacent hills—the *Columbian garden* —and the *mound,* at the *west end,* are the principal resorts.

CHAPTER IV, POLITICAL TOPOGRAPHY

Some apology is perhaps necessary for imposing on the people of the Miami country the items comprehended in the following chapter; with nearly the whole of which they must be already acquainted. It is hoped, however, that they will excuse it, from the consideration that persons at a distance, who may contemplate an emigration hither, will be gratified to know something of our political, as well as our social institutions.

POPULATION OF THE MIAMI COUNTRY

In the year 1790, this did not exceed 2000. In 1800, it was about 15,000. In 1810, the single county of Hamilton, not embracing more than 500 square miles, had 15,204; and the Miami country, excluding that part which lies beyond the state line on the west, had about 70,000, or one-fourth of the population of the state. At present (August, 1815) it cannot be less in this district than 100,000; which is spread over 4000 square miles, giving 25 for each mile. In 1810, the township of Springfield, in the interior of this county, had nearly 58 to each square mile; and could certainly support many more;—and that the density of population over the whole tract, in ten years, will equal 50 to each mile, is an expectation warranted by the general rate of increase since 1790; by the uniform fertility of our soil; and by the subdivision and sale of our lands in tracts of 160, and even 80 acres—a regulation, which in the United States, is indispensable to a thick population.

POPULATION OF THE TOWN

I have not been able to ascertain this, at an earlier period than 1810. It was then 2320. In the latter part of 1813, the Select Council made a census, which gave about 4000. From various estimates, it appears certain, that at the present time it is 6000—nearly 10, on an average, to each dwelling house; a number, which no one, who examines the town, will pronounce to exceed the reality; although it greatly transcend the limits which health and comfort would prescribe. In 1810, the males were to the females as one hundred to eighty-two and a half; and at the present time the disproportion is perhaps still greater—a striking contrast with Rhode-Island, where the former are to the latter, as one hundred to nearly one hundred and five.

NEGROES

By the ordinance of Congress, passed July 13, 1787, providing for the government and defining the principles on which the people of the North-western

Territory, when divided into states, should form their constitutions, it is expressly declared that there shall be neither slavery nor involuntary servitude, except for the punishment of crimes, unless with the consent of both the General Government and the people of the Territory. When the constitution of Ohio was formed, the prohibitory language of the ordinance was adopted, and slavery is forever excluded from this state. That the other Territories north-west of the Ohio will pursue the same course, there can be no doubt; and hence this fine river will acquire additional distinction in future, from being made the northern barrier to this execrable practice.

Both the ordinance of Congress and the constitution of Ohio, guarantee the recovery of fugitive slaves; but by the decision of our courts, those brought hither are free from the moment of their arrival. By our constitution, *white* male inhabitants *only,* enjoy the right of political suffrage: negroes are of course excluded from that privilege. By a statute enacted in 1804, and amended in 1807, free negroes are prohibited from settling in this state, without giving bond and security that neither they nor their children shall become public charges; but as this provision is considered unconstitutional, it has, I believe, in no instance, been enforced, and we have all the black population which an unopposed immigration could give. By the same laws, negroes and mulattoes are prohibited from giving testimony against white persons. Whether this be not unconstitutional, as well as the other, may be doubted; but it is generally carried into effect throughout the state.

At the time of adopting our state constitution, it was predicted that we should be degraded by the free negroes of other states, and infested with their runaway slaves—neither of which has yet been realized. The *political* distinction between the blacks and whites being abolished, the *social,* it was asserted, would suffer the same fate; but experience has shown, that the contaminating influence of slavery itself is most favorable to that dark effect. In no town of the state is there so great a proportion of black population, as in Cincinnati, where in 1810, it amounted only to 79, making about one-thirtieth of the whole. At present the number of blacks and mulattoes does not exceed 200, counting all shades and ages. They are a thoughtless and good humored community, garrulous and profligate; generally disinclined to laborious occupations, and prone to the performance of light and menial drudgery. A few exercise the humbler trades, and some appear to have formed a correct conception of the objects and value of property, and are both industrious and economical. A large proportion are reputed, and perhaps correctly, to practice petty thefts; but no more than one individual has been punished corporally, by the courts of justice, since the settlement of the town.

MILITIA

The militia of Ohio are organized in divisions, brigades, regiments, battalions and companies. Those of Cincinnati compose an odd battalion, in the first brigade of the first division. They number about 800; and are divided into five companies, one of which is light infantry. The days for mustering and training are only two in spring, and four in autumn; two of which are for officers alone—the discipline of the whole is of course imperfect, without any prospect of amendment.

SUPPORT OF THE POOR

No pauper is by law entitled to support from the township, without a residence of one year. The common mode of maintaining those who are permanent charges, is to offer them annually to the lowest bidder. The funds for defraying this expense, and for the support of poor generally, are raised by an annual tax on the same species of property which is taxed for county purposes.

With the design of extending charity to the needy, who in consequence of their recent arrival here can demand nothing from the overseers of the poor; and to those citizens who are, through misfortune, in want of temporary assistance, a number of charitable persons associated themselves in 1814, under the name of the *Cincinnati Benevolent Society*. They appointed two managers in each ward of the town, and by the voluntary contribution of a respectable portion of the inhabitants, a sum was obtained that has enabled the Society to dispense relief to a number of suffering immigrants. A part of the design, which will perhaps be hereafter executed, is the erection of a *work house;* where those who are unable *entirely* to support themselves, will find assistance, and be compelled to labor according to their abilities. Another important establishment by this Society, would be a *Dispensary,* for the relief in sickness, of those families who in health do not require gratuitous assistance.

CORPORATION

On the 1st of January 1802, Cincinnati was incorporated by the Territorial Legislature, with the following limits: viz. Mill-creek on the west; the township line, which lies about one mile from the river, on the north; and the eastern boundary of fractional section No. 12, which extends nearly half a mile above the town plat, on the east. On the 10th of January 1815, this law was superseded by another, which retained the same boundaries. By the latter, the town is divided by straight lines into four wards, in each of which three Trustees are elected for two years. When assembled for the first time, they appoint from their own body, out of the different wards, a Mayor, Recorder, Clerk and Treasurer. The powers delegated to the Town Council are, to pass and enforce such ordinances as may be necessary and proper for the health, safety, cleanliness, convenience, morals and good government of the town and its inhabitants. The tax which they have power to assess on real estate, cannot exceed one half per cent. annually, without a vote of their constituents. On all violations of the ordinances of the corporation, it is exclusively the duty of the Mayor to decide; an appeal being had either to the Town Council or Court of Common Pleas, at the option of the person considering himself aggrieved. The Mayor exercises, moreover, the principal duties of a Justice of the Peace, within the limits of the corporation.

TOWNSHIP OFFICERS

The boundaries of Cincinnati township are, on the east and north, the same with the corporation; on the west it extends a few miles beyond Mill-creek, until the northern boundary line touches the Ohio. In each township of the state, there are annually elected three Trustees, and several subordinate officers; whose duty it is to assess and collect taxes for the support of the poor, repair and improve the roads and streets, select jurors, and generally to superintend the affairs of the township.

COUNTY COMMISSIONERS

These are three in number, and are elected every third year. It is their duty to levy taxes for county purposes, to superintend the erection of public buildings, and generally to manage the revenues, property and concerns of the county.

RECORDER'S OFFICE

In each county of our state, there is an office for recording deeds, mortgages, leases, town plats, and such other written articles as it is important to preserve. A certified transcript of any of these, is received in evidence the same as the original. The Recorder is appointed for seven years, by the court of Common Pleas. He receives no salary; and his fees are determined by law.

JUSTICES OF THE PEACE

These officers are elected for three years. They vary in number in each township, according to the decision of the court of Common Pleas. In this township, they are generally three. In civil cases, the jurisdiction of a Justice extends to 70 dollars; and by consent of parties, to $200. In criminal cases, it is co-extensive with the county; but his only power, except in a few trivial offences, is to recognize the culprit to appear before a higher tribunal.

COURT OF COMMON PLEAS

In Cincinnati, which is the seat of justice for Hamilton county, there is a session of this court every four months. It is composed of a President and three Associates, elected by the General Assembly for seven years. It has cognizance of all violations of the statutes of the state, whether civil or criminal, which are not punishable with death. In the last cases, the offender has his choice between this and the Supreme Court. It has also unlimited appellate jurisdiction from the Justice's court, and may be selected as the court of appeals from the Mayor's decision. In these cases, its sentences are not liable to reversal. In all others, they may be set aside by the following tribunal.

SUPREME COURT

This is held annually, and is composed of three Judges, who visit every county in the state. They are elected for the same period with the last. The causes in this judicature are generally appeals from the court of Common Pleas; but it has original jurisdiction in all capital offences; and in civil cases, where the matter in dispute exceeds $1000. It is a tribunal from which there is no appeal.

The court of Common Pleas has jurisdiction in all cases cognizable by a court of Chancery, in which complete remedy cannot be had at law. The Supreme Court has original concurrent jurisdiction with the court of Common Pleas, where the title of land is in question, or the sum in controversy exceeds $1000; and appellate chancery jurisdiction in all other cases cognizable by the court of Common Pleas.

ATTORNIES AND COUNSELLORS AT LAW

By our statutes, these are licensed only by the Supreme Court; before which they undergo an examination. Certificates of moral character, and of a regular course of law studies, or of admission to practice elsewhere, are indispensable.

No previous residence is necessary; but the applicant must satisfy the Court, by affidavit or oath, that he intends to reside in the state.

CAPITAL PUNISHMENTS

A penitentiary having been lately erected in COLUMBUS, the capital of our state, the whole code of criminal law has undergone revision. Heretofore, the number of capital offences was five. At present it is but two, murder and treason. At this place there have never been but two convictions of this kind. They were both for murder, and within five years after the settlement of the town. One of the felons was pardoned, and the other executed. They were foreigners by birth, and the latter was attached to the army; but not in such a manner as to be tried by a military tribunal.

POLITICAL IMPORTANCE

Cincinnati was the residence of the Governor of the North-western Territory from 1790 to 1800. In that year the seat of government was removed to Chillicothe, as being more central.

In 1788 a wooden fort was erected here, which was garrisoned till 1802, and soon after erased. This was the key to a line of similar forts, extending quite to the Rapids of the Maumee; the whole of which, except Fort Wayne, were long since evacuated and burnt. For many years, therefore, Cincinnati has not been the site of any political or military establishment, and its position does not favor a prospect of any such distinction in future. Of course, no part of its unexampled progress in population and improvement can be ascribed to political aids, which might hereafter be withdrawn; but the whole has resulted from such natural and commercial advantages, as cannot easily be transferred or destroyed.

CHAPTER V, MEDICAL TOPOGRAPHY

Under this head it is proposed to communicate, as fully as possible, such information concerning our diseases, and such notices of the mineral springs within our reach, as a person about to emigrate to the Western country would desire.

SECTION I. PREVAILING DISEASES

Of the diseases connected with climate, we have most of those which are common in the same latitudes, east of the Alleghenies. Some of them, however, are less violent and frequent here than there. Of this kind is the *Pulmonary Consumption;* which, in the Atlantic cities, destroys from a fourth to a sixth of all who die; while in this town, it produces not more than one-twentieth of the deaths. So favorable, indeed, is this place to those who are threatened with Consumption, that a migration to it from the Northern states might be advantageously recommended, when this complaint is about commencing, or not very far advanced. The *Pleurisy* and *Peripneumony* occur every winter; but seldom prevail to any great extent. They are generally complicated with bilious affections; which renders the treatment difficult, and makes the use of calomel, in most cases, absolutely necessary to a successful issue. The *Croup* is a formidable disease in this place, annually carrying off a number of children. Like the preceding complaints, it is frequently attended with bilious symptoms; and occasionally shows itself in connexion with Cholera Infantum, forming a very dangerous combination. In general, it does not seem to be a worse malady here than in the East; and I have never seen it of that malignant and epidemic character at Cincinnati, which it exhibited in Virginia in 1799.* *Colds* and *Catarrhs, swelled tonsils, and other affections of the throat,* produced by sudden changes of weather, occur here in the same manner as in the maritime states; but do not appear to be so often followed by consumption. The *premature decay of teeth, pains in the jaw, and tooth ache,* frequent in all variable climates, are, it would seem, much less common here, than in some parts of New-England; as Dr. Hazletine informs us, that they make about an eighth part of all diseases incident to the people of the Province of Maine. *Rheumatism* occurs; but is not so frequent and formidable as in the Northern states.

Of the diseases ascribed to the exhalations from putrefying animal and vegetable substances; from alluvial ground, and from ponds and marshes, we have perhaps the whole catalogue, with the exception of the Yellow Fever of the Eastern cities. In the country, especially along the water courses, *Remitting* and *Intermitting Fevers,* including *Ague,* prevail every autumn; but are seldom malignant, and generally yield to the treatment elsewhere employed, if resorted to at an early period. In Cincinnati, the annual prevalence of these diseases is less certain, and the *mild* and *malignant Typhus Fevers* frequently supply their places. In the years 1809, '10 and '11, these complaints were prevalent here, without much intermission; but since that time they have been rare.

The diseases to which immigrants are most liable, are bilious and typhous fevers. This is especially the case with the natives of New-England and

---

* See Medical and Physical Journal, vol. ii.

New-York, who in coming here undergo a change of climate greater than they seem generally to suppose. They should, therefore, endeavor to arrive in the Miami country late in the autumn; and before the ensuing summer, place themselves in the most healthy situations which can be found. Those who intend to reside in the country, should get on upland farms at an early period: those who prefer the town, should choose the eastern and northern portions, which are more exempt from noxious effluvia; and, in the heat of summer and early autumn, expose themselves as little as possible, either to the evening air, or the noon day sun. With these precautions, and a strict regard to the prevention of what is denominated a bilious habit, very few will suffer an attack; but without such attention, *a seasoning,* as it is termed, will most likely be experienced the first summer after an arrival from the North. In the second, whether the first be sickly or not, there is but little danger.

Next to our fevers, are the different complaints of the stomach and bowels. These prevail chiefly in the summer, as in other parts of the United States, and precede the fevers which have been enumerated. The *Cholera Infantum* is commonly the first which occurs, and sets in with the earliest intense heat. Its greatest prevalence is in June and July, when it frequently proves fatal, particularly in town. It sometimes destroys life in a few days; at other times the unfortunate little sufferer pines for several weeks, when he either dies, or is restored by the frosts of autumn. In the country this disease is less frequent, and so mild as not often to prove fatal. The *Cholera Morbus* occasionally presents itself, at the seasons in which it is more or less prevalent over all the States. A few cases of *Dysentery* occur every summer; and once in two or three years, it is epidemic. When this is the case, its prevalence is sometimes very general, but not often mortal. Now and then it assumes a malignant character; when it is, for the most part, confined to a single family. Upon the whole, this disease appears to be less formidable in this country, than in the Atlantic states. On the head waters of the Great Miami, and in some of the adjoining parts of Kentucky, a disease called by the people the *Sick-stomach,* has prevailed more or less for several years. Its prominent symptoms are, a vomiting upon taking exercise, with chronic debility, lassitude and soreness of the extremities. Sometimes it continues for months, in the same individual; and frequently affects whole families. It is supposed to extend to horses, cows, sheep and dogs, varying in several of its symptoms. It does not often prove fatal, and the people, where it is endemic, seem to have learned by experience an efficacious method of treatment. It has been ascribed to some noxious impregnation of the water; to the use, by the animals whose milk and flesh are eaten, of some deleterious plant, and to marsh exhalation—the last of which is the most plausible. For two or three years past, its occurrence has been more infrequent, and it cannot be regarded as constituting any serious objection to the districts in which it prevails. The *Jaundice* is a pretty common disease in this country; but it seldom destroys life. *Inflammation of the liver* is met with occasionally, but not oftener than in the same latitudes of the maritime states. Sore-eyes (*Ophthalmia*) is a disease which now and then becomes epidemic over the whole of this country. It prevails most in the same situations where the ague, and other forms of bilious fever abound; and has therefore been referred to the same cause. It does not arise from heat or dust, as it occurs oftenest in shady vallies; nor from the smoke of autumn, as it precedes that phenomenon. It is less frequent than formerly, and will perhaps cease with those diseases which are acknowledged to depend on marsh exhalation. The *Periodical head ach* is a disorder which in this country is ascribed to the same cause, and can be cured in the same manner as ague and fever.

Of the diseases termed *epidemic,* the most frequent in the Miami country are the *Measles* and *Hooping Cough,* both of which have prevailed in Cincinnati every year or two, since 1800. They seldom affect a great number at once, but make their attacks successively for many months, and do not often terminate fatally. The *Mumps* now and then occur, with no unusual symptoms. The *Small pox* has not prevailed here to any extent for a dozen years. There is no institution for preserving and disseminating the vaccine virus; but a great number are annually vaccinated. The *Scarlet Fever* and *Putrid Sore Throat* have been of rare occurrence. About the year 1792, they were prevalent in all the infant settlements of the West, and produced many deaths. From that time till 1809, but few cases were observed at Cincinnati. In this and the two subsequent years, they appeared in an epidemic form, and destroyed a number of children. Since that period, but few cases have been seen, and those were of the mildest kind. The *Influenza,* so extensively prevalent in 1807, attacked the people of Cincinnati about the 1st of October, and disappeared in five weeks, leaving the town unusually healthy. Very few adults of either sex, but many children, escaped it. The number of deaths produced by it was inconsiderable. The Consumption, however, followed in its train, and carried off several persons in the two ensuing years. Since this visitation, we have more than once experienced wide spreading *Catarrhs,* which were ascribed to changes of the weather; but it seems probable that they arose from the same causes with the *Influenza.* The *Spotted Fever* of the Northern states has never prevailed here; but its successor, the *Typhoid Pneumony* (vulgarly called in this country the cold plague) affected a very considerable number in the winters of 1812-13 and 1813-14. In that of 1814-15, but few cases were met with. More men, in proportion, than women or children, suffered; and it generally attacked those who were most exposed to cold and moisture. It proved fatal in a number of cases; but was, on the whole, productive of much less mortality than in the North.

Eruptive diseases of the skin are common in the Miami country, and frequently prove obstinate. The *Itch,* and a breaking-out which nearly resembles that complaint, are the most common. These eruptions, however, exhibit a great variety of appearance, and are by the people ascribed to as many different causes. They seem to be more prevalent in the country, than the town. *Worms* are common, and affect children of every age, from one to fifteen years. They seldom prove fatal, unless combined with some other disease. The *Goitre* is an endemic of the western portions of Pennsylvania, and the eastern part of this state; but is unknown here, except in persons who have immigrated while laboring under it. The *Scrophula, Rickets* and *Scurvy,* especially the two latter, are rare diseases. *Hysteria, Hypochondria* and *Insanity,* are not uncommon. *Dropsy of the brain* is met with occasionally. *Locked jaw* is so rare, that but a single case has occurred here for many years. *Apoplexy* is scarcely ever seen; but *Epilepsy* is more frequent. *Dropsies* occur pretty often, but generally as the consequence of intermitting fever. The *Gout* and *Calculus* are seldom seen, and *Palsies* are infrequent. *Cancers* are uncommon; and no case of *Hydrophobia* has occurred since the settlement of the town. *Canine madness* has not been epidemic for many years. The venomous snakes are so few, that even in the newer settlements a *snake-bite* is uncommon; and in the neighborhood of Cincinnati, almost unknown. The *Coup de soliel,* or stroke of the sun; and death from the use of *cold water,* are not more frequent. *Drowning* in the Ohio, is an accident which often happens, and one which we are entirely unprepared to remedy, not having the instruments necessary, either for the recovery of the immersed body, or the restoration of life.

As no bills of mortality are kept in this place, it is not known what proportion die annually; what diseases carry off the largest number; or which of the seasons is attended with the greatest mortality—tho' the two latter may be estimated and expressed in general terms. The Cholera Infantum is more fatal to children than any other complaint. It is most destructive in the second summer; aggravated, no doubt, by teething, and the miscellaneous food with which children begin to be indulged at that age. Convulsions, in the first month after birth, carry off many; and should perhaps rank next to the Cholera Infantum in the number of their victims. After this follows the Croup, which for the most part attacks those between the ages of six months and two years. Of adults, the greatest number die with bilious and typhous fevers; with pulmonary inflammation, and with affections of the liver, stomach and bowels. In the months of June and July, more children die than in any others. The greatest mortality among adults is generally in August, September and October. When epidemics prevail, this however is otherwise, and the midst of winter is now and then attended with a greater number of deaths than any other part of the year.

SECTION II. CAUSES OF DISEASE

CLIMATE. Neither the extreme cold, nor the extreme heat of this climate, appears to produce many diseases, by its direct operation. If scurvy, goitre and chilblains arise from cold, that of our climate is not sufficient to produce them. The extremities of those who are much exposed in winter, are occasionally frozen; but there has been no instance of death from such exposure in this country. The most obvious effects of our hot weather are, oppression and lassitude in the muscles, with a diminution of appetite—all of which disappear upon the occurrence of a cool day, and are thereby distinguishable from similar affections produced by marsh exhalation. Few persons escape these complaints; but those who have emigrated from higher latitudes are of course the greatest sufferers. Some aged people, and a few valetudinarians, enjoy better health in our hot, than cold weather. Our children, during the great heats of summer, are liable to *rashes,* as they are popularly called—cutaneous efflorescences—which are troublesome, but not dangerous; and disappear upon the first occurrence of cool weather. There is even reason to believe these affections salutary, as they frequently appear on the healthiest children. Cholera Infantum is not produced by the direct action of heat on the system, but is so much aggravated by that cause, as to be generally incurable during the period in which the thermometer fluctuates between 76 and 96 degrees. The variations of atmospheric temperature are a more potent cause of disease, than either extreme. But they may in a great degree be rendered harmless, by a careful adaptation of clothing, lodging and fire, to the change. This cause usually produces pleurisy, rheumatism and other inflammations—colds, quinsies, croup, tooth ach, &c. uncombined with other complaints; but when the prevailing disease is a bilious or a typhous fever, it is commonly found, that the affections produced by changes of the weather, partake largely of the symptoms of the epidemic. The best examples of this combination are afforded by the pleurisy and croup. Variations of temperature, particularly changes from *heat* to *cold,* are sometimes the *exciting* causes of intermitting and other fevers, produced by marsh exhalation. In all these cases, the presence of moisture renders the depression of temperature more injurious. To water, indeed, in the form of dew and fog, it is fashionable to ascribe much deleterious power; but there is reason, perhaps, to doubt the correctness of this hypothesis in all cases, when the temperature of the atmosphere is *steady*.—Fogs and vapors are most abundant, where the decomposition of vegetable matter is greatest; and to this operation should perhaps be attributed most of the diseases which are vulgarly ascribed to moisture.

WATER. Throughout the Miami country, this is generally *hard;* from holding in solution *carbonate of lime, muriate of soda, muriate of lime,* and the other salts afforded by a calcareous region. It is apt, therefore, to disagree with emigrants from a country, such as that east of the Alleghenies, where most of the springs afford *soft* water. The complaints excited by this cause are for the most part transient; and to the natives of the country, its waters are as salutary and pleasant, as those of the Atlantic states are to the inhabitants of that quarter. Our springs and wells cannot, therefore, be regarded as affording a beverage absolutely prejudicial to health, though it may operate injuriously on strangers for a short period.

MIASMATA. The Miami country in general being level, ponds and morasses are frequent; especially in the northern part. Most of them might be drained, and certainly will be, at some future period. In the mean time, their environs must continue more or less infested with the diseases which spring from marsh effluvia; and therefore should not be selected for the residence of immigrants. Most of our vallies contain large quantities of alluvion, deposited at various antecedent periods; but whether from these tracts there be any exhalations still arising, which are noxious, is doubtful. The more obvious sources of miasmata, are the marshes formed in these tracts from the annual inundation of their lower portions; and the decaying remains of animals and vegetables deposited in the shores of the streams which flow through them. Whatever may be the truth on this point, it is certain that the vallies are less healthy than the uplands; but from clearing and cultivation, they are annually becoming more salubrious. With respect to Cincinnati, the sources of miasmata may be divided into those which are natural, and those which are artificial; or in other words, into such as are common to it, and other towns on the river, and such as are peculiar, and of our own creation. Of the former, we have but two—the drowned lands at the mouth of Mill-creek; and the river beach opposite the town. The former lie so far to the west, and are so much disconnected with the town by an intervening forest, that our summer winds but seldom blow their exhalations over us. Hence very little agency can be ascribed to this cause. The latter is, perhaps, more efficient. The great depressions of the Ohio, in August and September, expose to the sun a quantity of mud, with trees and some animal matter, in a state of decay: the exhalations from which are unquestionably prejudicial. The erection of the steam mill has augmented this cause; by producing, in high floods, an eddy, which annually deposits on the beach, for a thousand feet along the front of the town, a large quantity of filth and mud. Our *artificial* sources of disease are incomparably more deleterious. For many years the descent of gravel along the streets which run from the upper to the lower table, has kept several of the intermediate lots in a state of partial inundation, and caused them to accumulate large quantities of filth. Further west, in the same tract, nearly all the bricks hitherto used, were manufactured; and the pits whence the clay was dug, have been constantly receiving, through the gutter in Second-street, nearly all the wash of the town. Thus have we improvidently created, in the very midst of our population, the most offensive and destructive nuisances. Fortunately, the powers of the new Corporation enable them to compel the removal or abatement of the whole. The great purification has thus at last been commenced; and although its progress as yet has neither been creditable to the energy of the Corporation, honorable to the proprietors of those lots, nor beneficial to the public health, there is great reason to hope for relief at no remote period. When this salutary object is accomplished, our public sources of disease will be so few and inefficient, that we may without hesitation, expect to see Cincinnati approximating in healthiness, the driest and most elevated situations, remote from the river.

SECTION III. MINERAL SPRINGS

The Western country is abundantly supplied with *salines,* or salt springs. The richest and most copious are on the bank of Great Kenhawa, in the western part of Virginia. Along with the common salt, *muriate of soda,* there is a large portion of the *muriate of lime,* as I have found by examining the *bittern* or *mother water,* which seems to consist entirely of that salt. In various parts of Kentucky, salt springs were long since discovered, and are frequented by invalids. Several of them contain the *sulphates of soda,* or *magnesia,* and a few afford *sulphurated hydrogen gas.* In the Indiana and Illinois Territories, and in this state, near the Auglaize and Sciota rivers, springs of a similar kind are known to exist. *Chalybeate* waters, consisting generally of *oxide* of iron, dissolved by the agency of *carbonic acid,* are almost as numerous. On the present occasion, we must confine ourselves to those which are situated within such a distance from Cincinnati, as to be accessible to its valetudinarians.

In the bed of Licking, within a mile of its mouth, when the river is low, several copious veins of *chalybeate* water burst out, and have occasionally been resorted to by our citizens. In addition to the *carbonate of iron,* they contain the different salts common in the spring water of this region. They seem to be formed in the alluvial grounds which skirt the river, and may be mentioned as specimens of a numerous class of *chalybeate* springs, with which the alluvial formation abounds. The majority of them, however, are less copious than those under consideration.

About two miles above the town, on the declivity of the hill, a well has been dug in the loose clay and limestone, which have formerly been precipitated by the undermining action of the current. The water of this well is moderately charged with *sulphurated hydrogen gas,* common salt, epsom or glauber salt and iron, with some useless ingredients. Its effect on the system is that of a cathartic; and from its chalybeate properties in addition, it will unquestionably be found a valuable water. The proprietor intends, by the ensuing summer to make it a *watering place;* for which its topographical situation is highly agreeable. The road leading to it from Cincinnati, lies along the river bank, and its site is healthy, well ventilated, cool, and commands a view of the vallies of Licking and the Little Miami, which are seven miles asunder. In the vicinity of Northbend, marked *Cleves* on the map of the Miami country, there is a spring of a similar kind; but it is less highly charged with saline matter, and is without sulphur.

The most noted watering place in the Miami country, is the YELLOW SPRING, in Green county, 64 miles from Cincinnati, and two from the Falls of the Little Miami. It is a copious vein which bursts from a fissure in the silicious limestone rock; and is, at the distance of a few rods, precipitated into a ravine more than a hundred feet deep. On its passage thither, it has deposited an immense bank of brownish ochre, blended with leaves, twigs and other vegetable matter. The brook which flows along this wild and narrow valley, falls over many successive ledges, which adds much to the interest of the scene. Its margin is fringed with a variety of beautiful shrubs, whose broad & heavy foliage affords an agreeable contrast with the slender leaved cedars that adorn the rocks above. A quarter of a mile below the spring, this brook is joined by another, flowing in a similar valley. Along this, a number of excavations have been unsuccessfully made, in search of ores. Among these there is one, five or six feet deep and as many in diameter, which was dug at a period altogether antecedent to the settlement of this country by the Anglo-Americans; but whether by the French or the ancient inhabitants, is quite uncertain. The valley of these united streams exhibits to the geologist the transition from the common to the silicious limestone strata — and a visit to the Falls of the Little Miami will afford several charming prospects. Upon the whole, a tour to the Yellow Spring will amply repay the traveller, if not the invalid; and amuse those who are in health, if it do not in many cases heal the infirm. As to the fountain, it is transparent, emits no air bubbles, and has the temperature of 52 degrees; which is that of the springs in its vicinity. Its taste is that of slight chalybeate, and the examinations which have been made, indicate it to contain a portion of oxide of iron and carbonate of lime, dissolved by the agency of carbonic acid gas. In its other saline impregnations, it appears to have no excess over the springs of the Miami country generally; it is used for domestic purposes, and its sensible effects on the human system appear to be inconsiderable. In those cases of chronic disease and debility, where a chalybeate is proper, it has however been used with advantage.

An attempt has been made to prepare a paint from the deposit below the spring, which has been attended with the most flattering success.

The springs most resorted to by the people of Cincinnati, are the *salines* at BIG BONE, 22 miles south-west of the town, in the state of Kentucky. They are several in number, and their waters were formerly employed in the manufacture of salt; until the discovery of stronger *salines* on the Great Kenhawa, reduced the price of that article below what it could be afforded when manufactured at these *licks*. The waters at Big Bone hold in solution, besides common salt, the *muriate of lime, sulphate of soda* or *magnesia,* and a few other salts of less activity, but no iron. They afford a great quantity of *sulphurated hydrogen gas,* which is constantly escaping in bubbles. From their effects on the sulphates of copper and iron, they appear obviously to contain a portion of *gallic acid,* that is no doubt furnished by the vegetable matter through which the waters rise. The springs are situated near the termination of the back-water of the Ohio, and consequently at a point where great quantities of twigs and leaves (most of which from the nature of the surrounding forest must be of oak) are brought down by the current, and deposited. The temperature of the springs is 57°. Their taste and smell are sulphurous, and offensive to strangers; but the impression made by the gas is transient, and the taste of the common salt afterwards predominates. They do not increase the pulse, but their sensible effects on the alimentary system, kidnies and skin, are great. The action of the two former are very much increased; and the latter is frequently affected in a few days with a violent itching, and an eruption of pimples or pustules, which are now and then connected with large *boils*. These waters are, however, neither serviceable nor safe to persons whose constitutions have been long and generally debilitated; whose digestion is bad, from permanent weakness of the stomach; who are affected with head ach, and a general reduction in the energy of the nervous system; or who labor under that species of pulmonary consumption which will not bear depletion. The disorders to which they seem peculiarly adapted, are the torpor, obstruction or chronic inflammation produced by acute diseases in the lungs, liver, spleen, kidnies, in short, any of the viscera; and which have not continued so long that the constitution is exhausted. In these cases, experience has shown them to possess all the efficacy which could be expected in any mineral water. From a pint to a gallon, may be taken daily, according to the strength of the patient, and its sensible effects on the system. The quantity drunk at first, should be small, especially by those of a reduced habit.

The valley in which these springs are situated, is of moderate width, and bounded by a waving and irregular rampart of elevated hills. The scenery is romantic, and not destitute of picturesque features; but the verdure in spring and summer is rather unvaried, and the enchantment of a distant perspective is wanting. These defects in the configuration of the vale are, however, amply compensated by the mighty relics which it entombs. It is now more than half a century since these first attracted the attention of European travellers; and so many have been borne off, that a few fragments only remain on the surface, to excite the associations and recollections which this consecrated spot is calculated to inspire. As no other place hitherto discovered in the Union has afforded such quantities of huge animal remains, and as the first ever transmitted to the philosophers of Europe, were collected here, the BIG BONE VALLEY deserves, among naturalists, a classical distinction. It is indeed well worthy a visit from those who can relish the sentiments and the speculations excited by contemplating the ruins of the largest animal species which have appeared on our globe. And if, according to Mr. Jefferson, the passage of the Potomac through the Blue Ridge, be a scene worth a voyage across the Atlantic—the tomb of the mammoths will certainly reward the traveller of taste and science, for a journey from Cincinnati.

An establishment for the preparation of *artificial mineral waters,* was made in the spring of the present year; and during the few weeks that it continued in operation, it attracted much attention. The proprietor [Drake himself] has made arrangements for opening a greater number of fountains the ensuing summer; and will be able, hereafter, to supply the citizens of Cincinnati with as fine a variety of these salutary waters, as any of the large cities can afford.

* * * * *

CHAPTER VII, CONCLUSION

SECTION I. PROJECTED IMPROVEMENTS

Under this head I do not propose to mention any other improvements than those which are calculated to facilitate the intercourse between the town and country.

BRIDGES. Some enthusiastic persons already speak of a bridge across the Ohio at Cincinnati; but the period at which this great project can be executed, is certainly remote. Mean while, in a steam ferry-boat, we might find nearly all the conveniencies of a bridge; and the communication between the opposite sides of the river is so great, even at the present time, that such an establishment would yield a good profit.

A new and permanent bridge across the mouth of Deer-creek is much wanted, and will probably be erected in the course of one or two years; as those to whom this important charge is confided, will undoubtedly be ashamed to neglect it much longer.

There was once a wooden bridge over Mill-creek, near its confluence with the Ohio; but in consequence of a high flood in that river, it was destroyed. In the session of 1814-15, our Legislature authorised the erection of a toll-bridge at the same place; which, it is understood, will be commenced the ensuing spring.

ROADS. By the law of Congress which provided for the admission of this state into the Union, it was stipulated, that three per cent. of the nett proceeds of the United States' lands within the limits of Ohio, should be applied by its Legislature to the laying out, opening and improving of its roads. The policy pursued by the trustees of this valuable fund has been, to appropriate it on a great number; and, of course, to have not a single good road in the state.* The project of constructing, between the Miamies, from Cincinnati towards the sources of these rivers, a great road, which should at all seasons be equally passable, has been for some time in agitation. It will perhaps be undertaken in 1816, and pass by the nearest route from this town to Dayton. The benefits which an execution of this plan would confer, cannot be fully estimated, except by those who have travelled through the Miami country in the winter season, and have studied the connexions in business between that district and Cincinnati. The salt, the iron, the castings, the glass, the cotton and the foreign merchandise for at least eight counties, would be transported on this road; which would immediately become one of the most important in the state.

An improved road to Columbia is a great desideratum. The present one, for several years past, in the winter season, has been nearly impassable for carriages and loaded waggons; while all the materials for the best turnpike have been at hand, and even constituted one of the greatest obstructions on the route. Two years ago some efforts were made to form a company for this purpose, but they seem to have been ineffectual; and we must patiently wait for an accession of wealth and enterprise. The delay of this undertaking is the

---

* In the year 1809, the Legislature passed a law directing $9000 of this three per cent fund to be appropriated to other purposes. Against this the officers of the General Government remonstrated; and in the ensuing session a law was passed directing the money to be refunded, with interest.

more to be regretted, as we are thereby in a great degree precluded from the most agreeable airing which the vicinity of the town is calculated to afford.

CANALS. The points of near approximation between the waters of the Mississippi and the Lakes, appear to be six; not including those which may exist in the vicinity of lake Superior, and have not yet been examined. The *first* of these is in the neighborhood of Presq' Isle, where the highest navigable point of French-creek, one of the branches of the Allegheny, is found within 12 or 15 miles of the Lake. But whether a canal could be dug through the portage, has not been publicly stated. The *second* is between the Cayahoga and Tuscarawa, one of the upper streams of the Muskingum. The portage at this place is not more than a dozen miles; and so certain is it that the two waters may be connected by a canal, that in the law of Congress appropriating a portion of the public lands to the improvement of inland navigation, 100,000 acres were assigned for defraying the expense of this project; but the work has not yet been commenced. The *third* is betwixt the St. Mary and Auglaize, branches of the Maumee; and Loramies-creek, one of the most navigable waters of the Great Miami. The relative position of these small rivers may be seen by a reference to the map. The St. Mary is remarkably serpentine, with a general direction towards the north-west; which makes the voyage to the Lake circuitous and protracted. It is said to have an earthen channel, with low banks, and to be deep and narrow. In the course of the year, there are generally five or six floods, when its navigation would be perfectly safe, were it not for the bayous which are then formed. Its junction with the St. Joseph, at Fort Wayne, composes the Maumee. The Auglaize is a shorter river than the St. Mary, and entering the Maumee 60 miles below that stream, affords a much quicker passage into the Lake. It is also a larger river than the one first described; but has a stony channel and a rapid current. In the opinion of gentlemen who have descended both, the navigation of the Auglaize is generally not so safe as that of the St. Mary; tho' at certain seasons it affords more water. The highest navigable points on those rivers, are not more than 20 miles asunder; and between 12 and 18 from the head of navigation in Loramies-creek. The intervening tract is nearly level, and composed of a deep stratum of loam and clay. Which of these streams could be most easily and advantageously connected with our waters, remains to be determined; as does also, the more important question, whether the portage would afford sufficient water to feed a canal. The *fourth* connexion is between sources of the Wabash, and the St. Mary, eight miles above Fort Wayne. When very high, these rivers overflow the intervening lands to such a depth, that loaded boats pass over with facility. Of the practicability, therefore, of connecting them by a canal, there can be no doubt; and in the law of Congress just quoted, an appropriation of land equal to that for the Muskingum and Cayahoga canal, was made for this. The *fifth* point of intercommunication is between the Illinois, and the Chicago a southern river of lake Michigan, which I am informed are so connected, that in freshets boats can pass readily from one to the other. For encouraging the improvement of this navigation, the General Government have made the same appropriation as in the cases before mentioned. The *sixth* connecting waters, are the Ouisconsing and Fox rivers. The former runs into the Mississippi—the latter into Green Bay, an arm of lake Michigan. The portage at this point is said to be short.

Which of these connexions offers the greatest facilities to commercial intercourse, cannot at this time be determined. That between the Chicago and Illinois will, it is probable, be the least expensive; but as vessels in reaching it must pass through the straits of Michilimackinac, it is not likely to be used until the banks of lake Michigan and the Illinois shall become thickly inhabited. The canal between the Cayahoga and Muskingum will be first opened; and must greatly benefit the country watered by those rivers. But in the improvement of the connexion between the Great Miami and Maumee, the people of the western part of the state are most interested. Its utility to the inhabitants of Cincinnati and its vicinity will, however, in a great degree, depend on the execution of another and more difficult project; on which some general remarks will close this article.

To discharge a portion of the waters of the Great Miami into the Ohio, at this town, would, confessedly, be a great public benefit; but no proposition on this subject has yet appeared; nor does it seem to have attracted much attention. In the whole course of the Miami, there is perhaps but one point where a canal could be opened; and that is near Hamilton, 25 miles from the mouth of the river, and about the same distance from Cincinnati. In the valley, five miles south of the former town, there is a large pond, which is replenished by the Miami, when that river is high; and out of which, at the same time, arises one of the principal branches of Mill-creek. From this place to Cincinnati, following the meanders of the stream, there is nothing to prevent the opening of a canal. The valley it is true, contains great quantities of pebbles and gravel covered with soil, but by keeping near the hills that bound it, an argillaceous bottom could be had. The difference in level, at low water, between the Ohio at this town, and the Miami at Hamilton, has not been ascertained; but it may be estimated at 60 feet. About four miles from Cincinnati, the canal would have to be carried over Mill-creek, after which it might be conducted along the base of the high lands which border the site of the town on the north, to the valley of Deer-creek, through which it would reach the Ohio. The time when the enterprise and resources of the citizens of the Miami country will be adequate to the execution of this project, cannot be foretold; but when we consider the ratio of our progression in strength and numbers within the last fifteen years, there is much reason to hope that the era of this improvement is not remote. The transportation on this canal and the Miami above (if its navigation were somewhat improved) would, in less than half a century, be great indeed. The country on each side, for the average distance of 25 miles, and as far north as the navigable waters of the Maumee, about 110, would be dependent on it. In this parallelogram of 5500 square miles, there is no spot which is not susceptible of cultivation; and by far the greater part is equal to any land in the United States. It only, therefore, requires facilities for the exportation of its surplus produce, and the importation of foreign articles, to ensure for it a very dense population; and such facilities would be afforded by the canal. In addition to this, should the difficulties connected with the navigation of the Maumee and its branches, be removed at the same time, the skins and peltry, the fish, and perhaps the copper of the north, would reach the Ohio; and the cotton, sugar, tobacco and other productions of the south, would pass into the Lakes through the same channel.

SECTION II. FUTURE CONSEQUENCE

It will perhaps, to many persons at a distance, and particularly to those who have not studied our natural and commercial geography, appear altogether visionary, if not boastful, to speak of *cities* on these western waters. Yet it is certain, that those who have contemplated this country with most attention, are strongest in the belief, that many of the villages which have sprung up within 30 years, on the banks of the Ohio and Mississippi, are destined, before the termination of the present century, to attain the rank of populous and magnificent cities. The grounds which support this prediction are too broad to be travelled over at this time; but it may be rendered plausible in a high degree, merely by a reference to the Mississippi. If we consider the quantity of water discharged by this great river—the vast extent and number of its branches, many of which exceed in length the largest rivers of Europe—the general direction of the main trunk, nearly from north to south, passing through more than 15 degrees of latitude, in the temperate zone—the diversities of aspect, and inexhaustible fertility, of the region which it irrigates—the boundless and perennial forests, which in the east, and in the north, overshadow its sources—the numerous beds of coal and iron which enrich its banks—the reciprocal ties and dependencies, which can never cease to operate, between the inhabitants of its upper and lower portions—the numerous states which will possess in its navigation, a common interest, that must forever constitute a bond of political and commercial amity—we must be convinced, that there is no river on earth of equal importance; or at least none on whose countless tributary streams so many millions can subsist.

Of all the ramifications which enter into the composition of this majestic river, the Ohio will unquestionably retain, for ages, the highest rank. What comparison the countries dependent on it will ultimately bear to the Hudson, the Delaware or Potomac, cannot at this time be determined; but any hypothesis that assigns to the former a decreasing ratio of improvement will be seen to have no foundation; the opinion that these states cannot support even a denser population than any in the East, is altogether groundless; the associations of wildness and ferocity—ignorance and vice, which the mention of this distant land has hitherto excited, must ere long be dissolved; and our Atlantic brethren will behold with astonishment, in the green and untutored states of the West, an equipoise for their own. Debarred, by their locality, from an inordinate participation in foreign luxuries, and consequently secured from the greatest corruption introduced by commerce—secluded from foreign intercourse, and thereby rendered patriotic—compelled to engage in manufactures, which must render them independent—secure from conquest, or even invasion, and therefore without the apprehensions which prevent the expenditure of money in solid improvements—possessed of a greater proportion of freehold estates than any people on earth, and of course made industrious, independent and proud;—the inhabitants of this region are obviously destined to an unrivalled excellence in agriculture, manufactures and internal commerce; in literature and the arts; in public virtue, and in national strength.

Where will be erected the chief cities of this promising land? It may be answered with certainty—on the borders of the Ohio river. They are not likely to become places of political importance, for these must lie towards the centres of the states which this river will divide; but the commercial and manufactural advantages that exist in lieu of the political, are so much superior, as to justify, in this enquiry, the omission of every town not situated on the Ohio. Pittsburgh, Cincinnati and Louisville, are the places which at present have the fairest prospects of future greatness. The age of Cincinnati is intermediate to the others. Their population and business correspond at present with the order of their enumeration; but the time is apparently not remote, when a different comparative rank will be assigned them. Both Cincinnati and Louisville seem destined to surpass Pittsburgh. To this prediction the inhabitants of that town—for thirty years the *entrepot* of all the Ohio countries—are not expected to assent. It will even be regarded by them, as groundless and arrogant; but without stopping to anticipate and repel the charges of self interest and vain glory, I shall proceed to a brief exposition of the relative advantages of that town and this. It is well known to all the people of the United States, that for twenty years, both foreign and Atlantic goods, to the amount of several millions of dollars, have been annually waggoned to Pittsburgh, deposited in its warehouses, and shipped in its boats for the country below. The expense of these operations has, of course, been defrayed by the consumers in Kentucky, Tennessee, Ohio, and the adjoining Territories, who have thus made to the prosperity of Pittsburgh a yearly contribution of great value. Hundreds of our merchants were passing, moreover, through this town; and it was early discovered, that if manufactures were established, it would be possible to dispose of many articles required in the newer settlements below. Hence founderies, glass houses, breweries, and iron manufactories of various kinds, were erected; and the wares of this "Birmingham of America" superadded to the merchandise of the East, soon spread extensively over our country. During such a period of commercial prosperity, the borough could not but flourish; and were the causes of its growth as permanent as they have been efficient, it would unquestionably retain an enviable superiority. But a change in the current of our importations—such a change as has already begun—must inevitably reduce the ratio of improvement in that place, just as much as it will be increased by the same cause, in Cincinnati, Louisville and the other towns below. The waggoners employed in the transportation of our merchandise from Philadelphia; the boat builders and commission merchants; the freighters, and those who manufacture for these populous young states, will no longer receive our specie for their services; and must of course find other employments, or emigrate to other towns. The coal and iron of that place will indeed long continue abundant; but these are easily floated with the current to the towns below; which can thus establish the manufactures dependent on these important articles, with nearly as much facility as they are set up in Pittsburgh—while that town must obtain its cotton and sugar, its hemp and lead, at an expense of freightage, taking these articles together, more than twice as great as that paid by us. The country around that place, is moreover, rugged and sterile, in comparison with that about either Cincinnati or Louisville; and the greatest population it can support, will have a correspondent rarity. Pittsburgh, therefore, has not so high a destination as its younger rivals to the westward; but it must forever maintain a very important and respectable rank.

The chief advantage which Louisville possesses over Cincinnati, is the partial interruption of commerce at that place by the *Falls* of the Ohio. The cargoes of boats, when the water is low, are waggoned for two miles round those rapids. This not only gives employment to a great number of hands, but it makes the town one of the heads of navigation—a place of debarkation and deposit—where, of course, an active mercantile business may be done. If these obstructions to the navigation were irremoveable, Louisville would certainly arrive at a very exalted degree of commercial greatness. But the opinion of professional engineers is such as to dissipate much of this interesting prospect. The desired improvement was actually commenced more than a year ago; and altho' the prosecution of it has been for some time suspended—by causes not necessarily connected with the undertaking—there can be no doubt of its being resumed, and finished before the lapse of many years. When this is done, the commercial importance of that town must receive a signal reduction; but still it will possess the peculiar advantage of a site for great water works. It will, moreover, be the emporium of an extensive and fruitful district in Kentucky; for which its situation on a southern bend of the Ohio gives it a number of advantages. Still there are reasons for believing that CINCINNATI IS TO BE THE FUTURE METROPOLIS OF THE OHIO. Its *site* is more eligible than that of most towns on the river. It is susceptible of being rendered healthier than Louisville, and is extensive enough for a large city. The Ohio bounds it on the south-east, south, and south-west, so that all the streets, if extended, would, at one or both ends, intersect the river within the limits of the corporation. It has, therefore, a great extent of shore, along the whole of which there is not a reef nor shoal to prevent the landing of boats.—Opposite to Broadway, is the mouth of Licking; a river whose navigation will certainly be much improved. —Over the town plat, as we have seen in the preceding article, a canal at some future period may be conducted from the Great Miami; whose waters can, by another canal, be connected with those of the Maumee, and thus secure to us a new and profitable trade with the Lakes.—A survey of the Ohio will exhibit to us the important fact, that between Pittsburgh and Louisville there is not a single spot, where a future rival to Cincinnati can be raised up. Finally, by a reference to the map of the Miami country, it may be seen, that the river, in approaching Cincinnati from Maysville, which is 60 miles above, runs generally to the north-west; that after passing the town, it soon alters its course, and flows nearly to the south for more than 40 miles; and consequently, that Cincinnati lies in a situation to command the trade of the eastern and western, as well as the interior portions of the Miami country. This is the case for more than 30 miles in those directions; and when the improvement of the roads shall be such as to facilitate intercourse with this place, the power it must exercise over these opposite districts will be still greater. The adjoining parts of Kentucky, altho' politically disconnected, must long continue to acknowledge their commercial dependence on Cincinnati. Thus, it is the permanent mart and trading capital, of a tract whose area equals the cultivable portion of New-Hampshire, New-Jersey or Maryland; surpasses the state of Connecticut, and doubles the states of Rhode-Island and Delaware taken together—with a greater quantity of fertile and productive soil, than the whole combined.

These are some of the local advantages of Cincinnati; and if improved with a spirit corresponding to their magnitude, its inhabitants cannot fail to realise their most glowing anticipations of future greatness.

Valedictory to the Class at Lexington in the Transylvania University* (1818)

The justification of scientific study unalloyed by any combination with literary or theological interests was a constant concern of Drake's prior to the 1830s, when the vogue of "physiology," especially in connection with American reform movements, provided science with a new respectability as an essentially useful pursuit.[1] As early as 1818, in a valedictory address to his class in materia medica and botany at Transylvania University, he articulated his own conviction of the essential role which specialized scientific training played in the education of a physician. It was a theme he would repeat regularly thereafter, once ironically in his own defense when his colleagues at the Medical College of Ohio charged him with being ambitious, quarrelsome, and cultivating "other branches of science than [his] profession."[2]

As a Baconian in method, rejecting the certainty of rationalist science, Drake believed, as he said at that time, "that the science of medicine consists of something more than a collection of *infallable* [sic] *receipts*"[3] and that only systematic observation could determine the utility of particular measures in practice. For physicians to reject the information of the auxiliary sciences was to reject potentially useful tools for the effective pursuit of their profession, and to deny the essential unity of human knowledge, of which the science of medicine was but one "wide spreading branch," was to deny the unity of the universe to which such knowledge pertained.

Drake was appointed professor of materia medica and botany in the Medical Department of Transylvania University in December 1816, to begin the following autumn. Early in January 1817 he formed a partnership with his former teacher, Dr. William Goforth, and upon Goforth's death in May, with Dr. Coleman Rogers, in

---

* "Valedictory to the Class at Lexington in the Transylvania University," March 1, 1818, MS in Cincinnati General Hospital Library.

[1] Cf. Richard H. Shyock, "Sylvester Graham and the Popular Health Movement, 1830–1870," in *Medicine in America: Historical Essays* (Baltimore, 1966), pp. 111–25, and Drake's *Oration on the Causes, Evils, and Preventives of Intemperance* (q.v.).

[2] Drake, *A Narrative of the Rise and Fall of the Medical College of Ohio* (Cincinnati, 1822), pp. 36–40.

[3] Ibid.

order to maintain his practice during his absence in Lexington. In early March 1818, at the conclusion of the spring term, Drake returned to Cincinnati amid charges of violating an agreement to remain at Transylvania for two years and sabotaging the operations of the first medical school west of the Alleghenies. In two pamphlets published during the summer of that year Drake defended his actions, without mentioning however that he and Rogers had already begun to prepare for the inauguration of formal instruction in medical subjects in Cincinnati, and hence for a challenge to Lexington's monopoly of the field.[4]

[4] *An Appeal to the Justice of Intelligent and Respectable People of Lexington* (Cincinnati, 1818); *A Second Appeal to the Justice of the Intelligent and Respectable People of Lexington* (Cincinnati, 1818).

WITH THESE REMARKS, Gentlemen, I shall terminate tho' not complete the course of lectures assigned me by the Trustees of our University.

We have assembled about eighty times for improvement in the *Materia Alimentaria,* in the *Materia Medica* and in *Medical Botany;* but so numerous and important are the facts and principles comprised in these departments of our profession, that to review them all it would be necessary to multiply our meetings to double that number. We have generally, however, selected for examination the most interesting subjects and notwithstanding my efforts have partaken largely of the unskilful and inefficient character of a first attempt, your ardour has been such that I know you have made, during the winter, some useful acquisitions. It is not my intention to detain you with a general exposition of the means of extending these in future; but I feel unwilling to leave you without a parting exhortation on the subject of our indigenous *Medical Botany.*

That among the native plants of the United States there are a great number more or less endowed with medical properties, seems to be fully ascertained. There is perhaps no class in the *Materia Medica* which could not be furnished in part from the swamps—the mountains—the arid plains—the blooming prairies or the exuberant forests of this great continent. But we must never for a moment suppose that our shops can be supplied in the *best* manner from these sources. Nothing is more common in the vegetable kingdom, than a moderate degree of medicinal virtue, and in general no property is more useless. The really useful medicines rise in their remedial powers above this *mediocre* line of excellence. They have as it were an accumulation—a monopoly of active qualities. Like those great men who possess no other faculties than the bland and unaspiring multitude, but who from the energy of their intellect, the vehemence of their passions and the commanding dignity of their virtues, in a great degree controul and govern the destinies of that multitude, these active substances contain only the properties common to a numerous catalogue, but in so concentrated a state as to give them wonderful efficiency. Our main object, gentlemen, in the study of the vegetable materia medica of this country, is to identify these important productions and seperate them from the great field of comparatively inert simples. To do this with success, it is necessary to study them *botanically, chemically,* and *medicinally.*

To accomplish the first, it is requisite to cultivate the science of Botany; for which during the present winter you have I hope had a taste excited. Do not, gentlemen suffer this taste to expire. Follow its suggestions and they will conduct you thro fruitful and flowery fields. None of the sciences collateral to medicine, are so proper for cultivation by the physicians of the *Backwoods* as Botany. Notwithstanding this it has been hitherto strangely neglected by us, and many curious mistakes have been the consequence. Thus the *Serratuta spicata* has been gathered and administered for the *Lobelia syphilitica,* and this is very commonly mistaken for the *Lobelia inflata:* the *Gillenia trifoliata* has been collected for the *Ipicacuan* of the shops, and the our [*sic*] *Geranium maculatum* is supposed by many to be the *Tormentitta erecta* of Europe. These errors are not wholly unpardonable in us, who grope amidst the dark umbrage of primeval forest. They are not however confined to the benighted inhabitants west. In parts of the eastern horizon which have been for at least an age illuminated by the rays of scientific light, some *standard* works have been produced which exhibit mistakes of equal magnitude. Thus in one of them we are assured that there are in the United States several distinct *species* of Eupatorium *perfoliatum;* that our *Valeriana puciflora* is the same with the *Valeriana officinalis;* that the *Curcuria longa* of Hindostan, and the *Daphne mezercum* of the north of Europe grow along the Ohio, when not a single species of either genus has been found in North America, and lastly that the *Frasera walteri* is a *species* of the oriental Colomba although the plant producing that medicine was confessedly *unknown* to the *author,* and belongs to a different class in the Systems of Botany!

These are a few specimens of a class of errors which are sometimes mischievous to our patients, and always disgraceful to the profession.

You may perhaps be told gentlemen that the study of Botany will interfere with the practice of physic. To this opinion you will I hope turn a deaf ear. Botany, it is true is an extensive science and might constitute the sole study of any man throughout his life. But it would then be a principal: to a physician it should only be an auxilliary study. It is one of the handmaids of practical medicine, and precisely that which is most wanted in a new and unexplored region. Nothing is easier, or more natural for *us* than to connect botanical researches with the practice of our profession and make every distant visit to the sick subservient to the study of this beautiful department of nature.

A knowledge of chemistry as applied to the analysis of plants into their proximate elements, is almost as essential to the great object of distinguishing the active from the inert productions of our soil, as a knowledge of Botany. Every physician should, therefore, have a sufficient acquaintance with this branch of chemistry to determine in which element of a plant its remedial properties reside; what are the proper menstrua for extracting them, and with what substances they may be combined without producing decompositions. This is not a very exalted attainment, an [*sic*] yet it is true that many of us have not made it. Our profession must, however, improve with the general advancement of the arts and sciences in the West, and I venture to hope that we shall all live to see the time when the humblest practitioners of this country will at least be able to distinguish the gums from the resins or to comprehend the difference between a genus and a species.

You are already, gentlemen, sufficiently apprized that the only certain and conclusive mode of ascertaining the medical properties of a plant, is to administer it, and mark its effects. The rules and precautions to be observed in doing this have been sufficiently enumerated in the progress of our course and need not be reiterated here.

Gentlemen!—The attention which you have bestowed on the subjects that it has been my duty to bring in review before you this winter, is a full earnest of the diligence with which you intend hereafter to cultivate your profession, and warrants the anticipation that you will become successful practitioners. You have at an early period manifested a conviction, that knowledge can only be acquired by application. Let me exhort you to follow ardently the suggestions of this conviction, for a professional man can never neglect them with impunity. When you leave the medical school your studies are merely begun. The germ of your future professional knowledge is yet a tender seedling, which neglected by you must inevitably perish. Watch over it then unceasingly—foster it with tenderness—supply it with liberality and you will elevate it in time to a magnificent tree. Its balmy exhalations will diffuse health and comfort among the wretched victims of disease;—the *golden fruit* of its wide spreading branches will supply your numerous wants, and, in the shade of its ever green foliage you will glide serenely down the vale of declining life. Your studies ought not therefore to terminate with your residence in college. They should be continued diligently through all the engagements, the vicissitudes and the trials of after life. I hope, gentlemen, you will not be appalled or discouraged at this injunction. The cares and perplexities, it need not be concealed, that must environ and distract you, will be neither few nor trivial; but in the midst of them all your studies are to be prosecuted. You must review the books which are now the objects of your attention,—you must read the works which appear hereafter, you must carefully observe and register whatever may present itself in your practice that can explode the numerous errors of our science or extend and fortify it with new truths. You may perhaps doubt whether amidst the labours and fatigues of an extensive practice it will be possible to find time for so much professional reading. You have observed that the greater number of physicians neither read nor write on medical subjects, and you will perhaps assert with them that the drudgery of professional life leaves no intervals for those important duties. This is saying that the practice of that profession, which requires above all others a vigorous and inquisitive mind, amply stored with facts and principles and perpetually intent on new acquisitions, is utterly inimical to the application necessary to make them. I hope, gentlemen, you will never subscribe to this absurd and dangerous doctrine. It is *not* impossible for you to be at once laborious practitioners and diligent students, provided that in the early stages of your career, you providently establish compound habits of business and study. These should never be estranged from each other. Their union is natural and necessary: their seperation destructive to all our prospects. It is the divorcement of professional reading from medical practice that so often produces the melancholy spectacle of an old physician and respectable man, whose philosophical knowledge in his profession is absolutely less than at the hour when he began to exercise it. There are in the course of every day a number of leisure moments, and these you should seize upon for application to your books. You will perhaps be interrupted by a professional call before you have finished a single page, but let not this discourage you. Return to the sentence that was left unfinished and complete it. Fortify yourselves against the delusion that no improvement can be made in this way; and that it is necessary to withdraw from the tumult of business, to situations of retirement and tranquility, before you can read with any advantage. This unfortunate opinion has ruined thousands of our profession, and may justly be regarded as one of the causes which have retarded its progress. It *is* in your power, suffer me to repeat gentlemen, to form a habit of reading professional books, in the midst of professional business; and if this habit be once completely established your fortune and your fame are secured. Having this conviction, I should be wanting in my duty to you, if I did not urge you to the course which has been proposed. I hope that each of you will make the experiment, and thus secure the prize of glory, which, in our profession, is the reward of those only, who unite theory with practice, enlightening and correcting both, by reading, observation and reflection.

The moment has now arrived, gentlemen, when we are to part; & although we have been united in the same pursuit but for a single winter, we shall not I trust separate from each other with feelings of indifference. It seems to be a law of our constitution that concerted efforts in a benevolent work never fail to inspire mutual respect and affection. Such is the character of the object which we have had in view, and such I hope has been the effect of our feeble but unceasing and harmonious exertions for its attainment. It is not then I trust extravagant, for me to suppose myself uttering the sentiments of every member of our little class, when I express the opinion, that while by close attention to an important branch of medical science we have contributed to render ourselves useful to society, our social enjoyments have been multiplied by the attachments and friendships that have sprung up among us.

On one account, gentlemen, I may be supposed to feel sensations more acute at leaving you, than you experience at seperating from each other:—You part with the intention of reassembling; I leave you with no very sanguine expectations of another meeting, . . . and in

[LAST PAGE MISSING]

Anniversary Discourse on the State and Prospects of the Western Museum Society*

(1820)

In his two major discourses of 1820, before the Western Museum Society and before the Medical College of Ohio, Drake sounded a note of hopefulness appropriate to the personal success he had achieved as an American scientist in the West during the years since his return from Philadelphia in 1806. His two books on the natural and medical history of the Cincinnati region had been well received in the East on their own merits as contributions to a greater understanding of the Western country, and praised because of their Western origin, while his essays on the climate and geology of the Ohio Valley had earned him admission to both the American Antiquarian and the American Philosophical societies. He had renewed his acquaintance with the intellectual community of Philadelphia during a second winter of study at the University of Pennsylvania in 1815–1816, becoming in the process the first person from the West to obtain a degree from that most distinguished of American medical schools. He corresponded widely with physicians and scientists in the West, as well as with some of the leading scientists of the East, including Day and Silliman at Yale, Cleveland at Bowdoin, Mitchill at the New York College of Medicine, and the Brazilian minister plenipotentiary to the United States, José Correa de Serra. Alexander Wilson and Thomas Nuttall, the one a friend and the other a student of Drake's own teacher, Benjamin Smith Barton, and probably the two most prominent naturalists working in the United States at that time, stopped as a matter of course to visit with him in Cincinnati during their western tours. The organization of the Western Museum Society and the Medical College of Ohio as independent but affiliated divisions of the Cincinnati College marked the final triumph of his early years, and in his addresses before both groups he developed the proud themes of success: the independence and equality of the West with the East and of America with Europe,

---

* *An Anniversary Discourse, on the State and Prospects of the Western Museum Society: Delivered by Appointment, in the Chapel of the Cincinnati College, June 10th, 1820, at the Opening of the Museum. By Daniel Drake, M.D., Secretary of the Society; Member of the American Philosophical and Geological Societies; Counsellor of the American Antiquarian Society, and Member of the Philadelphia Academy of Natural Sciences.* Cincinnati, Ohio: Printed for the Society, by Looker, Palmer and Reynolds. 1820.

the feasibility and desirability of sophisticated scientific and educational institutions in the Ohio Valley, and the unity of human knowledge and its availability to the systematic practitioner of empiricism.

The Western Museum Society was organized in the spring of 1818 by Drake, the Rev. Elijah Slack, William Steele, Jesse Embree, and James Findlay. In September, the managers of the Society announced plans for the establishment of a museum of natural history, mineralogy and geology, and anthropology. Quarters to house the Society's collections and its library were acquired in the building of the Cincinnati Lancaster Seminary (rechartered as the Cincinnati College in 1819), and a staff to supervise the operations of the museum was appointed. Robert Best was named curator and John James Audubon his assistant. Prominent scientists of the East were elected to corresponding membership and encouraged to send duplicates from their own collections to fill the Society's cabinet.[1] The course of public lectures it sponsored was probably the first formal scientific training available in the West outside a medical college and certainly the predecessor of the scientific departments of the Cincinnati College.

Success was short-lived, however. Economic depression in Cincinnati and a weakness in leadership created by Drake's dismissal from the Medical College of Ohio and his subsequent decision to accept an appointment at Transylvania University in Lexington resulted in debts which the Museum Society could not or would not meet, and in March 1823 the Society's collection was put up for auction.[2]

[1] E.g. Drake to Zaccheus Collins, July 30, 1820, ALS in Collins Papers, Philadelphia Academy of Natural Sciences.
[2] Louis Leonard Tucker, "'Ohio Show Shop': The Western Museum of Cincinnati, 1820–1867," in Whitfield Bell and others, *A Cabinet of Curiosities: Five Episodes in the Evolution of American Museums* (Charlottesville, 1967), pp. 73–105.

WE HAVE THIS evening assembled to commemorate the establishment of the WESTERN MUSEUM SOCIETY.

Among the numerous reasons for this measure, there are two, which exert an almost imperative influence. First: At the expiration of the two years which have been spent in the collection and arrangement of curiosities, when they are prepared for public inspection, and the doors of the Museum are about to be opened, it is important that we should review the design and labors of the Society, and inquire what benefits they are likely to produce. Secondly: As the arts and sciences have not hitherto been cultivated among us to any great extent, the influence they are capable of exerting on our happiness and dignity is not generally perceived, and they have consequently but few friends and admirers. It is therefore proper, that we should institute and continue to observe an annual festival in celebration of the origin of a Society established expressly for their promotion; that we may elevate their character with the mass of our people, and multiply the number of their devotees and patrons, by the infallible method of augmenting their consequence.

The plan of our establishment embraces nearly the whole of those parts of the great circle of knowledge, which require material objects, either natural or artificial, for their illustration. It has, of course, a variety of subdivisions, and in its execution will call for very different architects; as its consummation will afford instruction and delight, to persons of very opposite tastes. Already, indeed, in possession of many specimens in ZOOLOGY, MINERALOGY, ANTIQUITIES, and the FINE and USEFUL ARTS, we venture to indulge the hope, that even at this time, we can offer *something* to interest the naturalist, the antiquary and the mechanician.

To assemble without arrangement, such a great variety of substances, would neither gratify curiosity, nor inform the understanding. It has been well observed, by one of the ablest philosophers of the last century, that *method is the soul of science*. The Managers and Artists of the Museum have not been unmindful of this celebrated aphorism; and, as far as their knowledge would enable them, have arranged the articles, which have been collected, according to the most approved systems. They have thus, in obedience to a valuable precept, adopted by an enlightened reformer* of the present century, provided *a place for every thing, and disposed of every thing in its place*. Should they adhere to these principles, the most important results must be obtained. A regular groundwork being laid, the groups of objects, thus associated by natural affinities, will rise from it in order and beauty, like those which start from the prepared canvass into imitative life, under the creative pencil of the painter.

To establish in this new region a scientific cabinet, on a plan so varied and extensive, may be considered by some as premature and impracticable. It is not difficult to show, however, that this objection is rather specious than solid. For an obvious reason, it is a new country in which such a multifarious assemblage is most proper. Ancient communities, only, exhibit a perfect separation of kindred trades and occupations, and a divorcement of the extraneous branches of science from the learned professions, to which in young societies we find them closely united. Old communities therefore are the only ones which can *successfully* establish cabinets and museums for particular classes of objects,

---

* Joseph Lancaster.

and destined for the benefit and amusement of particular orders of men. Let no one, then, charge our Society with temerity for aiming at a general collection; nor regard as an evidence of vain glory and undisciplined ambition, what, in reality, is both the effect, and indication, of our recent settlement in a new region.

Having thus briefly sketched the outlines of our plan, and offered an apology for their extent, let us proceed to inquire in what manner they are to be filled up.

It would be difficult to exhibit, on a graduated scale, the comparative importance of the different arts and sciences. The circumstances under which we may happen to be placed, exert a strong, modifying influence; and it is chiefly to them that the necessity of sometimes varying to a great extent the nature of our intellectual pursuits, must be referred. The information which to one man is of no utility, may be extremely important to another; and branches of science, which one people might neglect without detriment, may, to another, be indispensable. I will not venture to say, that every species of knowledge included in our scheme is deserving of encouragement here; but I hazard nothing in asserting, that many of our dearest interests are involved in the principal branches, which it is the object of our Society to promote and to illustrate. I will briefly draw your attention to these, and endeavor to show in what manner and to what extent our labors may be rendered interesting to the philosopher, the patriot and the amateur.

The illustration of our Natural History is, of course, the first, as it is confessedly the most curious object. I will neither insult your understandings nor consume your time with arguments to establish the importance of this branch of science. I expect from you a sort of intuitive acquiescence in the proposition, that the inhabitants of every country should be acquainted with its natural history. I anticipate from you an eager assent to the supplementary proposition, that a people in our situation have special need of an acquaintance with their productions and resources. With this conviction, which is equally the result of experience and reflection, you will be prepared to listen patiently to the feeblest exposition of the views and prospects of our Society, so far as they involve the physical condition of the region in which our destinies are fixed. I shall begin, by calling your attention to our ZOOLOGY.

The Quadrupeds of the United States have not yet been fully described, and it is even uncertain whether they have been all enumerated. What proportion of them are indigenous to the Ohio countries remains to be ascertained. The number which has been assigned to these regions by the zoologists will, in all probability, hereafter be augmented. The determination of the species of our foxes, wild cats, wolves, squirrels, otters and deer, will probably show, that many, which are now regarded as mere varieties, are, in reality, distinct species. To these points the attention of our Society is already directed; and I cannot but hope, that they and the other *desiderata* connected with this branch of our zoology will, ere long, be supplied.

It would be an act of injustice to speak of our ORNITHOLOGY, without connecting with it the name of Alexander Wilson. To this selftaught, indefatigable and ingenious man we are indebted for most of what we know concerning the natural history of our Birds. His labors may have nearly completed the Ornithology of the middle Atlantic states, but must necessarily have left that of the Western imperfect. When we advert to the fact, that most birds are migratory, and that in their migrations they are not generally disposed to cross high mountains, but to follow the courses of rivers; when we contemplate the great basin of the Mississippi, quite open to the north and south, but bounded on the east and west by ranges of lofty mountains, while the river itself stretches through twenty degrees of latitude, connecting lake Superior and the Gulf of Mexico, it is reasonable to conjecture, that many birds annually migrate over this country which do not visit the Atlantic states, and might, therefore, have escaped the notice of their greatest ornithologist in the single excursion which he made to the Ohio. As a proof of this supposition, it may be stated, that Mr. Audubon, one of the excellent artists attached to the Museum, who has drawn, from nature, in colored crayons, several hundred species of American birds, has, in his port folio, a large number that are not figured in Mr. Wilson's work, and many which do not seem to have been recognized by any naturalist.

It is not, however, among these important classes, that the greatest number of novelties in the zoology of this region can be found. The obscure and imperfect animals that swim in our lakes and rivers, infest our morasses, dimple our pools, and swarm among the flowers of our fields; those which, like our white perch, are remarkable only for their nutricious qualities; or, like the great rattle snake of our Museum, for living whole months in captivity without the aid of any nutritious substance; or, like the minute insects, which sometimes overcloud our atmosphere, and exhibit an unwelcome example of the distribution of a small portion of life among a multitude of beings, apparently to augment the sum of its power; those animals, in short, which delight the historian, rather than the poet of nature, have been least studied by us, and at the same time are not only most numerous, but contain the greatest proportion of what are peculiar to this country. The more noble and perfect animals traverse extensive continents, and become citizens of the world; while the imperfect are frequently limited to a narrow range, and seldom extend their migrations beyond a single district. An abundant feast is, therefore, in reserve for those who delight to study animated nature in every form, and can equally admire her attributes, whether humble or exalted.

The problems offered by our FOSSIL ZOOLOGY are still more curious and difficult than those presented by the study of our existing animals.

It was formerly supposed, that no animal species is ever suffered to perish. The fossil bones of the Ohio seem to have contributed largely to the correction of this error, and to the formation of juster views of the economy of nature. The first animal remains taken from the valley of Big-Bone were sent to Europe, about the middle of the last century. At that time the study of extraneous fossils had not been prosecuted to any considerable extent, and these bones were regarded as very uncommon and wonderful relics. Since that period, the indefatigable researches of Cuvier and other naturalists have led to the conclusion, as a general fact, that the alluvial tracts of every part of the earth contain the fossil bones of quadrupeds, which belonged to species that no longer exist. How many of these were inhabitants of this country remains to be ascertained. Judging from the prodigious extent of our alluvial grounds, and from the bones that have been already disinterred, we may reasonably expect that future examinations will develop to us many extensive and curious deposits.

Of those which have been discovered, the bones of the great *Mastodon* are the most remarkable. This immense animal belonged to an extinct genus, nearly allied to the elephant. According to Cuvier, five other species, most of them of less magnitude, have been found in America and in Europe. Whether the remains of any of these are mingled with those of the great *mastodon* at Big-Bone, or any other part of the Western country, is a question for future decision.

To the mastodon *we* have applied the name of Mammoth; but this was originally given by the people of the old world to an extinct species of *elephant,* whose bones are accumulated in great quantities along the rivers of the north of Asia. The remains of the same animal have been found in the valleys of the Western country; but are not so abundant as those of the great mastodon. The grinders of these two animals are easily distinguished; but all their other bones have been hitherto confounded by us, and present a difficult subject for future investigation.

The morasses at Big-Bone afford the bones of an extinct species of *Bos,* different from the domestic ox, the buffaloe and the bison; and a species of *Cervus,* of the size of the elk, but distinct from the round-horned elk of the Mississippi, the moose of Canada, and the fossil elk of Ireland, as appears from the researches of our late distinguished countryman Dr. Wistar.

Along with the bones of these extinct species, those of many of the existing races of animals are known to be deposited, and the separation of them will constitute another problem for our determination.

Nothing could be more unfavorable to the solution of this and the other questions which have been proposed concerning these fossils, than the practice of sending them abroad as detached curiosities. It is by comparison alone, that correct results in this, as in the other departments of natural history can be obtained; and to compare these relics it is obviously necessary that they should be collected at one place. I know of no spot in the Western country, which has geographical relations more eminently fitting it for their reception, nor better situated to display them to a great number of persons, than Cincinnati. I cannot therefore but hope, that our collection, already embracing a variety of interesting specimens from different places, will be much increased; and that we shall, ultimately, be able to exhibit large portions, if not entire skeletons, of the greater number of those which have perished in this country.

A more curious subject of inquiry than the cause which annihilated these animals could scarcely be suggested. That the catastrophe by which they were destroyed was an inundation, the naturalists think extremely probable; not only from the situation in which the bones are deposited, but also from the extreme difficulty of conceiving what other cause could have produced that stupendous effect. Of the time, however, when it occurred; whether it was partial or universal, and to what immediate agency it should be ascribed, we are uncertain, and must collect a much greater number of facts, before we can speak with confidence.

We pass, by a natural transition, from this subject to our MINERALOGY and GEOLOGY. I have lately, on another occasion, had the honor of pointing out to you, some of the more obvious relations between the cultivation of these sciences and the promotion of our independence and happiness.* Very little reflection must convince us, that this connexion is neither slight nor transitory; and that while we neglect the resources which can only be developed by the study of our mineralogy, and the arts which grow out of it, we must of necessity remain tributary to the more discerning inhabitants of other countries. To a consideration so powerful, it will, I trust, be unnecessary to add any others, to incite you to the study of this branch of our natural history.

An inspection of our cabinet of geological specimens, will show you, that

---

* See "An Introductory Lecture on the Utility and Pleasures of the Study of Mineralogy and Geology, delivered in the Western Museum, December 18, 1819," published in the gazettes of this city, and obligingly republished (though under a different title and with some alterations by the Editor) in the first number of the quarterly series of the Port Folio.

The Author candidly confesses, that his chief object in noticing this production, is to apprize the readers of that respectable magazine, that the paper referred to was not an essay, but an introductory lecture before a mixed audience; which he hopes will account for the rhetorical and declamatory style in which it is written.

the rocks which form the crust of our globe, have been divided into two great classes: *Primitive* and *Secondary*. Each of these has numerous characteristics, and a variety of imbedded minerals, peculiar to itself. The Western country, from the Allegheny mountains to the arid and uninhabitable savannas beyond the Missouri, is composed, exclusively, of secondary *formations,* and we can, therefore, expect them to yield only those useful and curious minerals which belong to that class of rocks. But although limited to these, our mineral resources, when fully disclosed, will be found sufficiently numerous to compose an important element of public prosperity. For a knowledge of what have been already discovered, I must refer to the shelves of our cabinet; and be content with a few practical remarks.

Limestone, so important to agriculture, architecture and the arts, is the predominant rock of the Western country; and constitutes a more substantial and permanent source of national wealth, than the boasted mines of Mexico and Peru. A soil spread over limestone rocks can never be exhausted of its fertility; and the people who industriously cultivate it, must be rich, powerful and happy. In different parts of our country this great formation presents several interesting varieties, and contains beds and strata of other useful minerals. In the valleys of the Great Miami and Kentucky rivers, it affords quarries of excellent secondary marble. In the northeastern portions of this state, it exhibits many indications of the existence of gypsum. In Kentucky and Indiana it is cavernous, and the walls and floors of its caves afford vast quantities of nitre and epsom salt. In the district where the Missouri, Mississippi and Ohio unite their waters, it is metalliferous, yielding inexhaustible supplies of lead ore, in connexion with sulphate of barytes and beautiful crystals of fluor spar.

Our sandstone formation is extensive, and furnishes numerous quarries, whence building materials are conveyed to the towns and villages along our rivers. The strata of coal which it contains are of such extent, as to present an exhaustless supply of that important species of fuel, whenever the state of the country shall require them to be worked. Either this or the limestone formation, affords at least two valuable localities of buhrstone, sufficient, it is supposed, to supply the entire demand for that article, so important to an agricultural people. In both these formations, there are many salt springs, so strongly impregnated as to induce the belief, that rock salt exists abundantly at no very great depth.

Our districts of slate are numerous; and from what has been discovered, there is no doubt that they contain beds of pyrites and aluminous shale, fit for the preparation of sulphur, sulphuric acid, alum and copperas.

The alluvial formations of the Western country, and especially of the states north of the Ohio, are extensive, and will doubtless yield their peculiar minerals in abundance. Those which are called wet prairies, will probably afford peat and marl; the former of which may be regarded as indispensable to the growth of a dense population on tracts so destitute of wood. Beds of potter's clay are every where met with in our alluvial grounds, and red and yellow ochres will doubtless be discovered. But the most important mineral supplied by these formations is iron, great quantities of which, in the state of argillaceous oxide, unquestionably exist, and will be drawn from them by future industry.

Among the useful minerals which we may expect to find, are mercury, zinc, antimony, and silver; the last of which, in the state of sulphuret, seems indeed to have been already discovered.

Our Geology has only been studied enough, to convince us, that it presents many interesting problems. The great formations of limestone, sandstone, and slate, have been examined but superficially. To determine their boundaries, comparative ages, and relative positions; and enumerate fully and faithfully their imbedded minerals, will require the united labors of many enlightened geologists. The organic remains which they enclose—so numerous, so diversified, so curious in their forms and so mysterious in their natural history, present subjects of inquiry and contemplation, equally difficult and wonderful. Let us traverse the level summits of our highest ridges, clamber up their steeps, descend through the ravines which separate them, or wander in our valleys, these petrified habitations of the ancient tenants of the ocean, are offered to our admiration at every step. The collection which our Society has already made, exhibits a great variety; and placed in connexion with the cabinet of recent shells and corals, presents us with the impressive fact, that the existing seas are inhabited by animals specifically distinct from those which tenanted the ocean that once rolled its waves over the spot where we are now assembled.

The very surface of this region exhibits several remarkable phenomena. Our extensive alluvial grounds contain the water-worn wreck of a multitude of rocks, many of which are not found in this country. On the tops of our highest hills, moreover, we find large detached masses of primitive rock, which have no geological affinities with the secondary strata that rest undisturbed and unbroken below. That polished wreck of other strata, and those primitive masses, were brought hither from the north by water and ice. To ascertain the extent to which they have been carried, and the spots from which they were detached; to determine the causes, assign the era, and note the impresses of the mighty currents by which they were transported,—are labors that remain to be perfected.* To engage in these investigations, it is not necessary to travel from our own doors. The valleys which we inhabit, were the beds of those ancient and overwhelming rivers. The Illinois, the Wabash, the Miamies, the Scioto and Muskingum—even the little stream, that lags among the maples and sycamores which skirt the western border of our city—were once more copious and majestic, than the Ohio in all its modern grandeur. Wandering feebly through expanded valleys, where accumulated waters once swept along, they call to mind a caravan of Arabs encamped within the spacious ruins of Thebes or Palmyra, and display changes in the natural, analagous to those which occur in the moral world.

In our Museum there is a collection of the utensils, weapons and trinkets of our Indian tribes. Some of these were obtained from themselves: others were found in the vicinity of their deserted villages, or disinterred from the rude stone or earthen *tumuli,* which we occasionally find overgrown with weeds in the obscurity of our thick woods. I hope to see this department of our cabinet extended much further. I trust that we are not disposed to forget that the curiosities which it contains, are the memorials of a people, who were lately the highminded proprietors and sovereigns of the country which we now inhabit: that the valleys of the Scioto, the Miami and Ohio were for ages overspread with their encampments: that our hills were once vocal with their songs and orisons, re-echoed their fiery and figurative war-speeches, and resounded the tumult of their dance and chase: that these hills and valleys were the land of their fathers, and those scenes their hereditary devotions, pastimes and pursuits:—but that a succession of wars and treaties have dispossessed them of their domain, and driven them, with the elk and bison, to remoter solitudes in the Northwest. Until we are prepared to deny, or can cease to remember, these simple and affecting truths, we must commend the curiosity that would seek to preserve from oblivion some memento of a people that seem to be doomed to inevitable extinction.

With a reference not only to the preservation, but to the acquisition of curiosities from among that people, I cannot forego the pleasure of adverting to the Union Osage Mission, which has recently passed through our city. The wooden foundations of Fort Washington, erected to protect the inhabitants of this place against incensed and hostile Indians, are not yet decayed, and this

---

* In the month of October, 1817, the Author of this discourse addressed to his excellency Joseph Correa de Serra, a letter containing facts and speculations relative to this subject, which was read before the American Philosophical Society, and ordered to be published. The vertical chart of the valley of the Ohio river, which was necessary to its illustration, was not, however, transmitted in due time to be engraved for the volume of Transactions that soon after appeared, and the publication was of necessity postponed.

Mission is destined to a permanent residence among tribes equally untamed, one thousand miles further west! The enlightened philanthropists that compose it, from that remote station, will shed upon civilized society all the rays of knowledge which they can collect; and we have received a promise that a few at least shall fall upon our institution. Independently of the obligation imposed by this pledge, I shall be excused by every friend of mankind, for offering an incidental tribute of respect to the holy and intrepid benevolence, which could detach for ever from the joys and comforts of home, so many intelligent and happy persons, and immure them in the depths of the wilderness, for the sole purpose of dispensing the blessings of civilization and christianity among its benighted inhabitants.

Our country exhibits older and nobler monuments than the recent vestiges of our Indian tribes. The number, extent and regularity of our mounds, and the implements of stone and copper which they contain, afford incontestible proofs that a people more numerous, enlightened and social, than the wandering hordes found on the discovery of this continent, had previously been its inhabitants. These monuments are our only antiquities; and although they may not, like the classical ruins of Asia and Europe, awaken inspiration, nor infuse melancholy, they will not, I hope, be thought altogether unworthy of our admiration. At what time were they erected, and deserted; have the people who formed them become extinct; did they emigrate to Mexico; or slowly degenerate into the existing hordes; and what were the causes of any of these events, are problems which can be solved only by researches into the relics which they have left. These should be vigilantly sought after and carefully preserved, that they may be compared with each other, and with the works of art which belong to the existing tribes. We should thus snatch from the grave a memorial of the former condition of our country, and gain at least a few materials for the portrait of a people, whose very name is blotted from the tablet of humanity.

To exhibit the connexion that often exists between things apparently remote from each other, as well as to demonstrate, that our Museum may be made an efficient means of inquiry into the aboriginal history of this country, I will here state one or two facts.

There are on the shelves appropriated to our Indian implements and ancient remains, several fragments of earthenware. The greater number of these were found in Kentucky and Ohio, about the deserted encampments of the present tribes: one piece was dug out from the center of the large mound, which imparts so much interest to the scenery of the western suburb of our city, and another, manufactured under the view of the person who presented it, was brought, a few years since, from the hordes which inhabit the banks of

Red River, in Louisiana. These different specimens have one remarkable character in common. A part of their composition is pounded river shells, an ingredient, which, not being like clay indispensable, seems strongly to imply a common origin of the art among the former and the latter inhabitants of this region; or a transmission of it from one to the other, and consequently a derivation of the existing tribes from the people whose monuments overspread our country.

The other fact serves, perhaps, to illustrate some of the migrations, if not the origin, of the same people.

In the Museum there are three large marine shells, which were taken from an elliptical mound near the center of the city. They had been deposited with many other utensils and trinkets around the bodies that were buried in that *tumulus*. The lip and internal parts of these shells had been removed, so as to convert them into vessels. The most interesting question which could be proposed concerning them,—from whence were they brought? we are already able, in part, to answer. The two larger are of the same species, and belong to a genus denominated Buccinum, by the naturalists. Among the shells hitherto obtained by us from the Atlantic states, there is not one of the same species with these; but the *desideratum* has been supplied by a gentleman from the West Indies, who has deposited in the Museum a shell, which is manifestly of the same kind. We are therefore at liberty to suppose, that those in question were brought from the Florida coast, or perhaps from the shores of Cuba.

The other shell found here, belongs to the genus Murex. It is strongly characterized by having its spire reversed, or turned from right to left; a conformation which belongs to no other in our collection. In the opinion of a late ingenious writer,* it is the same kind of shell, that is employed by the Hindus in certain religious rites; and from this and other facts, he has inferred, that the former inhabitants of this country were of Hindu origin. There is a reversed murex, however, in the northern European seas; and until it is ascertained that they, or some of our own waters have not supplied this, as well as the buccina which were found with it, such a bold speculation will not be received without hesitation.

I have already announced, that the promotion of the useful and ornamental arts, is among the objects of our Society. Drawings, models, and products of the former, whether mechanical or chemical, will find a conspicuous place; and every exertion will be made to acquire good specimens of the latter. Having been transplanted from countries where the fine arts were flourishing in vigor and beauty, our people in general are not without a relish for them.

---

* The amiable and lamented Mr. J. D. Clifford, of Lexington, Ky.

Those, indeed, who have once gazed with admiration on the sublime historical combinations of a West, or the faithful portraits of individual greatness by a Stewart, can never lose the taste for that enjoyment. Should any person object to the gratification of this taste, as a luxury that ought to be prohibited in a young republic, I would reply, that the cultivation of the fine arts should be regarded merely as a concomitant, and not as the cause, or the consequence, of a luxurious state of society; and that a love for the chaste and elegant labors of the painter, the architect and the sculptor, should not be ranked with a relish for the pleasures of the table, an admiration for personal ornaments, and a passion for public shows, and dissolute amusements. The former originates in sentiment, and its gratification imparts dignity and elevation; the latter are rooted in sensuality, vanity and vice, and their indulgence leads to ruin and disgrace. One springs from

"Faculties that walk the range of heaven,"

The others from

"Appetites that grovel on the earth."

It may be said, however, that we are too poor to encourage the fine arts. I will admit that but few of our citizens have sufficient wealth to become their individual patrons; but this very circumstance constitutes a strong argument for confiding to a collective body, the means and the duty of promoting their introduction into this country. This object has been assigned to our Society, and I hope to see it executed in a manner that will both delight and refine the public taste.

Among the variety of objects which it is designed to embrace in the Museum, are several kinds of philosophical instruments, calculated to illustrate the principles of magnetism, electricity, galvanism, mechanics, hydrostatics, optics, and the mechanism of the solar system. The whole of these can be fabricated by our ingenious *Curator,* Mr. BEST; and the acquisition of them will not only facilitate the progress of the solitary student, but enable the Society to institute public lectures on the different branches of natural philosophy, as well as of zoology, mineralogy, geology, American antiquities, and the fine arts. Popular courses on these subjects, delivered from time to time, as the means of illustration become adequate, and competent lecturers can be engaged, would accord with the spirit that suggested the establishment of the Museum, and could not fail to multiply the benefits which it is expected to confer.

In connexion with this subject, it is my duty to call your attention to the

Library of the Society. I do this, not to give a catalogue of our little collection, but to express an earnest hope, that in the various branches of science, it will be speedily augmented. One of the most painful deprivations experienced by the student of nature in these new and remote settlements, is the want of books to direct his researches. To employ arguments in support of this assertion, would be almost as superfluous, as an attempt to prove, that but for the invention of letters and the art of printing, we should, at this moment, be as debased as the untutored savages which prowl around the shores of lake Superior. Tradition is their only record; and in its archives no aggregation of the experience and observations of ages can take place. Their generations are insulated from each other; and, like the grasses of the prairies on which they roam, one passes away after another, leaving neither vestige nor monument behind. They have no ascending progression on the scale of excellence; but move for ever in a labyrinth, on the same degraded level. We have the happiness to be placed on a nobler eminence, but can never hope to rise from it, if we neglect the aid of books. Without their assistance, indeed, we can neither comprehend nor enjoy the wider and brighter prospects, which our superior elevation affords; but must resemble navigators becalmed in sight of new lands, abounding in all that could reward them for the perils and privations of the voyage; or guests bidden to a feast, and when seated, prohibited from tasting the luxuries which tantalize their longing appetites. We may be encompassed by a thousand interesting objects—our feet may press the rarest productions of the mineral world, and on every side the most beautiful forms of living nature may smile upon us and invite our scrutiny;—but the whole will be unavailing. It is not given to us, to penetrate by a single glance, the veil which the Creator has thrown over the relations, that bind into one beautiful and admirable system, the myriad of parts which compose the mighty fabric of this globe. Thousands of years have elapsed since the students of Nature began to unfold her mysteries. Books are the great repository of their discoveries, and he who neglects them, begins, like the first observer, unaided and alone. He may be compared to the astronomer, who, rejecting the records of the science, and even refusing to employ the telescope, might continue at the base of the observatory, and contemplate the heavens with his unassisted vision. Surrounded, then, as we are by a multitude of curious and important productions, we cannot, I think, refrain from introducing and consulting the oracles which might aid us in assigning their names, and acquiring a knowledge of their history and qualities.

These reflections naturally suggest the propriety of adverting to the introduction of the Museum into the College edifice, where we are now assembled. This connexion will, in all probability, be made permanent, and may be regarded as auspicious for both institutions. In some degree they are necessary to the success of one another, and the interests of both would, therefore suffer by a separation. They afford, in succession, all the aids that are essential to a liberal education. The College is principally a school of literature, the Museum of science, and the arts. The knowledge imparted by one is elementary, by the other practical. Without the former, our sons would be illiterate; without the latter, they would be scholars merely—by the help of both, they may become scholars and philosophers.

Dismissing the consideration of particular topics, I shall pass to a few general remarks. From the preceding review, it appears to be among the leading objects of the Western Museum Society, to collect and preserve the natural and artificial curiosities of the United States, and especially of that portion which we inhabit. If any enlargement of mind can result from the examination of them when exhibited in the Museum, the same effect would be produced, in a much higher degree, by inspecting and contemplating them in their natural situations. I cannot, therefore, but regret, that we do not attach more importance to journeys of observation through our own country. Travels of this kind were eloquently recommended, almost a century ago, by the celebrated Linnæus, and ought to make a part of the education of every young man. After having completed his scholastic, academic or collegiate course, and acquired the rudiments of his trade or profession, he could do nothing so well calculated to enrich his mind with useful knowledge, and qualify him for the practical duties of future life, as to travel through his native land. The objections that preclude the greater number of our young men from foreign travelling, cannot lie against domestic, which I do not hesitate to say would be equally serviceable. It is quite deplorable to observe, in what utter ignorance of the condition of their native country, they usually engage in the career of business that is allotted to them. Whether destined to be farmers, mechanics, or merchants; physicians or divines; soldiers, lawyers, or even politicians or statesmen; they, in general, enter upon their respective pursuits with equal ignorance of the geography, natural history, and statistics of their theatre of action; and of the character and genius of the people with whom all their future relations are to be formed. A few of our sons are sent to Europe; but this, even supposing them to derive some improvement from such a journey, does not make them acquainted with their own country, and cannot, therefore, supersede the necessity of exploring it. Such as rely on foreign journeys only, resemble scholars who spend their lives in the study of Greek and Roman literature, and die quite ignorant of their own. Accomplishing much which cannot be condemned, though but little that we can approve, they are to be admired rather than imitated. I would not, however, discourage foreign travelling, but it should be preceded by domestic. Until we are acquainted with the state of our own country, we must be wretchedly prepared to appreciate that of others. I may venture to add, that when we shall have acquired a full knowledge of our own land and nation, it will not often be necessary to an able performance of our civic duties, that we should visit any other. To indicate the multitude of curious objects, and depict the endless variety of sublime and beautiful scenery, which our beloved country presents to the inquisitive traveller at every step, philosophy should guide the pencil, and poetry infuse its inspirations. I shall not venture on the elevated theme, but hasten to conclude with some desultory observations on the influence of literature and science upon the complete establishment and future security of our national independence.

Not having commenced our career in barbarism, we are without the characteristics of an aboriginal nation. Our annals are recorded and uninterrupted. We were detached from civilized portions of the old world, and brought with us the habits of thought and action, the tastes, propensities and passions, which belong to a refined society. Although a young nation, our people are at maturity in much of what belongs to ancient communities. The edifice is new, but the materials of which it is constructed, were previously fashioned and employed elsewhere. We have, therefore, the wants and the desires of a highly civilized condition, while for our acknowledged deficiency in the means of gratifying them, we cannot offer as an apology, that we are a new people. With as little propriety can it be urged, that we suffered a premature alienation from the mother country; for the very acts of wisdom and heroism which disjoined us from the parent stock, were indubitable evidence of our being prepared for the separation. These, however, were the exploits of our fathers, and I fear that too much reliance has been placed upon them. Have we not admired the greatness of their achievements, when we should have been laboring to perpetuate the blessings of which they are the source? Have not our eyes been dazzled by the splendor of their virtues, until we have sometimes been rendered insensible to what we must perform to make us worthy of such distinguished ancestors? Have we not too frequently beguiled ourselves with the idea, that the formation and adoption of our federal constitution, was the full establishment of our independence? Like children, who subsist upon their patrimony, have we not drawn with prodigality upon a heritage of fame and glory, which it was our duty to augment?

It would be incorrect to deny, that many things have been undertaken: it would be untrue to affirm, that much has been accomplished. A variety of establishments and institutions have been organized; but their number and the labor bestowed upon them are inadequate to the objects to be attained. A simple enumeration of our remaining dependencies on Europe, would make a long and frightful catalogue; and to lessen their number should be the unceasing and anxious aim of every member of the republic. I do not mean to say, that we should have no foreign dependencies; but we certainly should have none that are not compensated by equivalent dependencies upon us. Such an equality with other nations, would be at once the sign and source of our permanent prosperity; and until that enviable condition is established, in vain will our political systems stand forth a proud monument of the wisdom of their authors; or our teeming earth send up its herbs, and fruits, and flowers, and our green fields display their richness and beauty—we shall neither be ennobled by the one, nor rendered comfortable and happy by the other. In the midst of Nature's choicest bounties, in the fulness of religious and political freedom, we shall remain the unhappy and ignoble dependents of the old world. To *us,* especially, who inhabit an interior region, and have our dwelling places among the sources of a mighty river; who cannot hold intercourse with foreign countries without an inland voyage of more than a thousand miles, or a difficult overland journey across rugged and lofty mountains, a dependence on Europe is equally disastrous and degrading. I trust that these opinions will, ere long, spread more widely through society, and inspire us to new and nobler efforts in the sacred cause of national independence. I will indulge the hope, that we shall, at no distant time, more fully perceive and acknowledge the momentous truths, that in a nation organized like ours, private and public prosperity are inseparable: That knowledge is the common basis of both: That efforts to promote it can neither exhaust nor impoverish: That expenditures for its cultivation, would not dry up our resources; but, like the exhalations which the earth sends forth, to fall, after a time, in fertilizing showers, would return upon us a rich and replenishing harvest: That periods of general pecuniary embarrassment should not be suffered to diminish these appropriations, as that would inevitably augment the evil: That the greater number of disasters, both public and private, originate in ignorance, and should not be allowed to perpetuate themselves by fortifying its empire: That under all the vicissitudes and trials of life, after a sincere invocation to Divine Providence, the safest reliance is on the dictates of learning and science; and that in the midst of the widest desolation, our exertions for their benefit should never relax.

An ample illustration of many of these propositions may be found in the difficulties produced during the revolutionary war by our ignorance of the arts and sciences. It is unnecessary that I should do more than advert to the sufferings which our patriot soldiers experienced in the field from a deficiency of clothing, and in our hospitals from a want of medicines and hospital stores;

to our defects in the science of engineering; and above all, to the inadequate supply of arms and ammunition, which of itself in the early stages of the war, seemed to threaten the cause of liberty with inevitable ruin. The young colonist saw his native land invaded, its soil crimsoned with the blood of his brethren, and the products of their united and peaceful labors consigned to the flames, or triumphantly borne away as plunder: His soul was fired with a noble resentment, but his weaponless hand hung palsied by his side: His agitations resembled those of the young eagle, when it sees its native haunts molested and defiled, without the power of punishing the aggressor: He was restrained, like those ancients, who, according to the fables of the poet, were metamorphosed into trees, retaining all their desires and aversions, but without the power of gratifying either: Like the kindling volcano, ere it has formed an outlet for its boiling lava, his breast heaved with the impulses of revenge; but the energies which might have redressed the wrongs of an injured country, were left to rend the bosom in which they had been nobly awakened.

If our beloved country were *now* to be invaded by a foreign enemy, not one of these agonies would be felt. And whence, let me ask, would be the exemption? I will reply, that since the perilous days of the revolution, Apollo and Vulcan have united in tempering the shield of Mars, and replenishing his quiver with arrows. The Genius of Science has shed his creative influence over our artisans and philosophers: by his beams they have explored the recesses of our primitive forests; pursued the meanders of our vast rivers; surveyed our spacious bays and harbors, and descended into our prolific earth. The fruits of this enterprise have been rich and abundant. Academies and Societies for military instruction have been formed; laboratories and manufactories have been erected; our magazines have been filled, and our weak places made strong. Thus, if not equal to Europe in the application of science to works of national defence, we have obviously passed the weakness of childhood; and might repel, almost without bloodshed, an invasion, to which our fathers could only oppose their dauntless and patriotic bosoms.

To specify all the connexions between the cultivation of science, and the increase and perpetuity of our happiness, would require more time and higher powers than have been appropriated to this occasion: for to what shall we ascribe the more admirable principles of our federal constitution, when compared with all others, but the profounder acquaintance of our statesmen with the science of government? To what shall we attribute the greater respectability and dignity of character which belong to the middle and lower ranks of our people, but the more general diffusion of information consequent upon our excellent political systems? To what shall we refer the extinction of many of the absurd prejudices and odious superstitions which, not half a century since, defaced our national physiognomy, but the influence of juster views of Nature and her laws? To what can we ascribe our better progress in the mechanical and chemical arts, but a closer study of the principles of natural science? To what shall we attribute the decreasing necessity and practice of sending our sons abroad for an education; but the improving state of our own institutions, the augmentation of their libraries, cabinets and philosophical apparatus, and the employment of more learned professors? To what, in short, shall we ascribe the decided amelioration of our national character, and its regular though tardy approaches towards refinement and elegance, but the cultivation of letters and science?

If we perceive, then, in the increase of useful knowledge the true secret of our permanent happiness; if literature can supply the talismanic agent of our prosperity and power, and philosophy, like 'a pillar of fire by night,' direct our wandering footsteps to the temple of glory, let us not ignobly stay our hands from the labors by which, only, philosophy and letters can be made to flourish. Let the architects of our national greatness conform to the dictates of science; and the monuments they construct will rise beautiful as our hills, imperishable as our mountains, and lofty as their summits, which tower sublimely above the clouds.

APPENDIX

The Managers of the Western Museum Society avail themselves of the opportunity afforded by the distribution of the preceding Discourse, to propose EXCHANGES with other Societies, and with individual collectors at a distance.

The principal articles, which they can send abroad with convenience, are—geological specimens of the secondary rocks of the basin of the Ohio, with their organic remains, and accompanying minerals; preserved specimens of the mammalia, birds, fishes, amphibia, crustacea, mollusca, insects and worms of the same region; grinders of the mastodon and arctic elephant; Indian implements, and aboriginal relics taken from mounds and tumuli.

In exchange for any of these, they will be glad to receive from the Eastern states and from Europe, foreign zoological, geological and mineralogical specimens of every kind; also manufactures and trinkets from the islands of the Pacific Ocean; coins and medals; paintings, casts from statues, and fragments of sculpture, of such excellence as to serve for models;—and finally, Books, in the various departments of physical science, whether written in the English, French or German languages.

As an Herbarium is contemplated among the future acquisitions of the

Society, and as the Managers are desirous of promoting the introduction and cultivation of exotic plants, they will be pleased to receive any valuable seeds and roots of other countries, and will, in return, transmit such of our indigenous vegetable productions as may be requested.

Foreign packages, intended for them, may be consigned to Professor SILLIMAN, Yale College; to Professor MITCHILL, New York; to Mr. JOHN VAUGHAN, Philadelphia; Mr. HORACE H. HAYDEN, Baltimore; or to Messrs. NOBLE and MILLER, New Orleans, who will reimburse the expenses which have been incurred.

From the authors of books on the various topics embraced in the plan of the Museum, whether published in the United States or in Europe, the Managers would be very happy to receive copies for the library of the Institution. By transmitting their works to this distant place, a wider dissemination of their fame would be effected, while they would experience the noble satisfaction of being instrumental, in naturalizing the sciences in a new country.

From their fellow citizens of the Backwoods, generally, the Managers would earnestly ask such curiosities, both natural and artificial,—whether as donations or otherwise would, of course, be left to themselves—as may fall into their hands. With this assistance, the labors and expenditures of the Society would be rendered far more efficient and productive than they otherwise could be; and the important but difficult design of making the Institution an extensive and useful School of Nature and Art, would be much sooner accomplished.

ELIJAH SLACK
WILLIAM STEELE
JESSE EMBREE } MANAGERS
PEYTON S. SYMMES
DANIEL DRAKE

Cincinnati, Ohio, June, 1820.

Inaugural Discourse on Medical Education*
(1820)

Daniel Drake concerned himself with the problems of medical education almost from the start of his own career as a practitioner. At least as early as 1809 he accepted students as apprentices and assistants to his medical practice, in the manner of his own teacher, Dr. William Goforth.[1] The organization of a partnership between Drake and Dr. Coleman Rogers in the summer of 1817 more properly marks the beginning of Drake's career as a medical educator, however; the doctors announced that they were prepared to receive medical students for instruction "in any number that may apply."[2]

Upon Drake's return from Lexington in the spring of 1818, it was announced that he, Rogers, and the Rev. Elijah Slack would inaugurate formal instruction in medical subjects at the Cincinnati Lancaster Seminary, of which Slack was president, beginning in the fall. Drake was to teach botany, materia medica, physiology, and the practice of physic; Rogers, surgery and anatomy, and obstetrics in conjunction with Drake; Slack, chemistry and pharmacy. Substantial enrollment in the fall of that year provided Drake with a rationale for requesting the Ohio legislature to charter a medical college at Cincinnati and to recognize and charter the Cincinnati Lancaster Seminary as an institution of higher education.

Quarrels among the members of the faculty, which were to reach a head in 1822 when his colleagues voted to dismiss Drake from his professorship and from the presidency of the college, prevented the Medical College of Ohio from opening earlier than the fall of 1820. It did open, however, with twenty-four students in attendance. In December Drake sought to complete the arrangements for successful medical education as he conceived it by obtaining from the Ohio legislature a charter for the Commercial Hospital and Lunatic Asylum in order to provide faculty and students an opportunity for the systematic observation of diseases in a kind of physician's "cabinet."

* *An Inaugural Discourse on Medical Education; Delivered at the Opening of the Medical College of Ohio, in Cincinnati, November 11th, 1820. By Daniel Drake, M.D. President of the Institution, and Professor of the Institutes and Practice of Medicine, and of Obstetrics.* Cincinnati, Ohio: Printed by Looker, Palmer and Reynolds. 1820.
[1] Drake to Return J. Meigs, Aug. 12, 1812, ALS in Horine Collection, University of Kentucky Library.
[2] *The Western Spy* (Cincinnati), July 11, 1817.

Disease is the appointed inheritance of Man. In that stage of his social existence which is called a state of nature, its forms are few and simple; but as he ascends from this, they increase in number, complication and mortality; and did not the progress of civilization open new sources of happiness, and afford additional means of counteracting both physical and moral evil, it would be deemed a curse instead of a blessing. Among the beneficent fruits of this condition, we may discover his ability to cure a variety of diseases, which, were they to occur in the uncivilized state, would inevitably prove fatal. This knowledge is, indeed, altogether indispensable to the welfare of his race, when associated into compact and refined societies; and may justly be regarded as at once a consequence of their nature and a cause of their continuance.

Among savages the cure of diseases can scarcely be said to be confided to particular and instructed members of the tribe; but in civilized life it has been found necessary to make a specific assignment of this important duty,—and hence the origin of professional practitioners of medicine. It was for the instruction of these, that medical schools were first instituted. To trace their origin and progress, and contemplate historically their influence on the profession, and on the happiness of society, might be highly interesting; but, on the present occasion, I shall pursue a different course, and proceed rather to call your attention to the principles on which they should be organized; to the nature and prospects of that which we are about to institute; and to the difficulties, importance and dignity, of the Medical Profession.

The end for which medical schools were established being the preparation of young men for the cure of diseases, we cannot adopt a better mode of ascertaining how they should be planned, than to enumerate the branches of science which seem indispensable to that important and difficult object. I shall devote a few moments to each of these.

An inspired writer has exclaimed, *I am fearfully and wonderfully made!* Subsequent examinations of the human frame have fully established the truth of this declaration, and proved that the preëminence of man over the other inhabitants of the globe, is not merely intellectual. His organs are more numerous, more complex and more intricately combined, than those of any other animal. The science which discloses to us this admirable structure, is Anatomy; and for its acquisition every medical school should furnish numerous facilities.

The anatomical study of the human body is the foundation of medical science. The materials of which the superstructure must be composed, are of a different kind. It is not sufficient to be acquainted with the forms and composition of the different organs. The student must inquire into the various functions which they are destined to perform. He must contemplate them in motion, and not at rest; as living, and not dead; as endowed with attributes, from which result an action among the organs themselves, as intimate and harmonious as their mechanical connexion; and a relation between our intellectual faculties and the surrounding world, more wonderful, perhaps, than any other phenomenon presented to our admiration. To the branch of science which teaches these laws of motion and sensation in the healthy body, the term Physiology, or the Institutes of Medicine has been applied. In the Medical College of Ohio, it has not been made the object of a distinct professorship, but considered as a proper introduction to the Theory and Practice of Physic, and, therefore, confided to the professor of that department.

Without a knowledge of Anatomy and Physiology, no pupil can prosecute the study of the profession; for diseases consist either in alterations of structure, or in disordered and irregular movements in the function of that structure; and in both cases, without an acquaintance with the *healthy* condition, no degree of genius can enable us to understand the *morbid*. The number of these morbid alterations is so great; the symptoms which distinguish and characterize them so equivocal; their causes so numerous and obscure; their progress so devious, and their terminations so uncertain,—that the study of them, even with the most ample preparation in the branches that have been mentioned, is as difficult as it is important. To this department of science, the term Pathology, or Theory of Physic, has been appropriated. The Practice of Physic is the application of the means of relief; and, to be conducted on scientific principles, it requires a knowledge, not only of Anatomy, Physiology and Pathology, but of several other sciences, among which the Materia Medica is the most important. The object of this branch of the profession is to teach the facts and principles which relate to the operation of the various medicinal agents on the human body, both in health and disease; together with their natural history and pharmaceutical preparation. In the college of medicine which we are about to establish, it is made the object of a distinct professorship, to which, when a hospital shall be properly organized, a subordinate branch, denominated Clinical Medicine, will be attached.

Those morbid conditions of the body,—mostly consisting of disorder in its structure—which require manual assistance for their removal, are the proper objects of Surgery. An accurate knowledge of Anatomy is especially necessary in the practice of Surgery, and hence the union which they sometimes exhibit in medical schools. At the present time they are thus associated in this college; but, constituting a professorship too extensive and difficult for one person, they will in all probability, before another session arrives, be confided to two.

Obstetrics is a branch of the profession for which there seems to be no necessity among savage nations; but its importance in civilized life is unquestionable. Its magnitude and difficulties, however, are not such as to demand for it a distinct professor in a school where it is considered an object not to multiply the teachers; and in that of Ohio it is attached to the professorship of the Institutes and Practice of Medicine.

Chemistry completes the list of sciences which it is essential for the student to learn, before he can attain the rank of an enlightened graduate. It is in some degree auxiliary to all the branches which have been named; but has the most important relations with Physiology, Materia Medica, and the Practice of Physic.

Medical Jurisprudence is the application of medical knowledge to the judicial inquiries which the law directs to be instituted, in cases of insanity, and of injury or death from wounds inflicted, or poisons administered, with a malicious intention.

We come now to sciences which lie on the outside of the group. Of these Botany is the most considerable. Its relations are chiefly with the Materia Medica, to which it contributes very largely, and in connexion with which it has for ages been cultivated by medical men. In the infancy of our college, it will not, like the branches that have been named, be made an essential study; but the means of prosecuting it will be provided, and, in the general state of the Western Country there is much to recommend its cultivation to every practitioner of medicine.

Mineralogy claims the attention of medical men for similar reasons. As contributing liberally to the Materia Medica, it is interesting to the profession in general; but to such as reside in these Mississippi states, where many valuable medicines undoubtedly lie concealed beneath the surface, it is peculiarly important. Our college has not yet appropriated a professorship to it, but the professors of Chemistry and Materia Medica propose to go largely into the natural history of the different medicines derived from the mineral kingdom; and the cabinet of the Western Museum Society will furnish those who are disposed to acquire further information, with every necessary facility.

Zoölogy, or the history of animals, ranks next to Botany and Mineralogy, in its relations to medicine. It contributes but little to the Materia Medica proper, though largely to that department called Materia Alimentaria. With Anatomy and Physiology, however, it has numerous connexions, and may be made to contribute so largely to their advancement, as to have strong claims on the attention of all the cultivators of medical science. Botany, Mineralogy and Zoölogy appear to me to occupy the first circle exterior to the sciences that are strictly medical, and to have with them a variety of obvious and intimate connexions.

The succeeding circle is occupied with various branches of natural and experimental philosophy, among which the most conspicuous and important are Electricity, Magnetism, Hydrostatics, Geography and Meteorology, with the principles of which every student of medicine should make himself acquainted.

Nor should he stop here. His researches must not be confined to the objects of the physical world. He must extend them to the mind itself, and to its operations and effects, as displayed in the civil history of our race. The endless variety of aspects which the derangements of intellect are known to assume; the insidiousness of their origin; the melancholy effects which designate their progress; and the dreadful and uncontrollable energy of their violent stages, render them of great interest both to society and the profession. The Philosophy of the Mind should, therefore, be made a part of the elementary acquirements of every physician.

The principal reason that should constrain him to the study of Civil History, is, that without it he must be unacquainted with the progress of his profession, and the manners, modes of living, and diseases of different ages and nations; a degree of ignorance unworthy of the dignity, and incompatible with the fulness of knowledge which should distinguish his official character.

The catalogue of studies is not yet completed. Literature and science are not the same; but a physician should acquire both, and the cultivation of the former ought to precede that of the latter. It is, however, a mortifying fact, that in the United States, and especially west of the mountains, the young men designed for the medical profession are in general destitute of this preparation in literature, so essential to their future acquisitions in science. Commencing the latter while ignorant of the former, their progress is comparatively slow and imperfect; and they learn, when too late, that a magnificent edifice cannot be erected on a narrow and badly constructed foundation. No young man should commence the study of medical science till he is at least sixteen years of age; and unless the preceding time have been devoted to the acquisition of language and the rudiments of general knowledge, he will neither possess that learning, nor those disciplined habits of application, that are essential to a successful prosecution of medical studies. While the standard of literary and professional excellence necessarily participated in the general imperfection which attended the institutions of our new country, this want of preparation in those who undertook the study of medicine was less striking, and had to be excused, from being unavoidable. The opportunities for prosecuting a better course of preliminary studies have been created, even in the western states, and no young man should hereafter be encouraged to become a student of medicine, who has not prepared himself in a manner corresponding with the vast extent and inherent dignity of that science. This preparation should not consist merely in detached knowledge of his own language. He should ascend to its ancient sources, and drink deeply at its pure and original fountains. If the principles of medical science, which are now taught, be not the same that prevailed in Greece and Rome, they are partly expressed in the language of those learned and polished nations; and to be thoroughly understood, the words in which they are conveyed must themselves be made an object of study.

So deeply impressed are the Faculty of this institution with the neglect of these studies, and the importance of them to the advancement and elevation of the profession, that they have offered an annual prize medal for the best inaugural thesis in the Latin language; and hope by this measure to excite among the students of the west an emulation for excellence in classical literature.

I have now made a rapid enumeration of the various branches of literature and science, which a physician should regard as indispensable, or important, objects of cultivation. By proper attention to these, the student may become a philosopher and scholar; but something more will be necessary to make him a successful practitioner of medicine. He must not only comprehend the principles of the profession; but, by acute and patient observation at the bedside of the sick, learn how to apply them in the cure of diseases. He has to make himself an artist; but his skill must consist in the practical application of the precepts and maxims of science. Clinical medicine is an unceasing employment of means for the accomplishment of specific or definite objects. Considered in relation to our knowledge of those means, the profession is a science— in relation to the application of them, it is an art. He who acquires the former only, is learned; he who relies on the latter alone, is ignorant, empirical and criminal; he who compasses both, reaches the highest attainable perfection.

I shall attempt to illustrate and enforce what has been said on the magnitude and difficulties of the profession, by a short history of the employments and responsibilities of a medical man in this country, where the various branches are united in the same person.

Let us suppose him to be called upon as an operator in surgery. He must be able to determine whether the age, health and condition of his patient will admit of the operation; he should have an accurate knowledge of the structure of the parts through which his incisions are to be made, and be acquainted with the variety of instruments that have been employed, so as to select those which are best adapted to the particular case; he should know the history of the operation, and be qualified to choose from among the various modes that have been proposed for performing it, the easiest and most successful. Should any anomaly of structure lead to the injury of parts which cannot be cut without danger, it becomes his duty, suddenly, to decide upon and employ the means which are necessary to the preservation of life; and this he must do in the midst of general consternation, when deafened with the cries, and drenched in the gushing blood, of his patient—circumstances, which requires an accurate knowledge of the surrounding parts and quick powers of invention, not less than a firm and unshaken hand. Finally, the operation being over, and every accidental difficulty removed, he must determine on the means which should be employed to avert the dangerous symptoms that may result from an extensive mechanical injury, and secure the final recovery of his patient.

Suppose at another time he is called to treat a malignant fever. He has then to encounter a case of general disease. All parts of the complicated machine have been invaded; every organ has its functions deranged, and every fibre vibrates with a morbid action. Prompt and energetic efforts must be made, but these according to the causes and character of the disease may be of opposite kinds. What an important and difficult decision is then devolved upon him. The remote causes of malignant fevers must be considered; but these are uncertain and obscure, and the study of them involves a knowledge of secret changes in the earth and the atmosphere, that can neither be detected nor understood by those who are ignorant of the science of Chemistry. The symptoms of the disease are its language; but they speak intelligibly to *him* only, who has, by a long and arduous course of study and observation, established in his mind the relations which exist between external appearances and internal changes of action and structure; and who, from his previous knowledge of the laws of the human body in health, can determine, from visible signs, how far, and in what manner they have been violated in disease. Having thus, by the lights of Anatomy, Physiology, Chemistry and Pathology, discovered the causes and character of the fever, it remains for him to decide upon its treatment. To fix upon the best, he should previously have acquired an accurate knowledge of the various plans of cure, which have been proposed in different ages and nations. He should be able to choose the most appropriate of these, and to modify it for the particular case under consideration. This will open to him a new duty, the selection, preparation and administration of medicines: the able performance of which requires a knowledge of the effects of the various remedies on the body, both in health and disease; in other words, an acquaintance with the Materia Medica, and certain parts, at least, of its subordinate and contributing sciences—Pharmacy, Botany, Mineralogy, and Zoölogy.

Again: Suppose he is consulted for a chronic malady of some internal organ, on which the powers of medicine have been exhausted ineffectually, and the important question of a change of climate and country is proposed. How can he determine in what region his patient may find a condition of soil, water and climate favorable to recovery, without a knowledge of Geology, Geography, and Meteorology?

Finally, he may be called to a maniac, and in the sad combination of mental and bodily disease, be presented with a case that requires him to possess a knowledge of the laws which regulate the mysterious influences of the mind and body upon each other. He will then perceive the necessity of an acquaintance with Metaphysics; for without it he must be utterly incapable of "administering to a mind diseased."

It would be easy to extend this exposition of the duties of a physician; but enough has been said, I trust, to prove, that no other profession requires in its practitioners so profound a knowledge of very different sciences, such a versatility of mental effort, such an association of ideas apparently the most distant from each other, such a power of calling up dormant facts, such nice discrimination in the selection of precedents, so collected a mind in moments of unexpected difficulty, and so ready a resort to expedients, when established means are unattainable, or not adapted to the end in view.

I have now completed a catalogue of the principal branches of science and art which should engage the attention of the student of medicine. To facilitate the acquisition of these, Medical Schools were originally designed, and continue to be encouraged in every civilized country. In the classical ages of Greece, under the guidance of particular masters, they shone forth with the lustre peculiar to that wonderful people; but lost their brilliancy as the flames of patriotism and public virtue declined, and were either extinguished, or carried away, by the copious torrent of barbarism that swept down from the north and east. In modern Europe, these institutions were among the first to emerge from the inundation; and for several centuries have been ranked with the most powerful means which have been invented for advancing the interests of science and humanity. Their establishment in this nation is of recent date; and till lately, like the other fruits of advanced civilization, they have been cultivated exclusively in the parent states. Whether they can be made to flourish in the new states of the west, can only be determined by trial. *The Medical College of Ohio,* which we are now assembled to institute, is an experiment of this kind. At the opening of its first session, the question whether it will ultimately succeed, forms a problem of many conditions. I shall briefly state the principal *data* from which a favorable prediction has been drawn.

The mountains that separate the western from the eastern states, and the declivity which extends from their base to the Mississippi, present an obstacle to free communication from the former to the latter, which requires from us an unceasing effort at self-dependence, in whatever relates to the support, the comfort and the preservation of life. On this broad and permanent foundation, we may confidently proceed to erect many important establishments, which, without it, like the rolling stone of Sysiphus, would fall whenever the immediate support of those who might sustain them, should be withdrawn.

More than half the states of the Union lie west of the great rampart, and are thus coerced into a union of interests, by an impediment to the east, while they are powerfully attracted into communion, by a common channel of foreign intercourse in the south. The population of these states, already one fourth of the national aggregate, is increasing at a ratio which must make it equal, within the present age, to such portions of the original states, as lie beyond the eastern sources of the Mississippi. In the United States, every ten thousand inhabitants, on an average, supply one medical student to our universities; and we may hence conclude, that the period is not remote, when one or more respectable schools of medicine may be supported in the west.

Of the numerous towns which have been built in this region, not one seems to me so proper for such a school, as Cincinnati. Its better geographical position, and more intimate relations with the western country generally; its more numerous population, and above all, its greater necessities and facilities for the establishment of a permanent Hospital, designate it as a spot peculiarly fitted for the erection of a medical institution. On the last of these local advantages, I feel it my duty to expatiate for a moment. Without the aid of an Infirmary, no school for the cultivation of practical medicine can possibly be made to flourish. The citizens of Cincinnati are, therefore, urged to the erection of a hospital by all the prospective advantages which a successful College of Medicine could bestow upon them. To enumerate these at large would be a work of supererogation. I shall state that only, which has an immediate connexion with the pauperism and sickness which prevail among the lower orders of our emigrants. *It is an unquestionable fact, that these wretched people, who at present subject us to the heaviest contributions, would, if an infirmary were provided, become a source of profit and prosperity to the city.* I make this declaration advisedly, and hope it will be remembered by all who participate in the weighty imposition of taxes which annually falls upon us, even if it should be thought doubtful or unworthy of notice by those who direct our municipal concerns.

It is in hospitals, that the lectures on practical or clinical medicine must be delivered. To hear these and witness the cases to which they relate, would be an object with every student who might attend the Medical College. The fees of admission for these purposes would go into the treasury of the hospital; and, as the professional attendance on the sick would, under this regulation, cost but little, the revenues thus accruing would, after a few years, become adequate to all the expenses of disease among this unfortunate and degraded class of our population. We should then make them do in sickness, what they did not perform in health—support themselves. The price of their exhibition, moreover, would be paid by persons from a distance, whose other disbursements during a residence here, would become a source of positive benefit to the city.

The Legislature of Louisiana, in the true spirit of benevolence, has proposed to the different Mississippi states the erection of hospitals for the sick boatmen on the various waters of that great river. On this subject the Governor of Ohio has received a communication from the Governor of Louisiana, which will be transmitted to the next legislature. I cannot for a moment doubt, that this honorable body will make an appropriation for an object involving so deeply the prosperity and reputation of our state; and Cincinnati, as its commercial metropolis, would, of course, be the spot where the establishment would be erected. It will not be necessary, however, that the State should maintain a distinct and independent hospital for this object. Their efforts might, with great propriety and advantage, be united with those of the guardians of our poor and the Faculty of our College—the State supplying the means of erecting a common edifice, the City maintaining its police and expenses, and the College supplying it gratuitously with medical assistance.

A Poor-house with shops and gardens might be made a part of the same establishment, and the whole confided to the care of a single board of managers. It is sufficiently apparent, however, that such a work cannot be accomplished without a general union of means; a hearty cooperation of efforts; a liberal and considerate course of legislative policy; and, above all, a deep and generous conviction of its necessity and benefits. If this happy communion of feeling and design could be effected, our city would soon be graced with a house of charity, in which the unfortunate, when diseased, would find refuge and relief; and the people of the whole state an asylum for the insane, that would wipe away the disgrace of confining them in the cells of our common jails;—while our students of medicine, enjoying more ample opportunities of improvement, would become the benefactors, instead of the scourges of society.

Conceding to Cincinnati all the advantages which have been enumerated, it may however be urged, that the establishment of a Medical College is premature; and that without endowment, the patronage which it can create will not be sufficient to sustain it. It cannot be supposed that the legislature of Ohio will long suffer an institution so beneficial to the people of the state, to struggle unsupported, with the difficulties that must beset its infancy. But should a different policy prevail, it will not therefore be abandoned. Its growth may be retarded, but like the pine on the stormy and barren summits of the Allegheny, what it loses in luxuriance, it will gain in vigor and hardiness. Its professors are determined on success, and hope to deserve it. To show more fully on what foundation their expectations rest, will occupy us but a few moments. The College to which they belong offers one advantage over all that have been hitherto established in the United States. Not one has a session exceeding four months—a period confessedly too short for the course of instruction which they are designated to impart. The lectures in this institution will continue five months, and there will be fewer of them daily. Thus the pupils will not be perplexed and oppressed by exuberance, nor hurried so rapidly on, as to be precluded from the necessary reading and reflection. To this single regulation, there is much reason to believe, our College will ultimately be indebted for no inconsiderable part of the support which is anticipated. The advantages of a protracted session may not however be perceived, until sufficient time has elapsed for the graduates of this school to be compared, in the extent, variety and perfection of their attainments, with those in which the pupil is revolved through the great circle of medical science in the short period of three or four months.

It is an axiom of the medical profession, that the same diseases in different ages and countries have many peculiarities. I shall not on this occasion attempt to assign the causes which operate to produce those modifications which are exhibited in this region. It is sufficient to know that they exist, and that they can be studied most successfully in a university established within the limits where they prevail, and supplied with professors whose daily occupation it is to investigate and cure them. A consideration so powerful will operate with the students of the west to counterbalance many of the advantages which the more ample collections of books, of anatomical preparations, and of philosophical apparatus, in the older schools of Europe, and the Atlantic states, may enable them to afford. To extend the comparison between old and young medical institutions, may appear preposterous to such as think they should be contrasted, and not compared. I shall not, however, be induced to desist from the task, by any apprehension of being charged with temerity or invidiousness. I venerate the founders of the medical institutions of the eastern states; I honor the liberality which has endowed them, and respect the eminent professors with which they are filled; but they have not reached an elevation to which *our* College may not ultimately attain, nor can I consent to regard those who conduct them as objects of imitation, rather than competition.

Every scientific institution is composed of teachers, and the machinery with which they operate; and the superiority of old over young ones, obviously consists in the latter, more than the former. Since the invention of printing, the discoveries and improvements of genius are no longer hoarded in the archives of the schools, but spread abroad and made the property of all. When a new professor, therefore, is inducted into the chair of an old institution, he does not acquire a heritage of useful knowledge, as none has been accumulated and transmitted with the office to which he has succeeded. With the single exception of being supplied with better instruments, he stands on a level with the professor of a new institution, and is superior or inferior according to the energy of his genius and application. Institutions, then, as it relates to those who conduct them, cannot be said to increase in knowledge as they increase in age. The professorships are perpetual, but their incumbents are successive. Hence we can understand how the professors of a new institution *may* equal, and even surpass those of an old; and how the former, with the greatest destitution of artificial aids, might sometimes possess an assemblage of talents, that would do more than the latter with all their accumulated machinery.

I hope not to be understood as arguing against the importance of books and philosophical apparatus to a scientific institution. On the contrary, I think they should be collected with the greatest diligence; but the possession of them should not be considered a positive evidence of superior excellence; nor a deficiency of them, a certain proof that it cannot afford many important advantages. One of the characteristics of genius is the capacity of attaining its end, independently of the means by which, only, inferior intellect can accomplish its objects. This quality of the mind might, indeed, be defined the power of *substitution*. It enables those who possess it, to contemplate abstractly the variety of agents by which an effect may be produced; and thus gives them the choice of a great number. The philosopher whose views of a subject are vivid and comprehensive, can render it obvious to ordinary apprehension, by means of the simplest apparatus:—with this he can instruct; with the multiplied instruments of an old university he can both instruct and delight.

In thus standing forth as the advocate of a young institution, it is not my design to draw a comparison between its teachers and those of the established and operative schools of the Atlantic cities. I may however be indulged in the hope, that they will be found to approach somewhat nearer to those eminent men, than our sparing collections and simple apparatus, approximate to their magnificent libraries, cabinets and laboratories. And, as more depends on the artist than the instruments which he employs, I shall flatter myself with the further expectation, that from its commencement, the Ohio school will be able to confer important benefits on those who visit it; and thus be found to have a self-dependent principle of growth. It is to this principle that the stimulus of public sentiment and patronage should be applied; and for the purpose of securing its influence, I shall proceed to display more fully the interest which the community has in the prosperity of medical institutions.

To enumerate the various advantages conferred upon society by well regulated medical schools, may be considered superfluous; as they fall upon us unceasingly, from the moment of our birth to that of our dissolution. Like the genial effects of heat and light, we cease, however, to observe them, because they are uninterrupted. Having become imperceptibly blended with the other elements of our happiness, we are unconscious of their influence, and without the aid of analysis, they continue through our whole lives, like secret benefactors, to administer to our comfort unobserved and unappreciated. There are occasions on which this analysis should be made; and in pleading the cause of an infant seminary, I should be recreant to its interests, did I not call upon the enlightened community, on whom it must rely for the aliment of its growth, to inquire how far their happiness is connected with the medical profession.

Most of the occupations in society reflect upon it merely the limited advantages of their immediate application to its necessities. The profession for which I have the honor to plead, is capable of dispensing a wider range of benefits and blessings. Its proper object is the cure of diseases, but in becoming qualified for this, its members are prepared to render many other important services. Imbibing in the course of their collegiate studies, a taste for the cultivation of letters and science, and being afterwards received into the bosom of every society, they contribute powerfully to infuse the same taste, where it might otherwise be wanting. Impelled by necessity, as well as inclination, to continue the prosecution of sciences which benefit mankind through other channels than the profession of medicine, they not unfrequently become the authors of discoveries and inventions which in their application to the common purposes of life, materially augment the sum of its enjoyments;—or, devoting themselves to the cultivation of *one* of the auxilliary sciences, increase the number of its facts; enlarge its boundaries; and elucidate its principles, or apply them with new success to useful and ornamental purposes. This is especially the case in a new country, where literature and philosophy are not yet self-existent; but must rely for protection and cultivation upon an alliance with the learned professions. It is in such a country, that the usefulness of a scientific physician spreads widest through society, and his character displays, comparatively, its broadest and brightest disk.

But the chief purpose of the life and labors of the physician, as already intimated, is the prevention and cure of disease; and this object is of the greatest magnitude, whether we consider it in reference to the preparation which it imposes on him; or to the countless multitude of blessings which it confers on society. In proposing to advert to these, I find them so intimately combined with all the scenes and situations of life, as to render selection difficult, and classification almost impossible.

Health, for the present purpose, may be defined that state of our sensibility, which enables the external agents that ordinarily act upon us, to produce effects of an agreeable kind. Disease is a condition, in which our sensibility is so changed, that those agents either produce no effects, or such as are painful. Now, if happiness do not *consist* in the former, that state is essential to its production; while the latter, of necessity, is attended with misery. And hence, either to be free from distress, or to be positively happy, it is equally indispensable to enjoy sound health.

I shall not dwell exclusively on the pains and sufferings of the sick; but call your attention, likewise, to the various distresses of which disease is the fruitful parent.

To estimate these, we must have approached its unfortunate subjects, and mingled with those who hung in anxiety and anguish over the sick bed. The united agonies of mind and body experienced even by a disgusting victim of prodigality and vice, affect us so sensibly that we cannot but desire his recovery, and feel inclined to unite with him in sentiments of gratitude to the physician who restores him to health, and perhaps to reformation and happiness. But what is there in the maladies of an insulated wretch to excite our pity, in comparison with the sufferings of the unfortunate, the useful and the good.

When a serious disease attacks an only son, on the threshold of manhood, and threatens the sudden extinction of genius and enterprise; or, when it fixes on a favorite daughter, in the midst of bloom and beauty, while every virtue is germinating in her youthful bosom, and the first fruits of taste and intelligence afford full sustenance to future hope—to what can we compare the agony and consternation of the afflicted parents? Or what language could express their gratitude to him who should preserve such beloved objects from a premature grave?

To accomplish this deliverance would be an enviable achievement; but a nobler triumph attends the conquest of the Tyrant, when his vengeful arrows are fixed on manhood in the zenith of its splendor.

Men of genius and beneficence are the brightest luminaries of the moral firmament: the choicest gifts of bounteous Heaven to our benighted race. They were the authors and architects of society; they decoyed the hunter out of the wilderness and weaned him from the chase; encircled him with the arts and sciences; inspired him with new and nobler propensities, and continue to furnish him with the means of gratification and happiness. In their preservation and prosperity then we should feel a deep and living interest. When they are assailed by disease, the very pillars of society are menaced with destruction; and their expiring struggles spread convulsion and disorder throughout whole communities. When a catastrophe of this kind impends; when our divines, philosophers and statesmen; our artisans, physicians, advocates, professors and philanthropists are selected as the victims of disease; when the fountains of benefaction begin to pour forth troubled waters, and we are even threatened with a diminished supply of these, where do we then look for relief, or on what can we rely but the medical profession?

But the apprehensions of society for the fate of a great and good man in disease, sink into insignificance, when compared with the forebodings and anguish of his friends and family. Who has ever cast his eyes upon the death-bed of such a man, without the conviction that it is a scene of the deepest anguish? Who has, at any time, gazed on the sad spectacle of weeping relatives, supplicating friends, and distracted children, and not been suddenly pervaded with horror? Who has ever contemplated the affectionate wife, immovably fixed at the head of her expiring husband, absorbed in unutterable grief, and silently rent with pangs of sorrow; and not turned in sympathetic dismay towards that profession, upon which, in these hours of emergency and distraction, the good as well as the bad, the wise as well as the foolish, are compelled to rely for hope and relief? When such a man is the prize, to rescue him from the grasp of death, and dissipate the portentous gloom that hangs over his family and friends, by the light of his renovated eye, is one of the happiest efforts of the medical profession. But although among the happiest, it is not the noblest triumph of medicine. There are periods of epidemic disease, in which the King of Terrors envelopes society in a pestilential cloud; when the salutation of the morning is not, who has expired, but who has survived through the night; when the stillness of our highways is interrupted only by the solemn rumblings of the hearse, and the silence of our apartments unbroken, except by the groans of the dying, and the more melancholy wailings of those who watch around; when Despair spreads her lurid mantle over the portals of every habitation; and Horror infuses his chilling influence into every vein, till the stoutest hearts are appalled; when Calamity sways his iron sceptre, and Terror, like a whirlwind, breaks asunder the bonds of society, and involves its members in anarchy and desolation:—then it is, when the ties of consanguinity and love have been dissolved, till the mother abandons her infected son, and the husband deserts his dying wife,—at this awful crisis, the good physician arises in a panoply of knowledge, as the champion of humanity. Deeply impressed with the sacred duties of his office, and nobly animated to their faithful performance, he sustains an aspect of serenity and confidence, and sublimely goes forth, like a ministering angel, to dispense health and hope and happiness.

With these dreadful visitations *we,* providentially, have no acquaintance but by report. Let us turn from the hideous picture, and resume the history of individual suffering.

The distresses of the sick consist, partly in the pains inseparable from disease, and partly in the seclusion which it imposes from all external sources of enjoyment. The efficacy of medicine in destroying the former is great; for, when the malady cannot be terminated until it has run a destined course, it may frequently be deprived of its virulence. The poison may be neutralized when it cannot be extracted; and wounds, which time only can cicatrize, may have their anguish assuaged, and be rendered supportable. The powers of the healing art in shortening the stages of a disease, are equally obvious; and, next to the mitigation of pain, nothing is more grateful to an invalid, than his early release from the chamber of sickness. Every convalescent has an exquisite relish for the objects and aspects of nature; and even in their coarsest and simplest dress they never fail to delight him. But there are moments when the great Artist arrays her in charms uncommonly fascinating. In every country the elements of this enchantment are in some degree peculiar. In this, we are presented with the most striking exhibitions of beauty and grandeur in an April morning, when the heavens have distilled their first and purest dews upon the tenderest flowers of the year; when the buds of our forest expand to the enlivening influence of the vernal sun, while the same influence has restored our migratory birds, and "made vocal every tree" with their songs of love; when the Ohio, swollen by the last snows of its parent mountains, wheels rapidly along its weight of waters, and reflects the brightened disk of that luminary, whose return has dissolved the spells of winter, and diffused new energy and action through every animate and inanimate form: or, on an evening of August, after the oppressive heats of the great fountain of light and life have been tempered by a thunder gust, and the freshened atmosphere wafts the sweet exhalations of our blossomed cornfields; when the green mantle of our hills assumes a livelier hue, and the rays of the setting sun illumine the departing clouds with the softest tints of red and yellow light. To languish in captivity till these evanescent glories have passed away, is the agony of privation: to be emancipated by medical skill, and set at large with renewed capacities for enjoyment, infuses gladness of heart, and inspires gratitude to the Great Physician above, as well as to the humble instrument by which his beneficence is administered.

But confinement in a sick room is rendered irksome by other causes, than a seclusion from the beauties of external nature. Occurring without our anticipation or consent, it produces an unexpected suspension of all the pursuits of business; exhausts the proceeds of those which had been made efficient; precludes the consummation of others; and restrains us from engaging in new ones, till the "golden moments of opportunity" have perhaps fleeted away for ever.

On our social plans and pleasures it exerts an influence equally unpropitious. A sick man is no longer a sociable, but a selfish being. He sinks to the state of a dependent on the community, and asks nothing from it, but sympathy and assistance;—and these afford him no other enjoyment than what arises from the removal of pain, or the dissipation of irksomeness. Their effects are negative, rather than positive. He wants the power to be an actor in the busy and bustling operations of society; and cannot even be a spectator of scenes in which he once performed a conspicuous part, and from which, in health, he unceasingly derived entertainment and happiness.

Finally, disease is a foe which invades us in as many forms as Proteus could assume. It is the great enemy of all enterprise and improvement: the sedative which paralyzes every faculty and passion: the poison which deranges every mental operation: the opposing power of patriotism, philanthropy and ambition—relaxing the arm of industry, subverting the schemes of benevolence, and extinguishing the lights of genius, to lead him captive through the mazes of error and dullness. It may be likened to the dark cloud which intercepts the sun beams till the germinating corn perishes in the earth; or the baleful mist that spreads mildew over the ripening harvest;—nay, its ravages are terrible as the volcano which breaks up the foundations of a country; prostrating as the tempest that lays waste its cultivated surface; overwhelming as the inundation which buries up its monuments, and "completes the work of devastation and ruin."

The struggle of the medical profession with this fell power, can only be compared to the holy but interminable contest of truth with error and falsehood; or the glorious warfare that liberty maintains against the black empire of despotism:—the magazines of science supply the shield and armour, philanthropy inspires the heroism, and the life of man is the prize of victory.

Introductory Lecture for the Second Session

of the Medical College of Ohio*

(1821)

Nowhere does Drake articulate his conviction of the universal availability of natural knowledge to the human understanding more clearly than in his "Introductory Lecture to the Second Session of the Medical College of Ohio," of November 1821. An essential assumption of modern science generally, and of the Enlightenment in particular, this notion had in fact been ignored more often than not by the practitioners of rationalism who saw themselves as priests of the truth and hence the natural leaders of those who had not learned the secrets of the natural world. The revival of Baconianism in the early nineteenth century involved a revitalization of this old idea, however, and an attempt at the democratization of knowledge to be achieved by ending the monopoly which the few had exercised over the tools of understanding.

For the Baconians, the means to this end was the creation of formal and informal educational institutions, colleges, seminaries, lyceums, mechanics' institutes, debating clubs, literary and scientific societies; the consequence was to be the creation of an equality of knowledge among men quite independent of their inequality of station, and a more effective utilization of the moral and intellectual resources of the community.[1] Ultimately, however, as Drake here points out, the very possibility of a democratization of knowledge depended upon the availability of the methods of induction—systematic observation and generalization—to all men, regardless of their backgrounds, and upon the effectiveness of induction in providing true knowledge of the operations of the universe.

The text of this extraordinary address was recently discovered attached to the manuscript of Drake's "Address to the Louisville Medical Society" of November 27, 1840, of which it formed the principal part. Drake was not one to waste good words. His Cincinnati temperance address of 1828, for example, was revised for delivery at

---

* "Introductory Lecture for the Second Session of the Medical College of Ohio, November, 1821," MS in Cincinnati General Hospital Library.

[1] Cf. "Dr. Drake's Remarks on Common Schools," Transactions of the Fifth Annual Meeting of the Western Literary Institute [1835] (Cincinnati, 1836), pp. 172–75; "[Address before a society for the promotion of general knowledge. Cincinnati,] June 6, 1842," MS in Cincinnati General Hospital Library.

Philadelphia and at Columbus in 1831. An address prepared for the Athens Literary Society in August 1841 received a new introduction and was presented at the Louisville Medical Institute on December 4. In the same way, the Introductory Lecture of 1821 was revised for delivery before the Ohio Mechanics' Institute on December 15, 1832, under the title "Gravitation, Affinity, Vitality," and subsequently underwent further changes for delivery to the Louisville Medical Institute. The version printed here consists of the original address, as well as Drake's additions and revisions, set off in brackets. These are included for their intrinsic interest as well as for the indication which they provide of Drake's continuing commitment to Baconian methods in science long after their allegedly peculiar utility in new and unexplored countries had passed.

IT IS ADMITTED that in choosing the subject of an Introductory lecture a public teacher is at liberty to exercise an unlimited discretion. Availing myself of this privilege and venturing to depart from the practice of my colleagues, whose classic eloquence exerted upon general and miscellaneous topics has given to every succeeding day a new charm, I have resolved upon a lecture that is purely physical. It may serve, therefore, not merely as an introduction to our course, but will constitute its beginning. In this way only can I hope to fix your attention, or to make upon you a new impression, after the high intellectual excitement with which you have been recently inspired.

Our object is to study the nature and cure of diseases; but as a preliminary exercise it may be useful to institute some general inquiries concerning the bodies in which they are observed to occur. These are the beings which we denominate living or animated, in contradistinction to those which are said to be inanimate or dead. From this proposition it would appear, that among the productions of nature there are classes of objects so unlike one another as to be designated by the most opposite epithets. Those which have just been employed stand in this relation to each other, and require to be well understood before we can proceed. These words are not the signs of simple, but of complex ideas; as we may satisfy ourselves by a moments reflection on the phenomena to which they are applied; and especially by comparing the properties and appearances of living objects with the characters of those which are dead. In making this comparison, one of the most obvious distinctions, and that which superficial observation would insist upon, is that the former are in motion and the latter at rest. Let us test the value of this distinction by attempting to apply it. The proposition then is, that bodies in which there is *motion* are *alive*—those in which there is *none* are *dead*. The dried seed, then, in the Herbarium of a botanist is *dead,* because it is without internal motion; but if subjected to heat and moisture it will germinate and produce a living plant. After this plant is cut down; it will ferment, and is then *alive,* because it has *inward* motion.—Again, the mountain is *dead,* because it is at *rest,* but the winds that beat around its brows are *alive* because they are *active*. These examples must be sufficient to show, that although absolute rest may in general imply death, actual motion by no means implies life. The position which has been assumed is, therefore, false, and it is quite apparent, that other ideas than those of *mere* motion and rest, should enter into our definitions of life and death. To arrive at a knowledge of these I know of no other method than analysis and shall therefore endeavour to employ it.

If by our senses we take cognizance of the objects that immediately surround us and of the earth and universe as far as it is allowed us to know them, we perceive that the masses of matter of which they are composed, exhibit an endless variety in *form* and *magnitude;* that their *motions* are equally diversified, and that they act and react on each other, in ceaseless operation. Now it is an axiom in philosophy, that matter of itself is inert; and hence we ascribe this multitude of determinate and indeterminate forms, and these complex and ever stirring movements to certain active powers or principles [or agents] with which we suppose it to be endowed. It is not my design [as already intimated], to institute a metaphysical inquiry into the essence of these attributes, for we can know nothing of them but by the effects they produce. These effects are the facts upon which we must reason, and from which all our inductions [conclusions] relative to the physical world should be made. The first and most independent business of philosophy is to observe and register them, the second to arrange them into natural orders, the third and highest to deduce from them the principles of science. I am not aiming, gentlemen, to reveal to you these principles. This, to the extent of my abilities and as far as they relate to the disordered functions of living bodies, will be done in the progress of our course. I would fain, however, reveal to you the means by which philosophy accomplishes her ends—they are acute observation, accurate comparison, judicious arrangement, and logical induction—these are the secrets of all her power, the engines with which she atchieves her mightiest conquests,—and if we expect success we must employ them in the analysis which we have undertaken.

[To Lord Bacon, the brightest, the profoundest, the most comprehensive genius, which the world has ever seen,—the human race are indebted, for this eternal truth, that acute observation, accurate comparison, natural classification, and logical induction, are the means by which philosophy *must* operate, in analyzing nature,—the engines, with which she has effected her past atchievements, and will continue to advance till every secret shall be revealed and all mysteries unveiled. In every age, man has been an observer of Nature, but he was long an observer only; and, indeed, in the mass, remains so still, even in countries the most civilized. He sees, and hears and feels, but never experiments, nor registers, compares, and classes his sensations. The mere observer, perceives only the surface of things—the philosopher looks through them, and beholds and records their inward beauties; like the Otakietan savage who pronounced the horse and his rider one animal, the simple observer, supposes the mass of matter which obstructs his path, to be uniform throughout,—but the philosopher finds it made up of many distinct parts: The *eye* of one travels over the exterior of creation, and stores the memory with independent facts—the *mind* of the others comprehends its laws. Hitherto, the philosophers have formed a distinct *caste* from the people; and like kings have been supposed to possess a divine right of superiority. But this delusion should be dispelled, is indeed fast disappearing, and the distinction between scientific and the unscientific, dissolved. Philosophers, like kings, are but men; and all men to a certain extent may become philosophers. Our Faculties are the same and if exercised in the same manner we should at length differ only in degree.

[Since the era of Bacon, Nature has been earnestly interrogated, and a vast body of facts has been amassed; they have been diligently compared, and, with a skill which philosophy never exhibited till she worked by his precepts, they have been classed according to the active agents by which they are produced. But much yet remains to be done, before we shall know the precise influence of each of these active powers, and be able, accurately, to assign to each, the phenomena which it generates. It is even doubtful whether we have, yet, ascertained the existence of all the principles of activity, with which matter is endowed.

[The objects and operations, which the *inanimate* portions of the earth and universe present, have suggested the existence and action of the following: 1. Gravitation, 2. Chemical attraction, 3. Caloric, 4. Electricity, 5. Magnetism. Let us proceed to consider the effects produced by each. First, of gravitation. This name has been given to that power, principle or attribute, by which masses of matter when at a distance attract each other. If they are not resisted, the effect of this attraction is to bring them nearer to one another—consequently it is a principle of motion. The laws of this motion were developed by Newton, whose name will be pronounced with reverence, as long as the noblest association of genius, virtue and religion, which the earth has ever known, can reach the human heart. Let us enumerate some of the phenomenon which are the offspring of this principle.

[When the apple is detached from its parent stem gravitation brings it to the earth: When the farmer chops off the lofty oak, gravitation lays it prostrate. The same power brought down upon Herculaneum and Pompaei the projected volcanic dust and ashes, which concealed them for 1800 years. When a fragment of rock is detached from the mountain cliff, gravitation precipitates it into the valley below. If the sailor boy aloft, let go his hold in the tempest, he dashes into the waves and finally rests on the bottom of the sea, from the influence of the same power.

[Indeed whatever descends to the earth in a straight line, or on an inclined plain, either fast or slow, is under the influence of gravitation—to fall is to gravitate; and to gravitate, is to tend to the centre of the earth, in virtue of its attraction. Seizing on this single and simple idea, you will perceive that many of the most sublime and beautiful natural phenomena, which give interest and expression to the air, the waters and the earth, are the products of this power. Thus the April Shower, and December snow storm—the gentle zephyr, the tempest and the trade winds—the tides the currents and the waves of the ocean, would have no existence without this power. On land its operations are equally striking; for it brings down the soil of the hills and mountains to fertilize the valleys below; maintains the currents of our rivers; and generates every variety of water fall, from the dashing little brook, in which the young engineer places his first, frail water wheel, to the cataract where the waters of a hundred lakes, fall headlong into, into the unfathomed gulph, dug out by their own weight.

[But the effects of gravitation are not limited to the fall of bodies from the atmosphere or their gradual descent from the mountains to the valleys;—they are found in the movements of the earth itself and in the motions of all its sister planets. Even the fixed stars themselves are kept in their proper places, by the same influences, acting reciprocally among them. The effect of destroying this attraction, by the sun and fixed stars, on the planets which revolve around them and of suspending it among themselves, would be, to leave them to the uncontrouled influence of the projectile power; Thus unbridled, they would instantly start off in straight lines.—The succession of seasons would then cease, and shooting in to new courses, across the Heavens, some of them might travel on forever through infinite space, while others would meet or strike against the fixed stars in dreadful concussion! Thus the quiet concert— the sublime harmony of motion, of the physical Heavens, (a most beautiful emblem of the moral Heaven, where divine love, sustains and regulates the whole) would be broken up; order fall into confusion, and the universe into chaos.

[But I must pass on to our second head—*Chemical* attraction or affinity. This differs from the attraction of gravitation, in at *least,* two important points—1st to exert itself the particles of matter must be in contact—2dly of the various sorts, some exert an attraction on this kind others on that. Chemical attraction is indeed essentially elective; so that we often meet with bodies which nature has never yet united, and which no art of man can cause to combine; while on the other hand we know of many, which the profoundest skill can scarcely decompose, so strong is the action which their elements exert on each other.

[While the attraction of gravitation draws and retains every thing as near the centre of the earth as possible, chemical attraction binds together the various elements which constitute the *parts* of which the planet is composed. While gravitation originates many *sensible* motions, on the surface of the earth, in the waters and in the air,—chemical attraction originate a class of motions among the ultimate particles of which the earth, the waters and the air are composed, which are *invisible,* and can be appreciated only by their effects.

[But let us look at some of these phenomena, beginning with those which are produced by art.

[If we mix sugar with water they combine by attraction, and the solution left in a warm place undergoes a fermentation in which the elements of both unite in a new mode and produce a fluid, which raised from the mixture by distillation, constitutes alcohol: Thus by chemical action one of the most palatable and nourishing substances in nature is transformed into a poison which has slain more than all the visitations of pestilence, which have ever scourged the Earth.

[Again. When sulphur, charcoal and nitre are intimately blended, into gun powder, and fire is applied to the mixture, an instantaneous action commences. The nitre is decomposed; its elements combine with the sulphur and charcoal; the new compounds are converted by the heat into gases, and an explosion, loud and terrible in proportion to the quantity is the result.

[Once more. If a plate of polished iron, be thrown into a vessel of diluted sulphuric acid it speedily disappears; but, if the fluid be slowly evaporated, we obtain crystals of copperas or green vitriol—a compound of the acid and metal. If this compound, called the sulphate of iron, be dissolved in a decoction of nutgals or oak bark, a different acid displaces, the sulphuric, and a new compound is formed of a black instead of a green colour and constitutes ink.

[Anecdote of Dr. Watson. *Mirabale dictu!* (Turn to the Laboratory of Nature)]

If we examine with minuteness the surface of our earth, we shall find that it is not a homogeneous and tranquil mass; but an aggregate exceedingly compounded whose parts are in perpetual action upon one another. Let us take an example. By the influence of heat and moisture certain bodies lose their bond of union; and the dissimilar atoms of which they were composed seek as it were the elementary particles of other compounds, and new ones are produced,—all of which implies motion. In these processes the water which facilitates them, is frequently separated into its constituent parts, and while one of them enters into almost every compound the other escapes; but at the moment of its liberation it seizes upon the sun-beams which have contributed to that effect, and forms a new compound of three elements hydrogen, light and caloric, to which the chemists have given the name hydrogen gas. By the agency of a fluid denominated electricity, this gas may be decomposed,—a great part of its light and heat disengaged, its hydrogen reunited, with one of the components of the atmosphere, and water reproduced, which falls to [and sinks into] the earth, to be again decomposed. In this new decomposition it meets [let us suppose that it meets] with the substance called carbon, for which both of its elements have an attraction; they are therefore [immediately] separated from each other, and form with that substance, two new bodies termed carbonic acid and carbonated hydrogen. The former of these continuing at [by its weight near] the surface of the earth, and meeting with lime, forms limestone; the latter ascending into the atmosphere gives us visible proofs of its escape in the burning springs which are met with in the various parts of the world. The limestone thus formed may not however continue for a single hour as it is instantly decomposed by either of the mineral acids. Suppose it meets with the sulphuric. A new compound is immediately formed by the union of that acid with the lime, which is denominated sulphate of lime or gypsum, an article of equal interest to the agriculturalist and the statuary. In the formation of this compound there is of course a liberation of the carbonic acid of the lime stone, which meeting with magnesia forms the carbonate of magnesia of the shops with which you are all familiar. These six may serve as specimens of the silent and unseen motions that are continually stirring the surface of our earth, and so altering the composition of its parts, that [These may serve as specimens of natural and artificial operations which depend on this power. To the latter we are indebted for innumerable useful and ornamental arts, not known till since the inductive philosophy of Bacon was applied to the study of chemistry. By the former, our earth near its surface experiences so many silent and unseen movements, and suffers such a continual change in the composition of its parts that] it may be truly said they are not the same at this moment, they ever were before; and, that should its duration even be infinite [eternal] they would never again be what they now are. All these operations are manifestly of a similar kind, governed by one system of laws and chiefly dependent on a single principle—*Chemical Affinity,* [Turn to p. 15 for the effect of a suspension of chemical affinity] aided, however, by heat, light, electricity and another principle to be mentioned presently.

If we now cast our eyes abroad for ampler views, we are presented with a class of motions essentially distinct from these. The fragment of rock that is detached from the summit of the mountain, owes that separation to a chemical affinity [operation], for when the water which had percolated into the fissures of the strata was converted into ice by winter and compelled to crystallize in such manner as to rend the rock assunder, it was affinity that gave it motion [the operation was chemical]. Was there, however, no other power in nature [than that which generates chemical phenomena] the detached fragment would forever remain where this cause had placed it, suspended between earth and heaven in eternal insulation. Neither Affinity, nor electricity could restore it to its lost situation nor bring it to the earth below. It is however, no sooner detached from its parent stratum than it descends with accelerated velocity into the plain below and would reach the centre of the earth if not intercepted. To the power which produced this descent the philosophers have applied the term gravitation.

I shall not dwell long on the effects of this power, they are less numerous than those of chemical affinity; but more conspicuous, and embrace many of the most sublime and beautiful appearances of nature. The april shower and december snow storm—the gentle zephyr, the tempest and the ceaseless trade wind, atmospheric phenomena that delight or terrify mankind owe their existence chiefly to gravitation; and the tides the currents and the waves of the ocean could have no existence without it. On land its operations are not less striking for it gradually reduces the mouldering hills by transporting their ruins to the low places; and to it the lover of nature is indebted for every variety of water fall, from the dashing rivulets that refresh our fields and forests to the deep and awful Cataract of Niagara. Finally, if we direct our eye to the visible Heavens we behold the mightier spectacle, of worlds sustained in harmonious and ceaseless revolution, by the influence of the same principle.

Here then we have two great classes of motions, embracing all that belong to this earth, considered either as a mass of inanimate matter, or a member of the solar system; and that we may prepare ourselves in the fullest manner to comprehend the nature of other motions of which we are about to speak let us dwell a little longer on the consideration of the powers which produce these. I ask you now to immagine [*sic*] a suspension of chemical affinity, and picture to yourselves the condition of our earth with this universal bond of union anihillated [*sic*]. Every compound in nature would instantly dissolve into its elementary atoms. The atmosphere would subside to the earth's surface in particles of this kind, and its splendid refractions of light, its purple clouds, its haloos [*sic*] and ascending mists, its lightnings, storms and rainbows would vanish in a moment. The seas, the lakes and rivers would undergo a similar change—water could no longer exist, its particles would fall to the bottoms of the present beds, and the whole earth present a surface more arid and inhospitable than the desert of Saharrah. It would not, however, like that sterile expanse present fragments of stone and other solid bodies. These would decay as speedily as those of the loosest texture; and the firmest rocks, the hardest metals, the diamond and the gems, would moulder like the atmosphere and ocean. The hills and mountains would consecutively disappear, their dust would gravitate to the valleys, and the whole earth acquire a smooth and barren mineral surface. Matter would then owe allegiance to the power of gravitation alone, and if the change were general the universe of worlds would exhibit nothing but a system of desolate revolving spheroids like that which has been described. Let us now carry on our immaginations [*sic*] a little farther and fancy gravitation as well as affinity to be destroyed. All remaining motion would then instantly cease, night and day and summer and winter would no more return—The music of the spheres no longer enliven the physical Heavens—creation would be still, and a silence more profound than the tomb would pervade the uttermost limits of the mighty empire of the Universe.

[3 Caloric. This is a principle or agent in many respects the opposite of those effects we have just reviewed. Caloric is a power of repulsion instead of attraction. It separates the particles of bodies from each other, melts down the solids, and exhales the fluids in gas or vapour. Thus it is a cause of motion, and contributes to form new compounds, by overcoming the attraction of those on which it acts and separating their constituents from each other. In the ordinary state of the earth, the influence of this agent is chiefly exercised on the air and water. Rarefying the former in certain places more than others, its equilibrium is destroyed; whereby winds are generated: when it acts on water, that liquid is converted into vapour, and rises in the air, which at all times contains an immense quantity, not visible to the eye. Abstracted from this vapour, the latter falls in refreshing dews or is condensed into clouds; which send down fertilizing showers: still further condensed, it falls in hail to desolate our fields in summer, or in snow bury them beneath a deeper covering in winter. Extracted from the waters on the surface of the earth, they are congealed and cease to flow. The little brooks are frozen into icicles, the rivers are bridged over, and the lakes and seas display floating fields, which compell the navigator to seek safety in the inlets of their shores. These phenomena, may serve to teach us what would be the state of the earth, if heat were destroyed, by him who created it. The motions which result from chemical affinity would almost cease among solid bodies; the streams would no longer ripple; the ocean itself must become solid to its profoundest depths; the last vapour of the air would fall in frost and snow, to lie forever unchanged on the bosom of the frozen earth; and the brightest glories of the atmosphere, would fade away,—for its ascending mists and the splendid refractions of morning and evening light, its haloos, rainbows and purple clouds would vanish forever.

[4 Electricity. Another principle of activity, is electricity or lightning. It is a popular opinion that we can see this fluid; but we only see the light which, is liberated by its action. Electricity appears to be present in all parts of the earth and atmosphere; for its phenomena may be excited everywhere. When its equilibrium is disturbed among different bodies, it causes them to attract or repel each other—establishes among them new affinities—promotes decomposition and recomposition, and abolishes old and promotes the formation of new compounds. Hence, it is a stirring and formidable agent—a generator of motion, and the cause of some striking phenomena both of art and nature. Thus, when made subservient to the genius of Sir Humphrey Davy, he was enabled to decompose bodies, which till then, had been considered as simple; and thereby set at liberty, certain elements, which from the era of the creation, had been cemented by an indissoluble bond of chemical affection.

[It is chiefly, however, in the atmosphere, that this agent delights to execute its wonderful powers. These are displayed in some of the most impressive and terrible phenomena of nature. By its influence invisible vapour gives out its caloric and is rendered visible in mist and clouds—or, still further condensed, falls in rain or hail. Its action on the air, is supposed to aid in the production of whirlwinds and tornados which open vistas thro' the loftiest forests; its passage from cloud to cloud, lights up the darkness of midnight with flashes and corruscations, which no human eye can look upon without quailing; while the thunders that follow its track reverberate along the valleys, and carry dismay into the stoutest heart. Were this agent struck out of existence, all these various exhibitions of sublimity and beauty, like the disolving cloud, would fade away and be seen no more.

[5 Magnetism. The essential nature of this active principle, is as little known as either of the others. The phenomena or effects which we ascribe to it, are comparatively few in number; and for anything we now know, if it were annihilated, no natural phenomenon, would cease. But employed in the service of philosophy its influence is precious to man, and indispensable to the present condition of society. It is known to all whom I have the honor to address, that to magnetism is ascribed that movement of the poised iron needle, which one of its extremities to the north—the other to the south poles, of the Earth. On this quality depends the construction of the compass; an instrument, which on the continent enables us to divide our land by straight lines, and determine the boundaries of our respective possessions; project the plans of towns and cities meander our rivers, survey our coasts, and ascertain the relative position of various places, without which the lines which now so accurately define the limits of every freehold, would be vague and uncertain; and descriptive geography have no existence as a science. But the most striking effects of the magnetic needle connect themselves with the ocean—the pathless, illimitable ocean. Relying on his compass, the mariner boldly unfurls his sails—and, throughout day and night, regardless of sunshine or rain with no cloud by day nor pillar of fire by night to direct his course, not even consulting the stars—he pushes onward through the mighty waste of waters—visits every continent, explores the remotest isles of the sea, circumnavigates the great globe, itself, and returns, to his native port, enriched with the productions of every land! Should this little instrument be lost to mankind, the condition of the world would speedily change, our race would fall back in the career of improvement and perhaps ultimately sink into barbarism.]

Gravitation, chemical affinity, caloric, electricity and magnetism, are the means, then, with which the "Most Highest" operates in our planet and produces—not matter itself—but all the varieties in its form and motion that have passed in review before us. What we have successively supposed to be lost, was produced and sustained by these powers; and as long as they continue the present order of inanimate things must remain. But I now call upon you, to consider what this globe would be, if it had no other active principles than what have been named. Place before your "minds eye" the assembled beauty and magnificence of the inanimate world, and you will still find it destitute of everything which now attaches us to existence! If no other attributes than these, I have named had been conceded to matter, the desert could not bloom and blossom; the landscape would not be cheered by the hum of insects and the song of birds; nor animated with flocks or herds or human face divine. Ere nature could give birth to these delightful objects, it was necessary to endow the matter of which they are formed with a new attribute. It did not supersede those which had been already granted but was added to them; it came from a benevolent Creator, as a higher principle of motion, a nobler grant of power.

Although it was not given to this attribute to annihilate the others, it was permitted within certain limits, to subject them to its sway, and use them as the instruments of its power; to modify their operations and to controul if I may so speak, their free will. Now it is to this power, imparted as we shall presently see to a small portion of objects on the surface of the globe, that the phenomena which we call living are to be ascribed:—All bodies having it, are said to be alive, those without it inanimate or dead. Our idea of life involves, then, a capacity for motions different from and higher in order than those of gravitation and affinity: and every movement of matter that does not spring from them we refer to it. We are as ignorant of the essence of all this new attribute as of the others; but its effects are obvious, and it is the business of philosophy to observe and deduce from them, the laws of animated or organized nature. This has already done for some of the other active powers. The laws of gravitation were developed by a single man! The laws of chemical affinity and caloric and magnetism and electricity have more lately been embodied by a succession of philosophers into an admirable system of Chemistry. Those of organic nature are but undergoing development and remain to be arranged into a natural method. It has been insultingly asked, why are physicians and physiologists behind the mechanical philosophers and chemists in the improvement of their science? We are now prepared, to answer this superficial question—There is in the nature of things, a reason for the relative perfection of these different sciences. The phenomena of gravitation do not involve those of affinity or life, and present, therefore, a simple, tho' stupendous problem. The phenomana [sic] of chemical affinity are more numerous, and involve in some cases those of gravitation, they present, therefore, a problem of many conditions and of more difficult solution. Those organic nature, or life, are still more diversified, and involve all the preceding classes; they offer therefore a complicated problem, with the conditions of which we are as yet but partially acquainted, and which in the present stage of our investigations, would defy a mightier genius than that of Newton.

I will draw your attention more closely to this new order of phenomena, and endeavour to give you a general notion of them, by contrasting them with the movements which result from gravitation affinity and the other principles of action which we have briefly reviewed. In the first place, vitality differs from those attributes, which belong equally to all matter, in being conceded to a small part of it only. The whole amount of substance which at the present moment is endowed with life, was every plant and animal, that inhabits the earth; the ocean, and the air, assembled into one aggregate, would be found to bear no assignable proportion to the remainder of the globe.

In the second place, while the principles of gravitation and affinity, as far as we can discover or even immagine [*sic*], are coexistent with matter itself; the principle of organic life, is an attribute which no particle can long retain. The age of the oldest animal or plant is not a measurable period in the march of eternity, and, therefore, did an atom continue in the animated being of which it is a part, from birth to death, it would enjoy, the privilege of vitality but for a single moment of its duration. The same atom however, does not long constitute a part of any living body. It is introduced from without, imbued with the peculiar properties of life, becomes for a time a part of the being which took it in, then looses the affinity of animation, if I may so speak, and must be ejected or it will destroy the body that vivified it. After having performed this cycle, it may be again endowed with the same high attribute, but this, to speak in metaphor, is a new benefaction, and endures no longer than the last.

In the third place the principle of animation, differs from affinity and gravitation, in this, that while *they* belong equally to every substance it can adhere to a part only. Thus the different kinds of matter that enter into the composition of plants and animals, are few, in comparison, with the whole; and there are many varieties, which have never yet been observed to make a part of any living body; as for example barytes, strontian, glucine and several other earths; copper, lead tin, arsenic, and, indeed, most of the metals.

A fourth point of difference between the living power and affinity, is, that the actions of which the former is the cause, cannot be long suspended, without ceasing forever. Both plants and animals have a constant intestine motion—a vibration of solids, and a circulation of fluids—and when these cease, death is the inevitable consequence. But the principles of gravitation and chemical affinity do not necessarily imply or sustain unceasing motion in the same body. Thus a stratum of rock beneath the surface of the earth remains forever in the same position by gravitation, and a crystal of quartz or a fragment of flint may continue for a thousand years without any intestine movement, and be as perfect at the end, as at the beginning of that long period. Indeed it is as necessary to *its* integrity and continuance, that no such motion should occur, as it is to the well being of animated bodies, that internal actions should perpetually take place.

In the fifth place, affinity and gravitation, have the greatest energy, when unfettered and unresisted, if I may so speak, by vitality—while that power would be utterly helpless, and inactive, without their aid. Thus the eagle whose vital energies, enabled it to defy for hours the power of gravitation, is brought quickly to the earth, by that power, when shot, and after death its body is as speedily decomposed and evaporated by the action of affinities that has been held in check by the living energies. These energies on the other hand would be absolutely inefficient, without the aid of chemical attraction. All the members and organs of living bodies, are composed of atoms, united in determinate proportions, by these attractions, and without their aid, the vital power could never have reared an organized fabric. In the creation of such a edifice, the materials are the elementary atoms of matter, the cement is affinity, and the living principle the artist. Affinity is then as necessary to the production of an animated body, as matter or vital force; but it is not the dominant agent, it is compelled to exert itself in a particular manner, and to do homage to laws enacted by the vital power.

In this servitude there is a perpetual tendency to revolt, so that whenever the power of the master, vitality, is weakened, the work of the servant affinity is performed indifferently; and no sooner is *it* extinguished, than, as if to avange [*sic*], himself for the captivity in which he had been held, he disjoins the whole fabric, and forms out of it many new ones fashioned according to his own taste. Thus, to lay aside metaphor, when the vital forces are reduced, the various parts of the body, and especially the fluids, suffer deterioration, in an exact proportion, and no sooner are those powers destroyed, in death, than a spontaneous decomposition, of all the organs ensues; and new bodies, assimilated in their properties to those of the inanimate world, are produced. These views appear to me to embrace the general principle of connection, between the vital power and chemical affinity; and will enable us to perceive, that while we should never attempt the explanation of a function of animate bodies either in health or disease upon chemical principles we cannot allow ourselves to remain ignorant of those principles in any researches of physiology.

In the sixth place the vital force not only compels the chemical to unite the particles of matter according to certain laws, as to their numbers and proportions, but likewise regulates their mechanical arrangement or juxtaposition. It is in consequence of this, that the shapes of animated bodies, differ so widely, from those which are inanimate and have resulted from an unsubdued and unbiased play of chemical affinities. This difference of form between these two classes of objects is so striking, as to constitute, in natural history, some of their most characteristic distinctions. It may, in general terms, be expressed, by saying, that chemical bodies are bounded by straight lines—living bodies by curves. It is, moreover, not merely the superfices of bodies that exhibit proofs of the overbearings of the vital influence. If we examine the internal structures of those which are chemical, we shall find it homogeneous, and assimulated to the eternal; but the internal structures of living bodies is highly complicated; exhibiting an infinite variety of parts, associated in a manner altogether unlike any thing to be found in the mineral world. It is to this structure that the term organization has been applied by the physiologists; and constituting all the *visible* diference between animate and inanimate beings, it has by some been considered as the sole cause of the phenomena of life. Sound philosophy, however, will teach us, that there is a vital power with which organization is coexistent. But for organized forms of matter we should have no conception of the existence of a vital power; and at the same time it is inconceivable that any structure, however, simple could have come into existence of itself, or, however, complicated could generate motion or perform functions.

In the seventh place, it is characteristic of every living body that it has derived its principles of animation and its structure from some other living body. Thus the germ of the young bird was detached from its parent, enveloped in a mass of nutricious matter, the egg; and, before its death, detaches others from itself, to perpetuate the race: And the young oak was secreted from the parent tree and deposited in the acorn as its defence and sustenance in the first stages of germination. It afterwards expands by attracting and appropriating mineral matter to its own use, and, finally, before its death produces by a peculiar action of its vessels, an embrio plant, which falling to the ground, takes root and continues the species. Bodies which belong to the mineral world never multiply in this manner; nor is chemical affinity like the vital power transformed from one to another. Hence, inanimate substances of the same kinds are produced *de novo* by the action of the affinities with which their atoms are equally endowed; while living bodies never come into existence of themselves—Life and organization can only be derived from beings which are organized and alive.

From the Western Journal of the Medical

and Physical Sciences*

(1828-1849)

Daniel Drake remained in Lexington as professor of materia medica and botany and, after 1825, as professor of the theory and practice of medicine and dean of the medical faculty, until the spring of 1827, when he returned to Cincinnati to join Dr. Guy W. Wright in editing the *Western Medical and Physical Journal.* Successor to the short-lived *Ohio Medical Repository* which Wright had begun with Dr. James M. Mason, the *Western Journal* within a year came under Drake's exclusive control as editor and publisher. As the *Western Journal of the Medical and Physical Sciences* it appeared monthly for a year, and then quarterly until the summer of 1838 when publication was suspended by agreement of the faculty of the Medical Department of the Cincinnati College, to whom Drake had transferred ownership and editorship during the previous year. Upon Drake's removal to Louisville in 1839 the *Western Journal* was revived and combined with the *Louisville Journal of Medicine and Surgery,* to form the *Western Journal of Medicine and Surgery,* edited by Drake and Lunsford P. Yandell, Sr. Until his withdrawal from active editorship in 1849 the *Journal* was Drake's principal vehicle for the publication of his continuing observations on the natural and medical phenomena of the West, as well as an important focus for the conduct of cooperative science. The brief selections which follow are characteristic of the tone and direction of the *Journal* during Drake's editorship.

* *Western Journal of the Medical and Physical Sciences* 2 (August 1828): i–vi [225–230]; 6 (July–Sept. 1832): 198–210; *Western Journal of Medicine and Surgery* 4 (August 1841): 403–407.

TO THE PHYSICIANS OF THE WESTERN STATES

At the close of the last number of The Western Journal of the Medical and Physical Sciences, the subscribers were informed that it has been transferred to the undersigned, who has, therefore, unexpectedly become at once Proprietor, Editor and Publisher.

It is not without regret, that he finds himself thus situated; as his editorial duties were, of themselves, sufficient to occupy most of the time, which professional engagements left at his disposal. Inured, however, to hard labour, and deeply interested, as he has become, in the success of the undertaking, he is far from cherishing a feeling of discouragement. Having, in 1819, issued proposals for a similar work, and obtained a considerable number of subscribers, he was the first to project such an enterprize in the Western States—although various causes, chiefly growing out of his connexion with Medical Schools, precluded him from entering upon its performance before the opening of the year 1827. Since that time he has encountered several unexpected difficulties; but indulges the hope that they have at length disappeared, and that the work will be rendered as permanent, as such publications generally are: whether it can be made as worthy of public patronage, must depend, not on his own exertions merely, but on the contribution to its pages, with which he may be favored, by the enlightened Physicians, Surgeons and Naturalists of the great region which it is designed to represent.

Such contributions he, again, respectfully, solicits, and would here take occasion to indicate certain subjects, to which he would, especially, invite the attention of his Western brethren.

1. Rare or singular cases of disease. To report all cases would be impracticable and absurd. To report the successful *only,* is uncandid. Cases which go to the suggestion of a new principle or rule of practice; which either weaken or sustain a disputed point; which, terminating unfavorably, may serve as a warning to the profession; and, finally, those which occur with exceeding rarity, may always be published with advantage to others, and credit to the reporter.

2. New forms of disease, depending on causes, either discovered or unknown.

3. Annual epidemics, in which the reporter should always omit, as far as

possible, whatever is common to the malady in all its visitations, and occupy himself with its peculiarities.

4. Brief or synoptic reports, consisting of a sort of account current with the weather and diseases, of the different seasons.

5. Accounts of new medicines; with a clear discrimination between their effects and the effects of those for which they are offered, either as auxiliaries or substitutes.

6. Experimental treatises on our *native* medicinal plants.

7. Chemical and therapeutick accounts of our mineral springs.

8. Comparative histories of the diseases of the negroes and whites.

9. Facts for an estimate of the peculiarities, which the diseases prevailing on board the steam boats of the Western and Southern waters may present.

10. Reports on the diseases in our penitentiaries.

11. Detached facts, in clinical medicine, which may be proper for insertion under the head of Original, Miscellaneous Intelligence.

12. Notices of the discovery in the West and South, of such minerals, as are in any way employed in pharmacy.

The Editor would not, however, absolutely restrict his correspondents to new facts; inasmuch as he believes that on many subjects a new arrangement of old ones, is imperiously required; he would be happy, therefore, to receive and publish essays and dissertations on all subjects in the profession, provided they develope, either new principles or new practical maxims.

In regard to the rules which should be observed in the preparation of papers, for a Medical Journal, he will take the liberty of offering to his junior brethren a few remarks.

In such papers, the first excellence is *truth*. Every thing that is published in a journal of medical science, should, as far as possible, be divested of error; for it may all, sooner or later, influence the treatment of the sick, whose lives might be jeopardized, in consequence of perversions or suppressions of the truth. The next great requisite is *perspicuity;* that what the writer has observed and thought, may be readily and clearly comprehended. The next is *conciseness;* which many readers would, in reality, place at the head of the column of good qualities:—And it cannot be denied, that they would have much reason on their side; for the quantity that is published, in the journals, and detached works, of the present age, is really enough to deter the timid and the indolent, from all reading. *Originality,* is another important property, which should never be wanting. Mere learning, and laborious rumination of that which has been masticated by others, will neither advance the interests of the profession, nor satisfy readers of good sense. Finally, those who write for medical journals, should not consider the minor rules of composition beneath their attention. If advanced in years and reputation, their example will be authoritative, and should, therefore, be a good one: If young, they should seize every occasion to improve themselves in the art of writing. Indeed no member of the profession, either high or low, should consider himself at liberty to disregard the laws of *orthography, etymology, syntax* and *collocation* in any thing which he presumes to address to his brethren. Lastly, in mercy to editors and compositors, he should write (or procure to be written) an unaffected and legible hand; without which, he has, in fact, no guaranty, that the printed will be a correct copy of his written communication.

The undersigned, regarding the "labourer as worthy of his hire," proposes to his brethren of the West and South:

I. A receipt for the year's subscription to the Journal, for every paper, of not less than four pages, which shall be deemed worthy of insertion.

II. One dollar a page, for every communication, exceeding four pages, which may, from its merits, be entitled to publication.

III. A premium of fifty dollars, for the best dissertation on each of the following topics:

1. The general principles of the pathology and treatment of the diseases, of the negroes of the Southern States; with a particular application of them to the malady generally denominated Negro Consumption; reference being constantly had to the constitutions and diseases of the whites of the same region, as standards of comparison.

2. The treatment of autumnal fever on the principles of Bronssais; with a comparative estimate of the success of that plan and the one previously pursued, illustrated by original cases.

3. The remote cause and morbid anatomy of the disease denominated in the Western country the *"Sick Stomach;"* with a successful discussion of the question, whether it is a new disorder.

4. The successful application of the process of Civiale, to the destruction of the *calculus vesicæ,* established by American cases.

The dissertations to be accompanied, as is usual in such cases, by private letters containing the names of the authors; which letters will not be opened until a decision is made on the respective merits of their papers.

It only remains for the undersigned to add, that, having divested himself of all connection with medical schools, he proposes to bestow upon the Western Journal, now the only publication of the kind in Cincinnati, all the time and attention that can be spared from professional duties; and that, in dedicating his humble abilities to its establishment and dissemination, he hopes to be honored with the countenance and co-operation of many of his senior brethren of the West; while he confidently anticipates the assistance of his former alumni, both private and public, wherever this appeal may find them.

DANIEL DRAKE, M.D.

Cincinnati, Ohio, August, 1, 1828.

NOTICES OF THE EPIDEMIC CONSTITUTION, OF THE SUMMER OF 1832, AT CINCINNATI

In the last number of this Journal, I gave some account of the Influenza and Measles which prevailed at Cincinnati, in the winter and spring of the present year. I propose now, at the end of August, to record a notice of the gastro-intestinal irritations, which have constituted the prevalent maladies of the summer. In any other than the present cholera times, such a history might, perhaps, as well not be written; but when the Epidemic is hourly expected, some account of those maladies which would seem to be its precursors, cannot be entirely useless.

The summer just ended, has been remarkable for nothing. It has, in fact, been temperate in all respects—in heat, in rain, in clouds, in wind, and in vegetable productions. The last have been neither more nor less abundant than usual, and in their properties have presented nothing indicative of the least peculiarity in the season. Yet, under this mean term in every quality, the past summer has been uncommonly productive of intestinal affections. These, as usual, began to show themselves about the end of the month of May, and have not yet entirely disappeared. Prevailing to the exclusion of almost every other form of disease, they have constituted a kind of reigning epidemic, and blending themselves with each other, in all their varieties have presented, in fact, but one malady; or at least have evinced themselves the offspring of a common cause. Some patients had vomiting only; others that symptom, with diarrhœa; a much greater number diarrhœa alone; a few real dysentery, and a number, diarrhœa with dysenteric complications; all of which varieties occurred simultaneously, though vomiting, on the whole, was more frequent in June, and dysentery in August.

Within the circle of my own practice, the cholera infantum was not quite as frequent and fatal as usual. Fewer cases than heretofore, were complicated with croup, or terminated in hydrocephalus. Indeed, I saw no instance of the former, and but one of the latter, throughout the whole season. As a general fact, it may be stated, that the warmer the summer, the greater is the prevalence and mortality of cholera infantum, and as the present was not one of our hottest summers, so it was by no means remarkable, for the number and severity of its cases of cholera infantum; nor was the cholera morbus of adults more prevalent or fatal than usual; while the number of dysenteric cases was far less, than I have often known them before. Diarrhœa, however, has been out of all measure, more prevalent than usual, especially among adults. It was not confined to any single class (if such can be said to exist in this country) of the community, but affected the rich and poor, idlers and laborers, the cleanly and the dirty, the old and the young of adult years, nearly in an equal degree. For many weeks this malady, (the *cholerine* of Paris, where, as in other places, it preceded Epidemic Cholera), attracted general attention, and must, I think, have affected nine-tenths of our adult population. Such a prevalence of diarrhœa was certainly never before known in Cincinnati; and suggested to all who knew the history of the dreaded Epidemic, that it was certainly impending. At present, however, this supposed precursor of pestilence has nearly disappeared, and with it the apprehension of a more serious visitation.

The following observations on the pathological character of this epidemic constitution, are, perhaps, worth recording.

1. In many cases the individual felt scarcely any indisposition, except a diarrhœa. The morbid action appeared limited to the intestinal canal, and was so mild, that he would eat and drink, and go abroad as usual.

2d. In a great number of cases, more or less fever was awakened, and this, in general, assumed an inflammatory character.

3. Abdominal pain was a common symptom.

4. In almost every case the secretion of bile was nearly or quite suspended. I have not, in any preceding summer, witnessed such an absence of that fluid in the dejections. They were generally thin, light colored, and serous—frequently described by the patient as white. Even when violent vomiting occurred, the matters thrown up, were rarely tinctured with bile.

5. The diarrhœa would cease for several days, and then return, and thus continue to afflict the patient for weeks.

Of the treatment of this epidemic, I shall say but little.

1. Bloodletting and abdominal cupping, were of great service. I saw many cases, which were so decidedly inflammatory, as to call for copious venesection, the omission of which in others, was, I believe, frequently the cause of a protracted disease.

2. Purging was attended with excellent effects. When the intestinal irritation was inflammatory, an oleo-saccharum with mucilage, was soothing and beneficial. In other cases, unattended with inflammatory diathesis, pills, composed of equal parts of calomel, aloes, extract of scammony and compound

extract of colocynth, given daily, were productive of the most beneficial results.

3. A pill of calomel and opium, or of those medicines in conjunction with ipecac, administered at night, generally gave great relief, and contributed to the cure. An increase of hepatic and cutaneous secretion, was a common consequence of this practice; and in proportion as these secretions were restored, the intestinal were diminished, and the diarrhœa abated.

4. When the diarrhœa was profuse without fever, the cretaceous mixture with narcotics, was followed by excellent effects; especially, if calomel or the blue pill, was exhibited at night with opium.

5. In all cases, a reduced and demulcent diet was indispensable. Violations in this respect, were often followed by relapses. In some instances, the first attack was, apparently, the consequence of excessive indulgence; in other cases, this cause did not appear to be operative. The prevailing opinion of the physicians and people of the city, was, that fruit and culinary vegetables, would induce the disease; and, in consequence of this sentiment, these articles were used more sparingly this summer than usual. Still, intestinal irritations were general; and hence, I am disposed to believe, that the influence of a vegetable diet in the production of these maladies, is greatly overrated. How, indeed, can succulent and mild acido-saccharine productions, excite inflammation in a surface perpetually accustomed to powerful stimulation by animal food and spices? We must, I think, either believe, that vegetables are not the morbific cause they are supposed to be, or else that the mucous membrane in the diarrhœa lately prevalent, was not inflamed.

So great was the resemblance of this diarrhœa to that which has every where preceded, or constituted the first stage of malignant cholera, that from the moment the physicians of Cincinnati heard of the latter, in the valley of the St. Lawrence, as I have already said, they constantly expected to see it here. Hence every violent case of cholera morbus was contemplated with suspicion, and became a source of popular alarm. The number of such cases was not, I think, as great, or at least, greater, than in common years; but they appeared to be marked by some deviations from their usual character, which gave them an affinity to the pestilential epidemic. Some of these fell under my own observation, and others were communicated to me by my medical friends. It may be well to record a notice of a few of them.

About the 8th of June, I had a female patient affected with obstinate vomiting of glairy matter, great epigastric pain and sinking, extreme reduction of temperature, and repeated cramps and spasms of the muscles of locomotion. She recovered, but had she resided in Montreal instead of Cincinnati, her case would have been reported as one of Epidemic Cholera.

Near the same time, a gentleman was attacked without any known cause, with pain *in epigastrio,* and diarrhœa, attended with considerable fever. The immediate loss of a pint of blood, and the use of a pill of calomel and opium, arrested the disease; but on the same night he awoke with *cramp* in one of his legs, an affection which had never before attacked him in his sleep; and which forcibly turned my mind to the affinity between his malady and the epidemic.

In both of these cases there was defective secretion of bile, which I found it difficult to remove.

* * * * *

Was the epidemic constitution of the present summer in the West, any thing else than the usual distemperature? I have already stated its characteristics. They certainly differ from those of our ordinary endemic to such a degree, as at least, to constitute it a variety. They assimilate it to, if they do not render it identical with *cholerine,* or the malady which precedes and accompanies Epidemic Cholera. But had this epidemic distemperature a different *cause,* from that which annually reproduces our *endemic* distemperature? If so, what was it? And what *is* the cause of our endemic constitution? Did the former supersede the latter? Or were they capable of combining? If, in reality, but *one* cause exists for both, what agent so far modified it, this summer, as to produce a modified disease? It certainly was not any peculiarity in the sensible meteoration of the atmosphere; for as I have set forth in the beginning of this paper, the summer exhibited no atmospheric peculiarities. I am far from being satisfied with any of the causes usually assigned for our endemic cholera morbus, cholera infantum, and diarrhœa. The theory of intense heat seems plausible, but still intense heat may only develope the real cause; insensible meteoration is but an assumption; and malaria a mere hypothesis.

Cincinnati, Aug. 31, 1832.

URTICARIA

In a late letter from a distant state, a medical friend says—"If you know of any means of cure for *chronic* urticaria, besides those usually laid down in the books, you will oblige me by stating them, either through the Journal or by letter. I have a patient who has had the disease from infancy, and been treated by the most skilful physicians in this part of the country—all to no effect. She is 19 years old, and apparently in good health in all her functions. The eruption makes its appearance more or less every day, and is very distressing. Can it be cured."

The demand of our friend is one which we cannot meet. We can only confess our own inability, and call upon our more skilful brethren to answer the question affirmatively and point out the means.

It has happened to us, to see many cases, both acute and chronic of this singular malady; the former have readily yielded—the latter for the most part have proved refractory. We have often observed a high degree of nervous irritation connected with *acute* urticaria, never a decided phlogistic diathesis, and we have found the administration of large doses of opium or its preparations, followed by an emetico-cathartic, of great efficacy. Bloodletting has not been of much benefit. We have, now and then, seen a compound of nitrate and bitartrate of potash do good. But in *chronic* urticaria, we have had little success. It has been said to depend on gastric or hepatic derangement, but we recollect the case of a gentleman, who labored under it for many months, during which his digestion was good, and the biliary secretion perfect, both in quality and quantity. He suffered most in winter; and the intolerable itching which arose under exposure to cold, was removed by the heat of a warm fire. Years have passed away, during which his general health has been good, but he is still more or less affected with this malady in winter.

We shall be happy to publish any thing that is valuable on this subject.

D.

CONSULTATIONS THROUGH THE JOURNAL

While writing the above, it occured to us, that many consultations and much reciprocal instruction might be conducted, through our monthly. We refer of course to chronic diseases. Cases are constantly occuring of an eccentric or anomalous character, which puzzle the medical attendant, and naturally turn his mind towards his professional cotemporaries. All such might be stated in the Journal, the readers of which could not fail to peruse them with interest, and whoever could, would no doubt bring to bear upon them, the results of his own experience. In this way much fresh and valuable information might be extensively communicated. Whenever cases are thus submitted, they should be drawn up with care. Fulness, clearness and brevity should characterize them; and while nothing of moment ought to be omitted, nothing unimportant should be introduced. All communications thus prepared will be promptly inserted. They must be subscribed by their authors, but the names will not be published unless requested.

D.

ARSENIC SOLD FOR ANTIMONY — EMPOISONING

Our friend Prof. J. Warder of Cincinnati College has communicated to us the following fact. A livery-stable keeper in Cincinnati applied to an apothecary for a quantity of what the farriers call antimony, (the well known sulphuret of that metal) and as he supposed obtained it. Administration was made in the usual way by mixing the powder with the provender of his horses. The consequence was that several of them were killed. The symptoms were not noted. A portion of the poison, not exhibited, was brought to the Professor who, on examination, ascertained it to consist of the impure arsenic, commonly called cobalt, or flystone.

Some of the horses were examined after death and exhibited evident traces of gastric inflammation.

We think this a favorable opportunity to put forth a few hints on the state of our establishments for compounding and selling medicines in the West. They are of three classes: 1st. The village and country merchant's dry goods' stores. 2nd. The Doctor's offices. 3rd. The Druggist's and Apothecaries' shops. The first have grown out of the necessities of a newly settled country; and are now seldom met with in any of our larger towns. The number of articles kept by merchants is generally few; and they are commonly sold in bulk, except tartar emetic and calomel; and therefore dangerous accidents are not likely to happen. The second class deserve a more extended notice. A student is generally the compounder of prescriptions in a Doctor's office, and no doubt he is *sometimes* qualified for the task, at the time it is confided to him. Such, however, is not always the case. The first study to which he is put, is the bones, when it *should* be the medicines he is practically to work among. Having gone through a tedious and too often unprofitable course of anatomy, he makes a superficial survey of physiology, and then takes up a work on therapeutics, or the practical application of medicines — sometimes even a treatise on the practice of physic! After this, there is a reluctant and unproductive recurrence to chemistry, pharmacy, and the natural history and sensible qualities of medicines — all of which should have been studied before any thing else. On this plan, or rather according to this practice without plan, the student passes through his private pupilage, engaged more or less in compounding the prescriptions of his preceptor, with a most incompetent knowledge of the properties and chemical composition of the articles which he is every day mingling together; and is of course liable to make mistakes. In illustration, we may mention one, which proved fatal. A physician of this state directed a certain quantity of ipecac., or some other medicine, to be mixed with a prescribed quantity of *aqua font.* for a small negro child. Instead of the latter, his student put up *aqua fortis*—the child took it and was killed. Had that young man understood pharmacy and the history and properties of medicinal agents, it is not probable that so serious a mistake could have occurred.

The third class, in most of our cities and larger towns, have, now, in their hands, the preparation and putting up of most of the medicines that are prescribed by physicians; and while the student of medicine generally has the recipe from the mouth of his preceptor and puts it up from memory, the apothecary's clerk has before him the written prescription. On the manner in which this is executed we must say a word. It should be in a legible, open hand, with the name of the patient, the date, and the name or private signature, of the physician, attached to it. How seldom it has these safe and essential requisites, any one may satisfy himself by examining the files of our shops, in Louisville or Cincinnati. If they always had the first, distinct legibility, the matter would not be so bad; but many are so obscurely written, that without a thorough knowledge of medical hieroglyphics, the compounder is liable to fall into egregious blunders. If the physician would arraign the apothecary before the tribunal of public opinion, he should enter it with clean hands. If, as he claims, he is the superior, let him set a good example to his inferior—let him show, on the face of his prescription, that he has thought, at the moment of writing it, of the great necessity of having it correctly understood. He will then have substantial ground to stand upon, in charging the druggist who has fallen into error. Many physicians, moreover, do not read their prescriptions after writing them, by which, mistakes of a formidable kind have passed uncorrected, till their effects were felt by the patient, unless the apothecary rectified them, which has not often happened. While on this prolific subject, we must add a paragraph relative to the language in which prescriptions are written. In the United States where the Latin language is so little cultivated, it is affectation and folly, to write, we should rather say, attempt to write, prescriptions in that language. Not one in a hundred (any more than ourselves) are capable of writing Latin prescriptions; and if they were thus composed, a still smaller proportion of our apothecarie's clerks would understand them. Moreover, we ask for the advantage of such a practice? We can see none, and wish it were universally abandoned. But we must return to the duties and qualifications of the compounding druggist.

It is the custom of our city apothecaries to take small boys as apprentices and clerks, who are often illiterate, and, always, unacquainted with chemistry and botany, while an elementary knowledge of those sciences, together with a tolerable education, and the age of 15 or 16, should be regarded as indispensable prerequisites. To place a little boy behind the counter of a druggist, as merchants are wont to place boys before their shelves of silks and gridirons, is homicidal; and no physician should sanction it. His own inevitable homicides are more than enough. If he cannot avoid being a principal, he may keep himself from being an accessory, by directing his prescriptions to be sent to shops where none but qualified and careful persons are permitted to handle and put up for use, articles which are the instruments of death not less than of life, when taken in excess. In Europe the apothecary is a man of learning and science—here, too often, he is a mere trader. There, he undergoes a long and laborious apprenticeship—here, he substitutes the probation of weeks, for that of years; and, from the beginning, works among poisons, with as much confidence and unconcern, as a confectioner's apprentice beats up eggs, or licks the syrup from the ends of his fingers.

D.

[*Louisville , August* 1841]

The People's Doctors;
A Review by 'The People's Friend'*
(1830)

Drake's conviction of the universal availability of natural knowledge to the human understanding was at once Baconianism's inheritance from the Enlightenment and an essential ideological element of the democratic revolution in the midst of which he lived. By the mid-1820s, as a consequence, Drake found himself increasingly caught up in ambiguity as the arguments which he and his contemporaries had used to oppose the elitism of rationalist science and Federalist polity were used in turn against the elitism of Baconian science and Whig polity. His response to this situation was of two kinds. He counterattacked from the somewhat uncertain stronghold of his own position, heaping scorn and ridicule upon herb doctors, mesmerists and toe-rappers, transcendentalists, demagogues who played on the credulity of the mob, and those who had not the social or educational credentials to merit his serious consideration and who in their turn denied the importance of credentials at all—the "empyrics" as he called them.

At the same time he sought to provide Baconianism and Whiggery with genuine viability by encouraging the establishment of educational institutions in which the methods of induction could be provided as an essential tool of the understanding to greater numbers of citizens; by urging the development of habits of industry, temperance, and social responsibility among citizens, and the creation of formal and informal institutions of social control to reinforce these habits; and by seeking scientific understanding of the behavior of men and of societies. The brief selection from *The People's Doctors* included in this collection, an example of the first kind of response, may be taken as introductory to Drake's new emphasis during the 1830s and 1840s on the scientific solution of the problems of society.

The People's Doctors; A Review appeared first in the *Western Journal of the Medical and Physical Sciences* and subsequently as a pseudonymous pamphlet signed "The People's Friend." Not a review in the modern sense, it consists principally of a series of long quotations from three recently published "doctor books" and from *The*

---

* *The People's Doctors; A Review, by 'The People's Friend.'* Cincinnati, Ohio: Printed and Published for the Use of the People. 1830. Pp. 7–12, 58–60.

Compleat English Physician, or, the Druggist's Shop Opened of 1693, interspersed with sarcastic comments by Drake on the attempt to erect into systems of treatment common sense practices known even to the American aborigines, or on the patent absurdity of medicines concocted from colts' heads, rubies, and dried scorpions. In addition Drake commented briefly on the appearance of *The Pulmist; or Introduction to the Art of Curing and Preventing the Consumption or Chronic Phthisis* by "the fishtaker, and ancient chronicler of Kentucky . . . who lately settled the limits, and *located* the wigwams of our Indian tribes for the last three thousand years," Constantine Rafinesque.[1]

One at least of Drake's sarcasms, directed against a common sense attack on the etiology of Benjamin Rush, has a peculiarly ironic quality in view of the rapid decline of Rush's reputation in the 1840s. "Orator Robinson informs us," Drake wrote, "that *doctor* Thompson's system made a very favorable impression on the mind of Dr. Rush. Now we wonder what that eminent advocate of the lancet must have thought of the following: 'The practice of bleeding for the purpose of curing disease, I consider most unnatural and injurious. Nature never furnishes the body with more blood than is necessary for the maintenance of health; to take away part of the blood, therefore, is taking away just so much of their life, and is as contrary to nature, as it would be to cut away part of their flesh. Many experiments have been tried by the use of the lancet in fevers: but I believe it will be allowed by all, that most of them have proved fatal; and several eminent physicans have died in consequence of trying the experiment on themselves. If the system is diseased, the blood becomes as much diseased as any other part; remove the cause of the disorder and the blood will recover and become healthy as soon as any other part; but how taking part of it away can help to cure what remains, can never be reconciled with common sense.' "[2]

[1] *The People's Doctors*, pp. 32–34.
[2] Ibid., p. 17.

"They shall have mysteries—aye precious stuff
"For knaves to thrive by—mysteries enough;
"Dark tangled doctrines, dark as fraud can weave,
"Which *simple* votaries shall on *trust* receive,
"While *craftier feign belief,* till they believe."

SUPERSTITION, says Robertson, the celebrated historian of America, 'was originally engrafted on medicine, not on religion.' However this may be, no one conversant with 'poor human nature,' can be unapprized of the close companionship which has always subsisted between them. In savage life, a belief in the healing efficacy of charms and incantations, is so universal, as to leave no doubt, that a principle of superstition is inherent in the human mind. It belongs, therefore, to man in every situation; though in civilized life its manifestations are comparatively few and feeble, for it is the tendency of all good education to limit its operations. Hence the most enlightened minds, display least propensity for the marvellous; but how can the intellect of a whole nation be cultivated on all subjects? This was never yet done, and we are sorry to add, never can be done. In despite of every exertion to illuminate the mass, many dark and impenetrable spots will remain; so that society, in its best composition, must continue to display enough of credulity to render it ridiculous. From the depths of ignorance, with its overshadowing superstitions,—when the hopes of the sick rest upon spells and coscinomancy,—the first step taken, is to blend with these *supernatural,* a variety of *natural* means, resting the efficacy of the latter, on the occult influence of the former. The next advance, leaves the mummeries of the sorcerer behind; but clings to amulets, seventh-sons, 'yarb-doctors,' and vagabonds. This brings us to our own age—than which, with all our boasted elevation in learning and philosophy, no other has ever presented a greater *variety* of barefaced and abominable quackeries. To eradicate them would be more difficult, than to root out the sour dock and Canada thistle of our fields, while the soil continues to favour their reproduction. Planted in the ignorance of the multitude, warmed by its credulity, and cherished by their artful and unblushing authors, these impostures are fixed upon us, as the 'poison oak' encircles the trunk of the noble tree, whose name it has prostituted. True it is, they are not always the same. The stupidest intellect at last comes to perceive their absurdity, and throws them off; but the impostors—

'New edge their dulness and new bronze their face,'

and speedily invent fresh draughts for the gaping and thirsty populace.

When one of these quackeries is inoculated into a community, nothing can arrest its spread, or limit its duration. Every dog has its day, and so has every nostrum. The gulping is universal; not extending, it is true, to every individual, but to all classes. The propensity to be cheated is not confined to men or women, the old or young, the poor or rich, the unlearned, or (we are sorry to add,) the learned; but displays its workings in the weak-minded and credulous of all. Like the small pox it prevails till all the susceptible are infected, and have gone through the disease. A moment of common sense may, perhaps, succeed to the period of suffering; as natural fools have sometimes spoken well from the shock of a violent blow. The desire to be cheated, however, returns apace; but not earlier than the desire to cheat—

"Then thick as locusts dark'ning all the ground,
"A tribe, with herbs and roots fantastic crown'd,

> "All with some wond'rous gift approach the people,
> "—Lobelia, pulmel, and steam kettle."

A new excitement now springs up. A blue light, such as the superstitious see rising from the church yard, spreads over the people, and reveals the *ecstacy* of every vacant and credulous countenance. Now is the time for dyspeptics and asthmaticks and hypochondriacks. Give them but a single draught. O how delightful! Perchance a ladle full from the chaldron of Macbeth; but no matter. Administered by wizzard hands it can do no harm. Down with the profession, *vive la Charlatanerie*. The world has been long enough duped by lawyers, and priests, and doctors. Let us rid ourselves of the last of them, if no more. If not the greatest impostors, they cheat us out of most money, and kill us to boot.—They bleed us to fainting, blister us to wincing, stupefy us with opium, vomit us with tartar instead of lobelia, salivate us with mercury, in place of the 'panacea,' or the 'stone mason's balsam,' and, purge off with calomel all kinds of phlegm, but that which encumbers our brain! Let no one be over nice. The end sometimes justifies the means. Suffering humanity cries aloud, and must be rescued from the keeping of science and skill and professional charity. The world has been in error four thousand years; and the path of medicine may be followed back by the carcasses of its victims. *Doctor* Thompson, and *doctor* Swaim, and *doctor* Rafinesque have received new 'gifts,' and are ready to distribute them. Push aside the 'riglar' Doctors!—Conceal all their cures, and publish all their failures! Go among their patients, and labour to overthrow a long established confidence! Brand them with ignorance of the human system! Stigmatize them with cruelty! Denounce them as mercenary! and Libel them as infamous! Break down the *aristocracy* of learning and science: give the people their rights: let the drunken and lazy among the tailors, and carpenters, and *lawyers,* and coblers, and *clergy,* and saddlers, and ostlers, now rise to the summit level, and go forth as ministering angels! Become their patrons, and snuff up in turn the *steams* of their incense: sustain them against the professed Doctors: lecture them into notoriety: mould them into form as the bear licks her shapeless pups into beauty: turn jackals and procurers lest they might want business: stand responsible for their success: newspaper abroad their pretended cures; and handbill away the proofs of their murders! This being done—

> "The dawn will break upon us, and bright day shall go forth and shine; when we may hope to live with the dear objects of our love, until ripe and full of years, we shall be gathered to our fathers."
> *Robinson's Lect.* ix.

But we must withdraw from the view of these ravishing prospects, to examine some of the objects which lie in the foreground of the picture.

We need not tell our readers, that the mountebanks who are to bring about the aforesaid millennium generally work on *patent* methods. This is indispensable to success—both in curing the sick, and getting the sick willing to be cured. It is edifying to see, how much of dignity and mystery are impressed on the vilest or the vapidest compound, by the great seal of state! Oh! to take medicine 'by authority,' and that, too, of the President, who would recommend nothing that he had not tried on himself, and found useful for the people. He is the people's friend. He is a good doctor (not from 'book-larnin') but common sense; he once made a *great speech* in Congress, and that shows that he is a great *doctor!* before that he was a great *lawyer,* and of course is a great *doctor!* he fought a great *battle,* and is therefore a great *doctor!* It is a crafty error of the 'riglar' doctors, that a man should study *their* books to know how to cure diseases; for this knowledge, like 'reading and writing, comes by natur.' Let us stick then to the patents—they are 'genuine,' and contain no 'marcury.' Such is the fanaticism inspired in the multitude by the diplomas issued by the office at Washington! To trace out all its consequences, would carry us to the charnel house, which at the present moment, we do not feel inclined to enter. But what will the fanatics think, when we tell them, that neither the president, nor the secretary, nor even good old Dr. Thornton himself, ever ventured to swallow one of these precious boluses or pills, the 'panacea,' 'catholicon,' or 'pulmel;' 'skunk cabbage,' 'number 6,' 'clover heads,' or 'cayenne'—except when they have added the last to their roast beef, or turtle soup? We fear they can scarcely believe us, credulous as they are; but still such is the mortifying fact. The whole process for the invention, patenting, selling, and killing off by means of these preparations, may be stated in a few words. Pick up a physician's recipe for 'antibilious pills,' 'stomachic bitters,' or 'diaphoretic powders'—or, if you should have been so fortunate as to be an apothecary's apprentice, or bookkeeper for a doctor, and have thus acquired the knowledge that will enable you to frame some compositions for yourself, in either case, apply to the patent office, and swear upon your *voire dire,* that you do *verily believe, that you are the true inventor or discoverer* of such compounds; whereupon a patent will issue, authorizing you to sell out for the 'benefit of the afflicted,' that of which the president knows about as much as Charles I. knew or possessed of the American continent, when he granted letters patent to Mr. Saybrook, for a slip of country from Long Island Sound to the Pacific ocean.

But let us turn more fully from the *patentees* to the *patentors;* and inquire by what authority they grant letters for the cure of diseases.

By the statute of 1793, it is provided that—

> "When any person or persons, being a citizen or citizens of the United States, shall allege that he or they have invented any new and useful art, machine, manufacture, or composition of matter, or any new and useful improvement on any art, machine, manufacture, or composition of matter, not known or used before the application, and shall present a petition to the secretary of state signifying a desire of obtaining an exclusive property in the same, and praying that a patent may be granted therefor, it shall and may be lawful for the said secretary of state to cause letters patent to be made out, in the name of the United States, bearing teste by the president of the United States."—*Ingersoll's Digest.* p. 656.

Now not one word is here said by the law, about drugs or medicines, or cures, or quack doctors; and we seriously doubt, whether the framers of the law ever contemplated an extension of its exclusive privileges, to the preparation and administration of remedies for the sick. The phrase 'composition of matter' evidently refers to the arts. Such obviously is and should be its meaning, for what can be more preposterous than a patent method of curing diseases; when not one of the whole catalogue, can be cured by the exhibition of any single medicine. A man would be entitled to the 'copy right,' a species of patent, for a chart of the Mississippi, by which a steam boat might be safely run from New Orleans to Louisville; but not a patent right for managing the boat on such a voyage. So, a chemist would be entitled to a patent for the discovery and preparation of Epsom salt or the sulphate of quina; but not for its exclusive application to the cure of diseases. If a man should discover the existence of aloes and rhubarb and ipecac, in plants which has not hitherto afforded them, he might fairly claim from the government the exclusive right to extract them from such plants. But these substances, by varying their proportions, may be formed into a hundred different kinds of pills, adapted in judicious hands, to a hundred different states of morbid action, though not *necessarily* a cure for any. How preposterous, then, to grant letters patent, for one of a series of compositions, that might be multiplied *ad infinitum!* But the patenting system, in its practical operation does not stop at this. A single composition will not do. This could not be offered as an infallible cure, for more than some twenty or thirty incurable maladies. Let the patents then be multiplied to 'No. 6' or upwards; print your directions accordingly, and let the retailers visit the sick, daily, and drench them *secundum artem, novem vel nigrum:* Thus under the *federal* patent you come to practise physic, the state laws to the contrary notwithstanding. Such is the origin of the Thomsonian or steam quackeries, which are the reigning epidemic of the day. That all who thus occupy themselves, do it in violation of the laws regulating the practice of medicine, and are obnoxious to their penalties, we have not a single doubt; but when our judges and legislators turn 'steam doctors,' and leave off butchering the constitution of the state, to butcher the constitutions of their fellow citizens, it would be credulous in the extreme, to believe that the statute against quackery is any thing more than a dead letter.

* * * * *

The further we look into the 'Druggist's Shop Opened' by the great people's *doctor* of the 17th century, the more we are convinced, that it is the magazine from which the people's *doctors,* of the present age, have drawn most of their learning. It contains some things, to be sure, which are not quite current at this time, such as witchcraft and spells; but these have still their advocates among us, and at the period when *doctor* Salmon lived, the majority of the people believed in their existence. On the whole, the 'Druggist's Shop Opened' of 1693, approached altogether as closely to the books of the profession at that time, as the 'Botanical Family Physician' of 1828, does to the medical writings of the present day.

In asserting the claim, of *doctor* Salmon, to the paternity of much which the people's doctors are now imposing on them as *new,* we perform, as reviewers, but an act of common justice; the more loudly called for, as *doctor* Salmon had no orator Henley, or orator Robinson, to trumpet forth his fame. He seems, however, not to have been altogether incapable of indicting an encomiastic paragraph on his own labours, which we are compelled to say is quite as applicable to them, as are the inflated periods of the *steam orator,* to the labors of the *steam-doctor.* The following is the closing sentence of his book, and will finish our extended notice of it:

> "Thus at length with much labour and pains, I have gone through and performed this so exceeding needful, so much desired, so long waited for work, wholly new in it its kind, and contrived in a practical method for the publick good, through the help of him, who works and none can let, who makes a way in the Sea, and a path in the mighty Waters, leading the Blind by the Hand, and guiding them in Paths they have not known."

Our extracts from the 'Compleat English Physician,' will, we think, be sufficient to rescue the reign of William and Mary, from any charge of defective originality; while they serve to show, that great as may be the fooleries of *doctor* Thomson, still greater had been previously invented. They differ, however, *only* in degree. In all ages, quackery has been essentially the same—a compound of ignorance, effrontery, falsehood, and libels upon science and virtue. Every quack is, indeed, a *demagogue;* and relies, for his success, on nearly the same arts, with his political and religious, or rather *irreligious,* brethren. He is one of the *people,* and pre-eminently the guardian of the *people;* while those who spend their lives, in acquiring the knowledge which has been handed down by the great physicians and benefactors of the world, are not *of* the *people,* but arrayed against the *people,* and bent on killing them off with rats bane, as if they were no better than so many Norway rats! Thus it is that the *people* allow themselves to be *charmed,* till they lose their senses, and crawl into the serpent's mouth. Would you arrest them, you thrust yourself between a snake and a simpleton—to be hissed into wonderment by the former, and brayed into silence by the latter—a predicament from which, of course, we beg to be preserved; and shall, therefore, conclude with a recollection, which, had it come up at the beginning, might have saved both ourselves and readers, a deal of trouble:—THOUGH THOU SHOULDEST BRAY A FOOL IN A MORTAR AMONG WHEAT WITH A PESTLE, YET WILL NOT HIS FOOLISHNESS DEPART FROM HIM.

EPIPHONEMA

In reviewing our 'Review' we do not find in it, an evidence, that like *doctor* Thompson we have 'had a call from God,' nor feel justified, like *doctor* Swaim, in recommending it as 'a *sacred* boon to the afflicted,' but we may say with *doctor* Salmon—*Its like not hitherto extaut!* It is, indeed, a *doctor-book* for the PEOPLE, on a *new plan;* and contains the *essence* of many others, to which we have humbly added, the results of our own personal experience; for which, of course, like other benefactors of the PEOPLE, we hope to receive the thanks of all who take our prescriptions, and outlive their effects.

As the FRIEND of the PEOPLE, we have long been concerned, to see how much of their money is squandered on hugh volumes, entitled Domestic Medicine, Family Physician, the Ladies Guide and so forth; *several* of which, are some times seen in the same parlour; where they act as amulets, and keep off diseases, with nearly as much effect, as a 'stran' of Job's Tears, or a rattle snakes tail, worn round the neck, prevents the nose-bleed or convulsion fits. However, we must admit, that they are *amusing* and *instructive* books to young ladies and gentlemen; and as most of them set forth the closing scenes of the *melo-drama amatoria* begun in Don Juan, they ought properly to lie on the round table of the drawing room, in communion with Anacreon Moore and 'my dear' Lord Byron. But as mere doctor-books to do good in a family, we must say, that we think them inferior to our own;—which costs but a small sum, is divested of technical terms and all manner of indelicacies, has a good adaptation to the distempers which are rife among the PEOPLE, at this present time, and deals, chiefly, in simples, which, to quote an old and safe maxim, *can do no good, if they do no harm,*—a recommendation that could be given to none of the books to which we have referred. Thus the PEOPLE are, at last, favored with such a *doctor-book,* as they have long wanted; and instead of consulting the 'riglar' Doctors, need, hereafter, only consult the 'PEOPLE'S FRIEND.'

Oration on the Causes, Evils, and Preventives

*of Intemperance**

(1831)

From the late 1820s Drake adopted as his own the cause of temperance reform. At no time an advocate of total abstinence, Drake differed from the moral absolutists of the movement in his insistence that the physiological basis for the use of alcoholic beverages must be understood if meaningful progress in the curtailment of intemperate drinking was to be achieved, and in his interpretation of intemperance as a physiological and social rather than as a moral and metaphysical problem. Intemperance, like gambling or gluttony, was the result of nothing more than the failure of the will to prevent the passions from dominating.[1] It was simple excess, and moderation was the remedy. Because drinking served the natural desire of the body for stimulants, more than a mere act of will was necessary to change an individual's habits, however. The physiological and emotional dependence gradually built up must gradually be broken down, either by reducing daily the quantity of stimulant imbibed, or by substituting minute quantities of an alternative and less destructive stimulant, like opium.[2]

Just as Drake advocated the interposition of an artificial mechanism between the will and the passions in order to eliminate intemperance in individuals who had allowed habits of excess to develop, so he urged the establishment of artificial means to encourage the development of temperate habits and the maintenance of temperate behavior once it had been established. Temperance societies to meet the influence of "fashion" by reinforcing reason, and instruction in the causes and consequences of

---

* *An oration on the Causes, Evils, and Preventives of Intemperance. Delivered and Published by Request, in the Town of Columbus, Ohio. February 12th, 1831. By Daniel Drake, M.D. Professor of the Institutes and Practice of Medicine in Miami University.* Columbus: Olmsted & Bailhache, Printers. 1831. This is a shortened and slightly revised version of *A Discourse on Intemperance; Delivered . . . before the Agricultural Society of Hamilton County* (Cincinnati, 1831).

[1] Cf. "Of Gambling," n.d., MS in Cincinnati General Hospital Library; "Breweries vs. Foundries," 1841, MS in Cincinnati Historical Society; *Strictures on Some of the Defects and Infirmities of Intellectual and Moral Character in Students of Medicine* (Louisville, 1847).

[2] Like many of his contemporaries, during the 1840s Drake turned against opium; cf. "Lecture before the Physiological Temperance Society of the [Louisville] Medical Institute on the Uses and Abuses of Opium. February 17, 1844," MS in Cincinnati Historical Society.

intemperance to meet the influence of the passions by reinforcing the will were the mechanisms he advocated. In 1841 he sought to combine the two by organizing the Physiological Temperance Society of the Louisville Medical Institute, probably an outgrowth of the course of lectures in popular physiology he had delivered during the previous winter.

Intemperance is not a special vice of the present day, or of our own country. On the contrary, there is reason to believe, that it prevailed more in the last than the present century; and while no people on earth ever had a greater abundance of Ardent Spirits, or obtained them at a lower price, perhaps, no nation, where they are a common beverage, ever abused them less. It must be admitted, however, that in America, as in Europe, Ardent Spirits are so often used to excess, as to justify the establishment of Temperance Societies. To infer from the existence of such associations, that Intemperance is peculiarly an American vice, would be fallacious. They more properly indicate a great energy of moral principle in our nation, and prove that it is not besotted. That drunkenness, at the present time, is less than formerly, is no argument against efforts for its further suppression, as philanthropy should aim at the desirable, though unattainable, result, of its total extermination.

It was feared at one time, that efforts to this end, would be unproductive; but experience has already shown this apprehension to be groundless. It is now generally admitted, that the use of Ardent Spirits, among the respectable classes of the eastern, middle and western states, has greatly diminished. The morning "bitter," and the glass of "grog" before dinner, are no longer taken, except in taverns; which are, truly, the citadels of this, as well as many other vices. But even there, it begins to require a certain degree of hardihood to be seen drinking; and, of course, many persons are deterred, who formerly indulged themselves without restraint. The drinking at the dinner tables of our Steam boats, has signally diminished, which, when we consider the multitudes who travel in them, is a most encouraging symptom; from our private dinner tables, the potation of whisky and water "when you are half done," is almost banished; our ships, occasionally, depart on long voyages without a supply of Ardent Spirits; the quantity imported from foreign countries is lessening, and the distillation at home has greatly abated,—although our population is rapidly increasing. A gentleman of observation, who has lately traveled over almost every part of Ohio, informed me to day, that the tavern keepers every where complain, that travelers no longer drink as they lately did; finally our canal commissioners and contractors, have shown, that labourers, and those too in situations the most exposed, can be found to work without that drinking, which was once considered essential.

These are ascertained facts, and they should encourage the benevolent to persevere. They demonstrate the practicability of greatly diminishing the amount of drunkenness in the land, and warrant the expenditure upon that object of still stronger efforts, than it has yet called forth.

In furtherance of this interesting object, tracts, sermons and discourses, which should present the causes, effects and remedies of Intemperance in every variety of aspect, have been considered as among the most efficient means; and it must be granted, that they are at least innocent: should they do but little good, they can do no injury.

Before proceeding to enumerate the exciting causes of this vice, I propose to say something of the desire for Ardent Spirits.

Man is endowed, by the Creator, with certain appetites, the regulated gratification of which, is necessary to his existence; and to the successful execution of the functions which he is required to perform in society. These desires are numerous; I shall, however, mention but three of them: *Hunger, Thirst* and the *Propensity* for *Stimulants*.

The first leads us to take the food, which is necessary to the nourishment of our bodies; the second, the water, which is requisite to the healthy constitution of the blood; the third, such stimulants, either solid or fluid, as impart activity to our systems. *Mere* hunger, seldom or never makes us gluttons;—thirst never makes us drunkards.

It is the desire for stimulation that makes us both gluttons and drunkards. Bread and meat, will satisfy the appetite for food, before we have taken them to excess; and water quenches thirst without disturbing the economy of our minds or bodies. It is the innate love of excitement, that constitutes the root of the evil, by tempting us into excesses in stimulation. To this distinct and original principle of our physical constitution, we should refer those abuses, which call for the associations, which distinguish the present age, above all which have preceded it.

In the savage state, the means of gratifying this desire, are few and feeble, in civilized life, they are diversified and abundant; and we find the desire for their use, correspondingly energetic. Their action upon us increases the appetite for them, and too often raises it to a state of morbid and ruinous importunity. This is conformable to an original law of our nature.

In savage life, man has but few functions to perform, and but little stimulation is necessary. In the civilized state, his duties, both physical and moral, are manifold and complicated. They demand the exertion of all his intellectual faculties and passions with his various bodily energies, and call for a sustained and diversified excitement, in all his organs. But this condition requires stimulation of various kinds; and without more of it than we find in savage life, the civilized state could neither have been created nor sustained. The more highly cultivated and intellectual portion of civilized men, are, it is true, sufficiently stimulated by moral causes and incentives: but these, in every country, constitute but a small part of the population; and the majority depend, *mainly,* on stimulants which act upon the body; and *must* have them, or the tone of society would fail, and its complicated operations begin to languish. Now the means of this adequate, moral and physical excitement, are created by the very civilization which they contribute to develop and advance; in which no one can fail to perceive an example of that wisdom and harmony, which are every where exhibited, when we attentively and philosophically, survey the works of God.

But, if these views be correct, on what can the friends of sobriety sustain themselves, in their warfare against Intemperance? I answer, on the same basis, upon which the moralist rests his efforts, against the inordinate indulgence of any other propensity. The ground is not broad, but firm and enduring as the laws of nature. As long as man stimulates himself moderately, and with such articles as do not impair his health, or pervert the faculties of his soul, he violates no moral or physical law, and suffers no immediate or prospective injury; but the moment he selects and indulges in such as do either, he is a transgressor, and must suffer the penalty of his violation. He does not raise in his system an excitement favorable to the duties and objects which lie before him; but an irritation utterly detrimental to their successful execution. Philosophy and ethicks do not, then, forbid all stimulation; but occupy themselves in regulating the selection and quantity of stimulants. They recognize an innate necessity for excitants; but distinguish between the *salutary* and the *pernicious*—prescribe the extent of indulgence in the former, and, labelling the latter as poisons, advertise the whole world to avoid them as destructive. They say, *"if you eat thereof you shall surely die,"*—and is not this enough?—*"if you drink thereof you will perish;"* and what deeper warning could be given?

I am aware, that these views are not consonant with those, which prohibit every kind and degree of stimulation. But the advocates of the latter system, are better moralists than physiologists. They do not understand, that the love of stimulation, is an original and necessary principle of our nature; and should, therefore, within proper bounds, be gratified. They who refrain from every kind of stimulants, if any such there be, would still more easily refrain from those, only, which are pernicious; while many persons might be induced to forbear from the latter, who would not consent to relinquish the whole. But he who abjures the pernicious, is out of danger; and *safety* is all that the friends of Temperance can desire. Why then should they insist on more? By excessive requirements, they pass into severity; and diminish their influence, by attempting to extend it too far. They become ascetics, rather than moralists: anchorites, instead of devotees in a good and great cause. They make proselytes, it is true; or rather, they are applauded, by those, who from peculiarities of constitution, or great elevation of moral feeling, are indifferent to physical stimulation: but the mass of mankind are not with them, nor ever will be. Nature interposes, and her power cannot be overcome. She calls for stimulation, but not for that which works out her destruction. Our errors, in selection and indulgence, are what make her importunate and reckless. If we supply her in moderation, with stimulants that do not vitiate her, she remains subordinate and harmless. It is our *improper* administrations, that rouse her into phrenzy, and place her on the throne of our intellect, a drunken and desolating fury.

As moral beings, we should oppose the motives of the soul, to the desires of the body; the spirit to the flesh; the pains which come from inordinate indulgence, to the pleasures of the indulgence. As rational beings, we should observe, what do us harm, and what do not: and proscribe the former, while we tolerate the latter; which, from being substitutes, become preventives. Here then is the spot, where reason and the moral sense should make their stand: the defile where the friends of Temperance should marshal their forces; the pass-Thermopylæ, where they should meet the conflict, and struggle for the victory—the triumph of the sentiments of the soul, over the propensities of the body!

Of the various salutary stimulants, I may briefly mention tea and coffee, cider, beer and the milder wines, most if not all of which, may be safely and conveniently employed, when stimulants are necessary to promote the activity of our systems, and will render more pernicious drinks unnecessary. Few constitutions, however, require the aid, even of these, much less of Ardent Spirits, which should be proscribed, outlawed and banished forever, from the catalogue of our daily drinks. He who excludes this, is in comparative safety, —he who drinks it, knows not the hour when his ruin commences. He *may*, it is true, escape its desolations; but he plays a deep and desperate game, on which he stakes his health, his fortune, his character, and the happiness of all who glory in his distinction, or hang upon his skirts for support and protection. And what does he gain, for these mighty and fearful risks? the vulgar and vanishing stimulation of a glass of grog!

The causes which give activity to the propensity for stimulants, are many and diversified. Some are moral, others physical. A part are universal, but the greater number are local and special; operating in particular places, or on certain groups of society. It is to these *causes,* that the friends of Temperance should direct their attention. Prevention should be the object: drunkenness is seldom cured, but has often been averted; and will continue to disappear, in the ratio in which its causes are laid open and rooted out.

Habitual drinking of Ardent Spirits is the first and greatest cause of Intemperance, which I present for your *condemnation*. While I assert the necessity for stimulation, I will equally assert, that except for aged persons, who have confirmed habits of daily drinking, the use of Ardent Spirits is superfluous and generally prejudicial, *even when taken in moderate quantities.* The ordinary stimulants, physical and moral, which act upon us in society, are sufficient, especially for boys and young men, whose systems are more excitable, than those of older persons. It has been said, however, that the daily but *moderate* use of Ardent Spirits, by young men, is at least safe, and may sometimes do good, by satisfying their curiosity, and generating the indifference, which comes from familiarity. All this is false and fatal. Physicians well know, that the repetition of a stimulant increases the desire for its action, and calls for augmentation of the dose. Moreover, the comparative absence of drunkenness in the respectable society of Friends, where daily drinking was never tolerated, is conclusive against the theory. A few weeks since, there died, in Cincinnati, a member of that society, who, for several years, had been the *only* intemperate person, born to such membership, in the city, although the society is considerably numerous, and much diversified as to the sources of emigration. This single fact is worth a volume of theories.

Dinner and supper parties promote Intemperance. I am aware, and admit the fact with pleasure, that the laws which prescribe drinking at these parties, have much relaxed; and that no one is now, as in former times, *compelled* to drink. But drinking is expected; and to go beyond the limits of what is called a puritanical sobriety, is no discredit. It is undeniable, therefore, that they encourage Intemperance; especially large evening parties of gentlemen only. I am far from wishing to propose their abolishment, but more reliance might safely be placed on intellectual stimuli, in literary communities; and Ardent Spirits should be banished, for the sake of example, not less, than the dignity and temperance of the distinguished men who generally compose those colloquial parties.

Gambling must not be overlooked, in scanning the causes of Intemperance. It is chiefly operative in towns and cities. Every gaming house is a centre of fluxion, for the idle and those who relish dissolute associations, not less than those who find a morbid delight in the chances of the game. Could the number of those who frequent gaming houses, as actors and spectators, in our towns and cities, be presented aggregately, it would make society shudder. They are all candidates for drunkenness. Drinking is the inseparable habit of every gaming table; and drinking to excess, at such a spot, is no discredit—but the reverse. It is the order of the day, the fashion of the time and place, the spirit of the age. The rule is, *"drink;"* the penalty of its violation, contempt and ridicule.

Idleness is a fruitful soil for habits of Intemperance. Man is an indolent animal. By nature he loves repose. Exertion is a forced state; the offspring of necessity, or the instigation of some passion, more powerful than the love of ease. Children, although constitutionally active, in the pursuit of amusements, are averse to labor, and require stimulation and discipline, to form habits of industry. I have been amazed to observe, how little fathers and mothers are aware of these truths; or, if aware of them, how little they are governed by the conviction. On this point, admonition is more necessary to the rich than the poor. Among the latter, children are often obliged to work for food and clothing—among the former, it is not uncommon, to see them grow up in ease and idleness. Youth is the era of life in which our habits are formed; and he who grows up in indolence and riches, may live and die in idleness and poverty. When extravagance and dissipation have squandered his inheritance, even the stimulus of want, may not break his established habits. This subject is of such deep interest, to all of us who are parents, that I cannot refrain from dwelling on it a moment longer.

INDUSTRY, promotes the health and bodily growth of children:
Indolence, impairs both.

INDUSTRY, renders their studies easy and pleasing:
Indolence, makes them truants.

INDUSTRY, is a substitute for genius:
Indolence, renders genius ineffective.

INDUSTRY, preserves our inheritance:
Indolence, squanders it away.

INDUSTRY, inspires society with confidence:
Indolence, repels its confidence.

INDUSTRY, provides for casualties:
Indolence, renders us helpless under them.

INDUSTRY, makes provision for old age:
Indolence, loads it with cares and embarrassments.

INDUSTRY, provides for our children:
Idleness, fails to do this, limits their opportunities, blasts their prospects, and, when we die, leaves them dependent on a heartless world.

INDUSTRY, gives us the means of charitable and patriotic donation:
Idleness, prevents our co-operating in works of beneficence, and inflicts on us the character of sordidness.

INDUSTRY, contributes to give us long life, while it condenses into a short one, the fruits of *many* years:
Idleness, abridges life, and renders the longest unproductive of happy results.
—Finally,

INDUSTRY, has transformed this vast and beautiful region, into a cultivated and populous country;—so abundant in comforts, and so noble in its public works, that when abroad, one is proud to say, in the manner of an ancient Roman, "I am a citizen of Ohio."
Idleness, would have left it a thinly peopled wilderness, without developed resources, destitute of the arts of civilized life, and inhabited by a few helpless adventurers; still grappling with Indians and beasts of prey, on the very spot where the eminent representatives of a million of freemen, are deliberating on the public good!

In no respect, can indolence be the parent of temperance, virtue or prosperity. All its tendencies are to vice. The idler is a prey to every folly. None is so much exposed to temptation: None yields himself up with so little resistance. He is the sport of circumstances. He walks into the snare, because he is too lazy to go round it: He suffers the net to descend upon him, rather than raise his finger to turn it aside. If any thing moves him, it is the love of dissolute pleasures; in the midst of which he luxuriates, and whence, having once entered, he seldom has the virtuous energy to return.

Fashion is a powerful cause of Intemperance. It is not limited to any particular class of the community, or state of society; though most operative in cities and in the highest circles. Fashion is rooted in that principle of human nature, which makes "man an imitative animal;" and involves that sentiment, which leads him to respect public opinion. Few persons, therefore, are raised *above* the influence of fashion, and that few, are none the better for their forced and unsocial elevation. It is one thing, however, to set fashion at *defiance;* another to become its *victim.* Fashion, to a greater or less extent, is, the taste and opinions of the world embodied. It is, therefore, always entitled to attention, if not to respect. It is characteristic of good sense and sound principle, to examine into the requirements of fashion, and conform to them, as far as they accord with nature, propriety and convenience. It is the vain and frivolous, only, that yield a blind submission. Good taste rejects all that is absurd or ridiculous: bad taste swallows the whole without examination. Fashion exerts the greatest domination over young minds; and in youth, acts upon both sexes, in nearly the same degree. Education being equal, the weakest minded are the greatest devotees of fashion; but in early life, it imparts delight to every grade of intellect, though in varying degrees. Young persons, are not aware of the delusions of fashion, and should be admonished, against yielding to its absurd demands. I do not know a harder master. It has no heart, no conscience, no stability. It governs without law, and sentences without a hearing. Its changes, like epidemic diseases, come and go, when we least expect them; and often with a social devastation which might carry out the metaphor. No perspicacity can foresee its caprices, or prepare to meet them. The edicts of the morning, are reversed before the evening lamp of pleasure and dissipation is extinguished. That which it lauds to-day it scorns to-morrow, and ridicules those who joined in the praise. Such is its character, and this character should be made known to our sons. They should be warned, never to deliver themselves into its power. If once reduced to servitude, they are on the road to ruin. If the fashion of the club or coterie to which they belong, says, *"drink!"* they cannot refuse; if fashion says, "prefer Ardent Spirits to tea or coffee, or fermented liquors," they acknowledge the preference; if fashion says, "pour into the fatal chalice, the sweets and spices, that honey over the poison," they comply; if fashion says, "drink again, and again repeat the draft, raise your spirits, elevate your soul, exalt your feelings, send abroad your heated fancy, become or believe yourself a genius, mount into the clouds and look down with smiles and contempt, on the plodders that walk the valleys of the earth;"—you drink, you rise, and you fall headlong, to grovel, in scorn and infamy, beneath the footsteps of those whom you despised. *Such* is the issue of a life of fashionable drinking.

Time was when fashion, on this point, governed our young men in the spirit of a tyrant. It was held, that drinking and riot are indications of talent, and a sentence of contempt, rested on those who declined to participate to excess. The spirit which presided over those convivial parties, pronounced all who held back, to be nothing more, than nature's down right common places —Drones or bigots. Dunces, if they would not drink to stupidity; smart fellows, if they did. I can recollect when this test of genius was more relied on than at present. It reminds one of the ancient ordeal for witchcraft.—If the young man would not drink freely, he was a fool—if he did, he became a brute:—Verily, a sad dilemma. I am happy to know, that nobler views of the character and destiny of youth are beginning to prevail; and trust that our sons of genius, will, soon, have invention enough, to manifest their superior endowments, in some other mode than scenes of dissipation and uproar.

Sunday drinking is a fruitful source of Intemperance. He who appropriates the Sabbath, to the society of taverns and coffee houses, is already vitiated in his moral taste, and ends his career a sot. He there dissipates a part of his fortune, or of the earnings of the week, and with it, go his habits of application, and his powers of self denial. Better were it for such an one, that he should be altogether denied the privilege of a day of rest; for he might, then, escape this deep contamination. As a general fact, the people of the United States, are pre-eminent in their observance of the Sabbath; and long may they continue to merit this distinction! The nation, which dedicates a seventh part of its time, to retirement from the cares and contentions of business—which recurs at stated periods to a sense of its moral accountability—which devotes itself, on the Sabbath, to religious exercices, and the study of books of sound morality—which assembles, at the end of every week, around the family fireside, and purifies the domestic relations, by imbuing them with appropriate devotional sentiments and moral feelings—is in the way of duty, which is the way of happiness. About such a people, there is an atmosphere of moral and social grandeur, which must repel a host of crimes and follies. Let me, then, exhort such of you as are guardians and masters, to look well to the conduct of your children and wards, on the Sabbath day. Let innocent amusements be invented—let attractive and suitable books be placed before them—let fathers remain at home, and instruct them in the first principles of religion, and the simple precepts of moral and social duty:—Above all, let SUNDAY SCHOOLS be encouraged, patronized and extended; not merely as scenes of religious exercise, but as seminaries of literature, religion and morality, united. If not sent to such schools, many poor children never learn to read, but grow up in ignorance; and before they attain to manhood, fall into most of the vices, which beset the footsteps of those who spend the Sabbath in idleness, and in wanderings among the haunts of dissipation and profligacy. But it is not the poor, only, who may be benefited by Sabbath Schools. To the rich, through a certain age, they are scarcely less beneficial. They diversify the existence of the child, and reconcile him to the salutary restraints, of the day of rest and meditation. They connect his literature with religion, and the principles of moral obligation; they civilize and soften his heart. I know of no institution, which might be made to exert so much power, in the great work of moral elevation—of none, so worthy the attention of those who labor in the mighty enterprise of ennobling a whole nation!

As a means of preventing Intemperance, Sunday Schools, indeed, deserve unlimited confidence. I am aware that *children* do not often contract habits of drinking; but when suffered to go at large, on the Sabbath, they form those habits of vice and vicious companionship, which, as they grow up, too often lead directly and powerfully, to dissipation and drunkenness.

Volumes would be necessary, to delineate the calamitous effects of Intemperance.

Ardent Spirits are a poison. A fit of drunkenness, is a paroxysm of acute disease, which, arising from any other cause, would be regarded with dismay. Habitual drinking generates chronic maladies, which, ultimately, extend to all the organs of the body. It inflames the stomach, the liver and the brain; which are, finally, disorganized. It poisons the whole nervous system; disorders the senses, and palsies the muscles. Thus the entire man, is at length transformed, from a condition of health and vigour, to a state of loathsome disease; and the grave is, at last, the only purifier.

In the mind, the sad effects of Intemperance are equally conspicuous. It impairs the power of observation, weakens attention, renders the memory treacherous, excites the imagination, and subverts the understanding. Neither the observations nor the judgments of one in this condition, are to be trusted; they *may* be correct, but are always liable to be false.

Even Madness may be the offspring of the habitual use of Ardent Spirits; although deep intoxication may have been seldom perpetrated. Incessant irritation of the brain, at last perverts the reason, and sets up the creations of fancy, for the realities of observation. The perceptions become disordered, and the individual is delivered over, to strange and terrific phantasies. In this condition, he is successively the victim of every kind of delusion, and exerts himself on those around him, as he would upon strangers and enemies. His friends are transformed into foes, and the dearest objects of his former love, become the prey of vindictive and murderous designs. Unable to distinguish between right and wrong, and, mistaking the creations of his own shattered intellect, for actual facts, he acts accordingly, and commits outrages the most shocking to humanity. In this melancholy condition, which bears but little resemblance to a fit of intoxication, and frequently occurs after a suspension of the practice of drinking, he is actually *insane;* and should no longer be held responsible, for his actions. This view of the case has not, however, been generally taken; and hence the history of our jurisprudence, furnishes examples of punishment, not compatible with the prevailing wisdom and mildness of our penal laws. Our criminal courts have confounded the *insanity* of drunkards, with their fits of *intoxication,* from which it is distinct; and punished the offences of both states, in the same manner. A more searching analysis, would have prevented such an error. The mental alienation of habitual drinkers, *is* of that kind, which brings them under the judicial maxim, that *he who is insane shall not be punished*. The proper place for such an one, is a hospital, instead of a prison; and the time *must* come, when he will find that destination. Our courts of justice are not at liberty, to sit in judgment on the remote causes of insanity, and discriminate among its varieties. The man who is *non compos mentis* from disease, *however* produced, is no longer an accountable being, and should be confined, but not punished.

Even the delirium of a *fit of drunkenness,* should be plead in mitigation of punishment; for the individual often does that, when intoxicated, from which he would recoil with horror, in his sober moments; and this should be the test. But drunkenness itself, not now recognized by the law, as a crime, should be punished. It *is* an offence against the peace and dignity of society; against the wife and children, who may, by this practice, be reduced to beggary, and thrown upon the public charity for support. The drunkard himself, may come to the same end; and, finally, subsist for years, on the earnings of the industrious and temperate. Hence it is, that society acquires the same right to punish drunkenness, that it can claim to punish any other outrage. It inflicts legal penalties on no one, who does it no injury. Blasphemy and irreligion, it leaves to a higher tribunal; while it punishes the slightest and every aggression, upon its interest and happiness. Drunkenness in all its stages, is one of these, and should be met with appropriate penalties. The personal rights of those who practice it, should be restricted; their political consequence abridged; their children placed under guardians, and their property transferred to trustees. By the fear of these penalties, thousands would be deterred from becoming intemperate; while the friends and families, of those who might still drink to excess, would be screened, in part at least, from the calamities, which, in the absence of all protecting legislation, never fail to overtake them.

The perverting effects of Intemperance on the heart, are not less, than on the head. It transforms equanimity into petulance; aggravates impatience into irascibility; engenders suspicion; blasts the domestic affections; and converts a good husband and father, into a capricious and cruel scourge. It generates a taste for dissolute society, with its diversified obscenities; vulgarizes the feelings; inflames every resentment; introduces the language of profanity, and ends, by establishing habits of falsehood and treachery.

On our actions and pursuits, the influence of Intemperance is equally deleterious. It speedily breeds an indifference to business, which at length rises to ruinous neglect. A total disregard of property not uncommonly ensues, and the earnings of former years are speedily dissipated. Economy is replaced by prodigality, and the maxims by which property is acquired and preserved, are trampled under foot.

In this reckless condition, the attractions of the gaming table, too often begin to exert their influence, and the victim of Intemperance, thus acquires another impulse on the road to ruin.

Gaming, as we have seen, is a cause of drunkenness, but in towns and cities, it is equally a consequence of that habit. Nothing, indeed is more common, than to see the drunkard become a gambler; and at last fall a prey to their united consequences.

He who adds gambling to drunkenness, renounces all the interesting objects of life. He no longer goes abroad to gaze on the beauty and loveliness of nature; to traverse the fields or forests, inhale their fragrance, and invigorate his mind by the contemplation of their ceaseless variety. When the setting sun fires the whole Heavens with beams of red and yellow light, which dazzle and delight the eye of taste, he is already in the 'den of thieves;' and feasts his *distempered* sight, on the colours of his *cards*. When the stars come forth in beauty, to illuminate the clear blue canopy, and elevate the lover of nature into feelings of poetry and devotion, *he* sits toiling with inflamed and watery eyes, amidst smoking lamps, whose oil is consumed, before his guilty passion is satisfied. When the morning dawns, he staggers forth, but not to refresh himself in its balmy breezes, or enjoy the songs of animated nature, that float upon them; for he is insensible to the whole. Even the purple splendors which clothe the east in glory, fall unheeded on his *stupid* eyeballs. Still less does he watch the rising sun, and, with the great poet, exclaim—

'Hail, holy light, offspring of Heaven, first born!'

No! ah, no! He delights to dwell in darkness; the light which cheered him once, cheers him no longer; it displays his shame: he skulks along narrow alleys, to avoid the companions of his virtuous days; and seeks his desolate home to, play the drunken despot, or prepare, by a few hours of disturbed and morbid slumber, for another night of debauch and knavery. Thus he sacrifices to his guilty pleasures every elevated enjoyment, arising from the view or the study of nature; and equally alienates his heart, from all communion with Nature's God.

In the same degree, he loses the gratifications which flow from the study of books. His mind is not enriched by the lessons of science: his language is not refined by works of literature: his feelings are never fired by the sublime and thrilling examples of history.

He is equally estranged, from the rational gratification, imparted by the knowledge and practice of the useful arts. He is ignorant, or neglectful, of every kind of professional skill, except that of his new and despicable calling;

for the debaucheries of which, he foolishly barters away the dignity and happiness, which flow so plenteously from a participation in the useful pursuits of human life.

Still further, he loses the enjoyments of virtuous society, and accepts for the companionship of the high minded and faithful friends of his youth, the treacherous and drunken associates of the gaming table.

Thus it is, that whatever may be his winnings at play, and however his constitution may bear up under habitual stimulation, the victim of drink and cards, inevitably relinquishes those enjoyments, which a man of unperverted taste and sound moral feeling, would never put at hazard, much less forever renounce.

These negative losses, however, are of but little moment, compared with the positive desolations of heart and character, which his indulgencies generate. Thus, it is well known that the drunken gamester comes, at length, to view the obligations of religion, and the attributes of the Deity, with indifference or disgust; and at last surrenders himself up, to habits of unmitigated profanity.

Cunning and knavery, are equally the offspring of his evil passions. No man plays with another, without having the conviction, that he is that other's equal. Whenever, therefore, he finds himself a loser, he naturally concludes, that his opponent is a cheat, and, forthwith resolves, himself, to cheat in his own defence. Thus, all who lose, are tempted to defraud; and beginning as men of honor, though not of temperance, they terminate their career as knaves and swindlers.

Broils, assassinations and duels, are other fruits of this tree of death. Drinking arouses the angry passions, and losses generate resentments and revenge. Hence personal combats, as fierce and furious, as those among wild beasts, suddenly spring up: The more sober and powerful grow violent, the drunken are overthrown; and the floor is drenched in blood: or, if revenge postpones its fatal blow, the parties at length meet, on what might be, ironically, called the field of honor, and society, perhaps, has the good fortune, to be rid, at the same moment, of two of its monsters.

But this happy result,—happy for the surviving; dreadful, indeed, for those who thus enter eternity, covered with unrepented crimes—does not often occur; and a more protracted catastrophe is in reserve, for the martyr of vice. His business being suspended, both his fortune and his good name, are at length destroyed. For a time he may supply his wants, by an encouraging course of success; but this only serves to determine his fate; for it feeds his cupidity, and deepens the awful fascination, which binds him to his wicked pursuits. At length, his tutelary goddess, capriciously, withdraws her smiles, and bestows them on his opponent. But his prudence is now annihilated, his understanding impaired, his appetites perverted, his passions inflamed, his will subjugated to his dreadful propensities; and with the glass in one hand and his cards in the other, he drinks and plays still deeper and deeper.

When the victim of drunkenness and gambling, is an insulated being, the ruin thus induced, is less affecting. But it too often happens, that he is both husband and father; and having expended the proceeds of his days of business, and sold, for the wages of iniquity, the venerated heritage of his fathers; having cheated his guilty companions; and, with lies and deceit reduced his credulous friends to poverty; he comes, at last, like a famishing beast, to fix his fangs on the hard and scanty earnings of his wife and children. Regardless of the vows of wedded love; dead to the sobs and entreaties of the beautiful, but faded form, that kneels before him; insensible to the fate or feelings of the innocent children that cling to his knees, and in voices of love and obedience, beseech him to remain at home; he seizes, without remorse, the little fund designed to purchase bread for him and them, and prepares to escape to the scene of his vices. In vain do the tears of anguish fall upon his robber-hand, or sighs and prayers ascend up to Heaven; unmoved by the cries of love and horror, he is intent on nothing but his booty, and looks not back, till he sees it lodged on the fatal board. But his days are numbered. His race is run. The hand of death is upon him. A raging fever kindles up in his corrupt and cankered stystem, and ends his mad career; or phrenzy seizes on his "burning brain," and his own arm raises the poison-bowl, or wields the dagger, that consigns him to the tomb, and leaves his family the heritage of his disgrace.

YOUNG MEN! shut not your eyes to the hideous aspects of drunkenness, here dimly shadowed out. Let them alarm you. Walk not in the paths which are beset with such spectres. Frequent only the abodes of Temperance. I have not described what has, but occasionally, befallen a young man in the lowest walks of life. Not one of you can say, can truly say, that *he* may not become the ridiculous, the humiliated, the scorned victim of drunkenness. Therefore, drink *no* Ardent Spirits. Make it the rule of your lives. If none of you drink—all will escape the drunkard's fate: *whoever* drinks, may sooner or later be lost. I warn you affectionately, in the midst of this respectable assembly—within these holy walls—I exhort you solemnly, to distrust your firmest resolves against drinking *too much: rely only on the resolution, that you will* NEVER *drink*. He who never drinks, has little temptation to resist, and is safe: the habitual drinker must combat a desire, which every day becomes more importunate, and combat it successfully, or he perishes. The struggle is for victory or death! the habit, or the gay and animated form of opening manhood, *must* be destroyed. If you drink from fashion, how unspeakable your folly: if from desire, how appalling your danger! A young man, perhaps an only son, loaded with the honors of the first seminaries of his country, and about to ascend the theatre of that country's noblest operations, engaged in the daily ingurgitation of gin or whisky! what a sorrowful spectacle! what a gross absurdity! Claiming the applause of the good and great—but trammelled in the habits of the degraded and sensual! Aspiring to fortune, influence and fame,—but yielding a voluntary submission to the tyranny of a vitiated appetite! In the proud consciousness of cultivated intellect, almost enrolling himself with the angels that never die,—but stooping to drink of that, which sinks him below the brutes that perish, and are no more!

FATHERS! permit one of your own number, to speak to you with freedom on this momentous subject. Look not with approbation or indifference, on the first departure of your sons, from the line of sobriety. Strive, both by precept and example, to inspire them with a horror of Intemperance. Wash your hands of their ruinous indulgences, by an earnest and affectionate protest. Keep your skirts unpolluted with their blood, by pointing out the destruction, which awaits their erring footsteps. You desire them to be good and great men, or at least, virtuous, respectable and happy men; let your desires lead to active efforts; urge them onward in the paths of Temperance, and frown, with paternal indignation, upon every deviation. You give up your days to labour and anxiety, your nights to watchfulness and meditation, that you may earn a fortune, and establish a name. Before either is acquired, you find the sun of your existence declining; and you turn your departing eyes, upon those who are destined to inherit the products of your toil. Would you not wish them to be worthy of the heritage? Would it not embitter your last, lingering hours, to know and feel that your estates would be speedily dissipated in hotels and gaming houses? that your very name would become a byword and a reproach! yet such will be the issue of your protracted labours, your deep schemes of gain or ambition, your bright anticipations, and your ravishing hopes, if the sons, who are to succeed you, sink into habits of Intemperance.

MOTHERS! *You* have a still deeper interest in this matter; for you suffer still heavier affliction, from the drunkenness of your sons. In what other way, short of committing robbery or murder—and drunkenness may lead to both—could your happiness be so mortally wounded? On whom, but them, do you rely, when their fathers are mingled with the dust? But what reliance can be placed on a son addicted to Intemperance—with its disgusting consequences —idleness, extravagance, disobedience and treachery! Better for you, far better, would it be, to stand alone on the earth, exposed, like the last tree of the mountain, to every tempest,—unallied, unnoticed, unpitied and desolate— than to rest under the calamity inflicted by a drunken and reckless son; with no husband to interpose the protecting hand of conjugal love, or wield the rod of paternal authority.

FATHERS and MOTHERS! You have daughters, the *tender* pledges of your virtuous love. They are flowers of the prairie; whose unfolding beauties, you have beheld with a delight which no compass of language could express. In the feeble hours of infancy, you have watched away the longest and the dreariest nights, over the cradle in which they lay scorched with fever, or writhing in convulsions. You have given them the first lessons of instruction —conducted and guarded their tottering footsteps in the open air—defended them from every assault of vice and violence—and sought for them the ablest teachers in all the branches of useful knowledge, and every accomplishment of mind and body. You have laboured to fashion their sentiments and manners, after the best models of the age. You have led them with pride, over the threshhold of society—and perhaps resumed your suspended relations with its gayer circles, to accompany them, to defend them from treachery, to guide them by your wisdom, and to drink deeply of a gratification, which, in the world's wide waste, flows not from a purer fountain. But to what good end have you done all this, if your daughter must be exiled from your arms, to the loathsome companionship of a sot? If she is doomed to leave the happy and cheerful paternal mansion, venerable by every early association—its books, its little decorations by the hand of domestic taste, its enlightened visitors, and its thrilling scenes of family affection,—for the dreary and echoing walls of the drunkard's house, to wither, in solitude, a transplanted and neglected flower?

PARENTS! As you value the happiness of your daughters, I call upon you to discourage the Intemperance of young men. As the number increases, the chances of consigning the blooming objects of your love, to the society and authority of drunkards, will likewise increase. Discourage Intemperance, not only in your own sons, but in the sons of your friends and neighbors; who, in the order of nature, must become the husbands and companions—good or bad —of the daughters, whose destiny is to fix the character of your declining years. Do all that you can, in this respect; and if fate should at last return upon you a broken hearted daughter, to die in the chamber which gave her birth, the consciousness of having performed your duty, will console you under this, the last dreadful calamity, which can fall upon old age.

To all who can realize the horrible consequences of Intemperance, it must be astonishing that there are men, and men of some influence too, who discourage by sneers, or more decorous means, the efforts of the present age, to repress drinking and drunkenness. I cannot but regret, that any should be so misguided; or so lukewarm, in a cause of such great magnitude. He who has the smallest influence on others, should feel his responsibility. No expression of his opinions, can ever be without some effect. He is the repository, however limited, of a moral power; and, should be held accountable to society for the manner in which he exerts it. Public sentiment should arraign him for every abuse, and mete out its indignation according to the measure of his transgressions. Can any man deny, that Intemperance is a vice? that it is a vice which brings ruin upon the individual, and wretchedness on those dependent on him? How, then, can any man justify himself, for dropping even a solitary drop of cold water, on the genial fires of benevolence which glow in the bosoms of those who devote their days and nights to the suppression of drunkenness? Should they not rather sustain the flame; and labor to render it perpetual. What would be said of a man, who might rail against the efforts of our Legislature to limit the number of thefts and murders which disgrace the land? He would at least be denounced, as a fool or a misanthrope! What is said of him, who looks with displeasure, on the laws against gambling? That he, himself, is a secret though not a sleeping partner, in the midnight abominations of the gaming table! Why then should society tolerate those, who array themselves against exertions to suppress Intemperance?—a vice, the effects of which are but feebly embodied, in gambling, robbery and murder. Either the head or the heart of such an one, must be wrong. If a *good* man, he is a *weak* man: if strong in intellect, he is perverted in moral feeling. But, perhaps, he may be perverted in his bodily feelings. Aye, he may, himself, possibly, be inclined to the very habit he thus indirectly encourages. He may, at least, be *suspected,* and should be listened to with caution.

Every age brings forth its carpers; every scheme of philanthropy or patriotism, rears up its own blind or interested opponents. They would retard that, in which they do not participate; not because it is bad, but because it is good, and they are too indolent, or too selfish, to lend a helping hand. They are, however, but drift upon the mighty current of benevolence, which they may agitate, but can never arrest.

The *great* men of the land, should look to their example. Our Legislators —the men who fill high offices—the distinguished of the learned professions —the aristocracy of wealth—the men of our chief cities—the community of self-styled gentlemen—the *magi* who wield the wand of fashion, at whose movements we see manners and customs, rise and fall, as if by enchantment; —these are they, who govern the destinies of the multitude—who wield 'a power greater than that of the throne.' From *their* lips proceed precepts, which all beneath them adopt as rules of conduct: by *their* example, will the actions of the nation be regulated. These are the men, among whom reformation should commence—where sobriety, and self denial, and purity of manners, like purity and propriety of language, should be cherished and perpetuated. *Their* precepts should fall upon the millions below them,

'Pure, as the fleeces of descending snow!'

They should stand forth, as bright examples of Temperance and virtue; as burning and shining lights in the firmament of society, to guide the benighted footsteps of those who have no light in themselves. When the wealth and knowledge of a people, lend themselves to the practice or countenance of vice, a moral overthrow is at hand. Another Phæton has ascended the chariot of the sun, and great social desolation may be expected. While the men of wealth and the men of letters, preserve the integrity of *their* manners, the national dignity is safe, and the virtue of the people uncontaminated. The stream which is pure and unpolluted in its fountains, can never afterwards be poisoned in its depths. Again, I say, let those who wield the sceptre of moral and social power, look well to themselves. They are models for imitation—their footsteps are trodden over, by long trains of followers—their conversations are rehearsed—their maxims of life spread abroad upon the breath of the people—they live not for themselves only; for their lives modify, if they do not mould, the destiny of the countless numbers, less favored than themselves.—If *their* example is bad, they inflict upon the age to which they belong, a curse, which descends to the third and fourth generations: if good, they exalt the nation, and perpetuate its happiness.

Nations like individuals have had their rise and fall. But why should they? The individual man has his day of bodily perfection, then declines, and descends to the tomb. Such is the law of his being. Human wisdom may prolong, but cannot perpetuate his existence. But nations are not under such a fiat; and, *still,* they rise and fall. To assign all the causes of these vicissitudes, would require the analysis of their whole history. It may however be averred, in general terms, that they rise by their virtues and sink by their vices and follies. Without wisdom and virtue, no nation ever rose: *with* them, no nation would ever sink. Every vice is an element of national decay. Multiply vices, and, at a greater ratio, you augment the tendency to decline. They are so many modes of diseased action, in the great social body; which may still remain sound in parts, but the hand of moral death is upon it. Its perpetual verdure begins to fade; its fruits fall, unripe and bitter, from the boughs; limb after limb, is blighted, and tumbles to the earth; the trunk itself ceases to grow, and becomes hollow at the heart; but it lives on, a perishing, though, never dying victim, of disease and desolation!

Such has been the growth and decline of nations; and such it will be, till they learn wisdom and walk no more in the paths of folly. Let no one presume to treat this subject with scorn or levity. I would ask such an one, if such there be, to say whether national degradation and downfall, would not come from multiplying to a great extent any single class of vicious men? The number of those who sacrifice every thing to the pleasures of a luxurious table, or the hazards of the gaming table;—of those who labor to repress the spread of intelligence and religion;—of those who employ unhallowed means, to encompass wealth or attain political power;—of those who encourage and indulge in idleness;—of those who drink themselves into sots and dumb brutes! What, I would, again, inquire, would befall the nation, in which either of these vices might become universal? Why, it would sink! Though raised so high in the moral firmament, as to attract the gaze, and guide the footsteps of the whole earth, it would fall, and fall to rise no more!

What then are the duties of patriotism? the dictates of beneficence? the requirements of religion? the demands of self interest, properly understood? To oppose wisdom to folly, and virtue to vice: To explore the fountains of crime, and dry them up: To throw across the pathway of every vice, a solid phalanx of virtuous men, who should say, at the beginning of its career, 'thus far shalt thou go, but no farther:' To look, like prudent physicians, to the forming stage of the moral disease, and arrest its development: To single out the infected, and brand them with a mark, or exclude them from society; that the sound may not be corrupted by their contact! By doing this, we shall rest the destinies of our young and beloved country, on its morals cemented by wisdom. Such a foundation will be imperishable. On it we should raise the pyramid of our liberties. Let us inscribe on its walls, the motto—

> TEMPERANCE!
> INDUSTRY!
> INTELLIGENCE!
> RELIGION!

It will then defy the revolutions which have prostrated those of other lands; and endure from generation to generation; a proud monument of that national grandeur, which passeth not away like a dream, but shines brighter and brighter, unto the perfect day!

Remarks on the Importance of Promoting Literary and Social Concert in the Valley of the Mississippi*

(1833)

During the second and third decades of the nineteenth century Daniel Drake was forced to revise his earlier notions of the West as an extension of the East with the potential to become identical with the older settled portions of the nations. It was during these years that he developed his conceptions of the role which geography and geology played in the determination of the natural boundaries of social and political community, thus anticipating by half a century or more the environmentalism of Ellen Churchill Semple and Orin Grant Libby. It was during these years also that he broadened his own vision of the West in which he lived to include the entire Mississippi Valley.

As a discrete region of the nation, this newly defined West possessed a natural unity as well as natural connections with the East and South. Experience had shown, however, that what was natural was not always actual, and Drake, Whig that he was, looked to the establishment of artificial institutions in order to realize the potential which geography had created. Literary conventions, journals and newspapers, professional societies, and mercantile and commercial establishments could bring citizens of the section together, and instruction in the history and character of the West could provide them with a sense of identity as members of a community; analogous institutions, cutting across sectional lines—and in particular Drake's proposed railroad from the Ohio to the Carolinas and Georgia—could create more binding ties than the arbitrary action of any central government; for in the one case the relationships were natural and the institutions only were artificial, while in the other the relationships as well as the institutions were wholly the creations of men.

* *Remarks on the Importance of Promoting Literary and Social Concert, in the Valley of the Mississippi, as a Means of Elevating Its Character, and Perpetuating the Union. Delivered in the Chapel of Transylvania University, to the Literary Convention of Kentucky, November 8, 1833. By Daniel Drake, M.D., of Cincinnati.* Published by Members of the Convention, at the Office of The Louisville Herald. 1833.

Drake's remarks before the Literary Convention of Kentucky contain his most explicit statement of the political implications of the West's status as a discrete geographic entity, an interpretation he had casually advanced as early as the *Natural and Statistical View, or Picture of Cincinnati and the Miami Country* of 1815, but one which he had just used as a basic tool in his analysis of the history and etiology of epidemic cholera during the spring of 1832.[1]

[1] *A Practical Treatise on the History, Prevention, and Treatment of Epidemic Cholera, designed both for the Profession and the People.* By Daniel Drake, M.D. (Cincinnati, 1832).

ABOUT THREE YEARS AGO, several respectable teachers in the valley of the Ohio River, most of whom reside in Cincinnati, projected and organized a society, which they denominated, "THE WESTERN LITERARY INSTITUTE, AND COLLEGE OF PROFESSIONAL TEACHERS." Its second annual meeting was held in Cincinnati, in the month of September last, and was attended by a number of teachers and professors of Ohio, Kentucky and Illinois. Several interesting topics connected with education, it is understood, were discussed by these gentlemen, and a number of public lectures, by themselves and others invited to that task, were delivered to large audiences of ladies and gentlemen.

Before the Institute adjourned, it was thought advisable to enlarge its limits, both as to the objects on which it should in future exert itself, and also to the qualifications of membership. Accordingly the following resolution was adopted:

"*Resolved,* That a Central Committee be appointed to devise a plan of a Society for the Improvement of Education and the diffusion of useful knowledge, which shall include the citizens of all classes, in the several Western States, and be calculated to exert an influence on the whole mass of the people; and that said committee shall make its report at a General Convention of the Citizens of the Western States and Territories, in this City, (Cincinnati,) on the second Monday in April next."

In the first week of November, by the efforts of the Rev. B. O. PEERS, acting President of Transylvania University, a similar Convention, for the State of Kentucky alone, was held at Lexington, and attended by several gentlemen from Cincinnati, invited thither by Mr. PEERS. The Author of this pamphlet was among the number, and was one of those whom the Convention honored with the request to address them. The subject which he chose, was the

Physical, Intellectual and Moral Education, *appropriate to the two sexes, respectively.*

The lecture being concluded, he deemed the occasion a suitable one for promoting the *general* literary meeting just referred to, and accordingly offered the following resolution, which was ultimately adopted by a unanimous vote.

"*Resolved, as the sense of this body,* That the State of Kentucky should be represented in the proposed meeting of delegates, from the different States of the Valley of the Mississippi, in April next."

After the Convention adjourned, a respectable number of its members, did him the honor to ask a copy of both the *Lecture* and the *Remarks* for publication. Expecting, at no distant time, to be able to present the former to the public, in a different way, he has complied with their flattering request, in reference to the latter only.

His remarks were chiefly extempore or from brief and hasty notes; and in writing them out for the press, he has extended them on certain points, so as to present the subject more fully than it was then displayed.

He is not a statesman, nor even a politician, but a naturalist; and has applied his geographical and geological observations, to the discussion of certain questions of patriotic and social duty. By this application he hopes in some degree to promote uniformity and elevation of character in the Valley of the Mississippi, and thereby contribute to the preservation of the Union; which, however, he regards in no present, and believes, by the West alone, may be preserved from all future danger.

Cincinnati, (Ohio,) December 15, 1833.

Our happy UNION enjoys unlimited sovereignty among the nations of the earth, but over its own people, and the different states to which they belong, its powers are restricted. On many points the states are sovereign, in relation to the confederacy, but they have few attributes of sovereignty in reference to each other, individually, and still fewer, in regard to foreign governments. This complex political organization, the only one perhaps that could enable the inhabitants of an *extensive* territory to establish union, and at the same time enjoy the blessings of laws adapted to their respective wants, like every thing complicated, is liable to decomposition. At a period when such a catastrophe is spoken of by all, and apprehended by many, it cannot be unprofitable for the people of the different states, to consider what they may do to avert it. As a citizen of the valley of the Mississippi, addressing those who dwell in the same region, I propose to say something on the means of prevention which lie within *our* reach; and hope to show, that the intellectual and moral elevation which it is our absolute duty to promote, is precisely that which would most effectually perpetuate the UNION.

In past ages of the world, such a union would, perhaps, have conferred but few benefits on those who might have formed it, and could not, in fact, have been sustained. To the discoveries and inventions of modern science—physical, mechanical, political and moral—applied to national objects, we are indebted for the means of its preservation in our own case. But even these might be ineffectual, if nature did not favor their application. Thus guided, it is, I think, the coldest sceptiscism, to doubt their perfect efficiency.

Before the means of diffusing knowledge, favoring personal intercourse, and facilitating an exchange of productions, between the remote sections of a great empire, were invented, the ties which bound them together, were woven and sustained by the hand of military power; and when it became convulsed or paralyzed, decomposition was the inevitable consequence. The natural objects and operations which might have promoted union, or which required to be controlled for the purpose of maintaining it, were too often overlooked. In the United States the case is far different. A profound policy of the people, exerted at the same time through the federal and state governments, has laid the foundations of union on the plan of nature. Where she favors intercourse between the different portions of the country, the hand of art lends its cultivation; and when she opposes it, the same hand is successfully raised against her power. Let them persevere in this policy and the UNION is perpetual.

To understand how the natural configuration of our country, under the influence of science, must of *necessity* give permanence to the UNION, we need but turn our eyes upon its map, and contemplate the different great valleys or basins into which it is naturally divisible.

The seaboard presents a range of states, the "Old Thirteen;" the whole of which, except Vermont, are connected with the Atlantic ocean. Each has its navigable rivers, its bays, harbors, and wharves, enabling it to establish and maintain an independent commercial intercourse with every other state, and with all the world. Most of their rivers originate in the Alleghany Mountains; which, commencing in the north of Georgia, terminate in the state of New York, traversing the states of North Carolina, Virginia, Maryland, and Pennsylvania. The average distance of these mountain ranges from the ocean, is about two hundred miles. Large portions of two of the states extend beyond them into the interior. Between the extremities of these mountains and the gulf of Mexico, towards the south and Lake Ontario, towards the north, the land is low and level. New England, separated from this Alpine range, by the valleys of Lake Champlain and the Hudson river, has its own mountains. Such is the maritime or Atlantic basin of the Union; and the states which it comprehends, extending from East Florida to Maine, form a sort of arch, of which New York is regarded, as the keystone, though nearer to one extremity than the other. New England is the northern buttress of this arch.

The original states, lying in this basin, were settled in a great degree, by separate colonies from Europe; and if *they* composed the *whole* union, it might, at any time be dissolved; for there is among them no physical tie of paramount influence. Indeed, I think it a fair presumption, that before this time, the Chesapeake bay would have politically divided them into a northern and a southern confederacy. But, happily there rests on the arch a weight, which, unremoved must forever preserve it. This weight is the superstructure of trans-alpine states and territories, which stretch from the western foot of the Alleghanies to the wilds of Missouri, in prospect even to the Chippewan mountains; and from the Lakes to the gulf of Mexico, in *natural* association. The waters of this extensive inland region, flow off to the sea in two opposite channels; the Niagara and the Mississippi, dividing it physically into two great valleys or basins. Let us consider them separately.

The southwest corner of New York, the adjacent parts of Pennsylvania, the northern portions of Ohio, Indiana, and Illinois, a part of the Northwest Territory, and the whole of Michigan, lie in the Niagara basin; and are, commercially, connected with the city of New York by the Clinton Canal, and the Hudson river. The connexion of the west with that city, is not, however, limited to the states just enumerated, for the Grand Canal of the enterprizing state of Ohio, has recently extended the water communication between New York and the West, quite into the valley of the Mississippi; and Indiana, and Illinois have similar works in contemplation or actual progress.

The connexion between the Niagara basin thus enlarged and the Atlantic states, is not limited to New York, but extends to New England, especially to Connecticut, Rhode Island, and Massachusetts. Thus the northern parts of the United States present a natural zone, which reaches from the eastern extremity of Maine to the Upper Mississippi, through nearly thirty degrees of longitude. This is our lake country, an interior maritime basin, of twice the length of the Atlantic, and four times its fertility. The states which it comprehends, form, like the thirteen, a kind of arch, of which New York again is the keystone, and New England the eastern abutment. The two lines of states, indeed, meet in New York, which is common to both and the "land of the pilgrims" constitutes the point of the angle which they form. The long chain of northern lakes with their connecting rivers and canals facilitate emigration from east to west; and, as man never migrates in numbers from a warmer to a colder climate, the predominant population of this great zone will be *Yankee*. The manners and customs, the literature, religion, arts, sciences, and institutions of New England, and its derivative Western New York, are destined to prevail throughout its whole extent. All this is the offspring of *natural* causes; which, whenever enterprize is left free, and laws are enacted for the public good, will be found to guide emigration, govern the investment of capital, and direct legislation.

Every friend of the Union must look with pleasure and confidence on the interest which the eastern and western halves of this zone must forever have, in maintaining their various mutual relations. The focus of these relations is the city of New York. In her resides the centripetal power, which can never cease to attract the whole. This power has increased a hundred fold within the last twenty years; and cannot be annihilated. Nature has decreed that she should be the commercial capital of the northern belt of the Union. All the states and parts of states, which it naturally comprehends, will be brought under her paramount influence; and she, on the other hand, will never cease to perceive, that her prosperity rests upon theirs, and that if her connection with them were dissolved, the gorgeous visions of future greatness which cheer her enterprising citizens, would vanish in an hour. Of the vast, and already populous region which administers to her wealth, that part which stretches from the Falls of Niagara into the wilderness, will soon be the most important, and, in her wisdom, will be especially cherished. In prosperity or adversity— in her days of pride and exultation, of conscious superiority over many of her humbler sisters of the Sea-coast—in the calm of political peace or amidst the schemes and ragings of faction—she will never be so mad as to disconnect herself, if indeed, she could, from her western resources.

Such is the northern girdle of the UNION, extending from the Bay of Fundy to the Lake of the Woods. New York is the link which connects its opposite parts, and until there arise a power strong enough to displace her, it cannot be broken. A fruitful fancy may conjure up undefined images of such a power, shooting forth from the midst of possible revolutions;—the speculative politician may have his reveries of the future, and the hypochondriac his forebodings—but the naturalist, who quietly contemplates the overruling influence of physical causes on the political and social relations of a free and enlightened people, will confide in their power, and continue a firm believer in the stability of the north.

The integrity of that commercial and social confederacy being preserved, the *Union* itself could scarcely be dissolved. The western portions of that zone, must forever exert an attraction on the northern parts of the valley of the Mississippi, while its eastern half, composed of New York and New England will act with equal power on Pennsylvania and New Jersey. Physically, the north of Pennsylvania, and the south of New York, are one region. The great rivers of the former originate in the latter, and there is between the two no natural line of separation. The north-west of Pennsylvania, moreover, has a direct interest in the Erie basin and the canal which connects it with the ocean. The connection between my native state and New York, is still closer. New Jersey, in truth, must forever remain in political, as she is in commercial, association with the great emporium. It is, however, East Jersey only, that feels this influence. West Jersey is allied in trade and social relations to Philadelphia, and hence, that little state must, at all times, constitute a link of union between the valleys of the Hudson and the Delaware. But dismissing the influence of the basin of the Lakes, let us turn to that of the Mississippi.

It is said by Dr. Goldsmith in his Natural History, that however large one fancies an Elephant to be, from reading the description, it always appears larger, still, when seen by him. I would apply a similar remark to the valley of the Mississippi. Whatever ideas may be formed of its extent and importance, from the ordinary notices of it, they will always be found too limited, when a profound examination is made. Compared with it, the maritime and lake basins dwindle into insignificance. They are but belts, and at many points narrow ones. Their united area does not greatly exceed that part of the valley lying east of the Mississippi; which is itself, much less than the portion situated beyond that river. This great region extends through thirty degrees of longitude and twenty of latitude, and no part is as far north as England. It is at least equal in area to Europe south of the Baltic and west of the Black Sea.—Bounded on the east by the Alleghanies; on the west by the Chippewan mountains, its numerous rivers meet in the channel of the Mississippi through a distance of more than two thousand miles. To the north, the sources of these rivers blend themselves, on an elevated plain, with the shorter streams which flow into the lakes. To the east, from New York to Georgia, originating on the slopes of the Alleghany mountains, they interlock with the Delaware, the Susquehanna, the Potomac, James River, the Roanoke, the Santee and the Savannah. It is no exaggeration to say, that considered in reference to area, soil, aspect, and climate, this valley is superior to any other on the globe. Its only natural highway to the sea is the river Mississippi;—New Orleans constitutes its mart; and between that city—the New York of the south—and the vast country above, there is and must ever be an action and reaction still more natural and powerful than that between the city just named and the basin of the lakes. Let us consider the civil divisions of the valley, with a reference to the influence it is exerting, and must continue to exert, on the whole union.

Several states, as Missouri, Kentucky, Tennessee, and Arkansas, lie wholly within its limits. Of Illinois, Indiana, Ohio, Mississippi and Louisiana, the greater portions are included in it. A part of New York, a larger one of Pennsylvania, and a still larger of Virginia, with a small portion of Maryland, North Carolina, Georgia, and Alabama, dip into the same basin. Thus twenty states and territories out of twenty-nine, the latter of which are of vast extent and destined to sub-division, are embraced wholly or in part in the Great Valley; the inhabitants of which already make one-third of the entire population of the Union, and are daily augmenting by emigration from every quarter.

As the extent to which the old states run into the Mississippi valley is very different, so there must be degrees in their influence on the stability of the union.

The participation of Pennsylvania and Virginia is most extensive, and to this we may look for a permanent effect. The Western portions of these great states, are, in truth, natural and unalienable elements of the Mississippi community; united to the true and proper *West,* by ties not to be dissolved; dependent on our great river; familiar with its banks; and proud of its name. They can never consent to become a distinct people from their brethren below, through whom they must forever wish to pass to the ocean; and among whom, it must always be their interest, to distribute their mountain forests of pine and cedar, their beds of iron ore, their banks of coal, and the products of their salt springs. So intimate, indeed, is the *physical* relation of the western declivities of these states, with the other parts of the Mississippi basin, that no influence of the maritime portions could ever draw them from the West. Sooner shall we see them, respectively, broken asunder along the spine of the Alleghenies, than their western extremities detached from the Ohio states.

With respect to Pennsylvania, especially, what motive can ever madden her into a desire to leave the West? Certainly none. She has the same interest in the West as New York. It has contributed largely to make her what she is. To facilitate intercourse with it, she has even anticipated the resources of generations to come: she is turning her rivers into artificial channels, reducing her mountains, and perforating her hills; in short, she is laboring to bind herself with the West and the West to her. Thus we see that the commonwealth of William Penn, populous, orderly, respectable, and situated in the centre of the maritime zone, is equally bound to the North and the West, and must forever maintain her position in the confederacy.

Her neighbor Maryland, united to her by many natural and artificial ties, will not consent to see them broken; and although, more remote from the

Great Valley than Pennsylvania, she is deeply impressed with the importance of participating in its trade, and has for years been stretching towards it her enterprizing arms.

The public sentiment of Virginia is moulding itself on the same plan. The Atlantic can never think of a separation from the Ohio portions of that state. I assume that the two halves will remain united; but the western, for the reasons I have just assigned, will adhere politically to the Valley with which it is naturally associated; and thus, the whole is permanently bound to the West by physical causes.

Would North Carolina leave Virginia? Contemplate the imaginary line which separates them. The waters of the same fountain may bubble up on the territories of the two states, and the fallen tree which lies across the stream below may serve as a bridge to connect them. Virginia, moreover, the greater in extent and physical resources, must forever be superior in political and moral power, and would be a dangerous rival. Would North Carolina withdraw from her daughter, Tennessee, estranged from her by no impassable barrier; and prepared, under a proper system of internal improvement to administer to her wealth and power? She can never willingly consent to such a separation.

What of our high-minded and palmy South Carolina; the brightest orb in our southern constellation, will she seek a new zodiac, and become the lost star of our political heavens? Of all the states she has the least of natural and commercial connexion with the valley of the Mississippi, and, as if to afford a negative evidence of the over-ruling influence of physical causes, she alone has shown symptoms of secession. But she will not wander off by herself, and none of the sisterhood *can* accompany her. The geographical cords by which she is united to her twin sister on the north, and her younger sister on the south, are too strong to be snapt assunder, and those states are bound to others by ties of equal strength and durability.

The participation of Georgia in the valley of the Mississippi, is small, but her natural relations with Tennessee are intimate and profitable, and those with Alabama and Florida permanently controlling. She cannot disconnect herself from the union, without the concurrence of Alabama, and the prospective co-operation of Florida. But Alabama is naturally associated with Louisiana, Tennessee and Mississippi, and is not therefore, politically separable from them. Moreover, New Orleans exerts on the entire region immediately east of it, an effect precisely similar to that of New York on the south of New England; and, therefore the maritime portions of Mississippi, Alabama, and Florida, are commercially under its control, and cannot become politically detached from the West, till Louisiana shall secede; an event that may happen when the Mississippi finds some other route to the ocean, or is swallowed up in the sands of its own delta.

Thus, in travelling along the Atlantic coast of the union from the St. Lawrence to the Balize, we find such natural connexions and dependencies of its states on the valleys of the Niagara and the Mississippi, as must forever set the spirit of disunion at defiance. Within these basins, together constituting the West, lies the centre of gravity of the union. Here dwells the conservative power. The cement of future adhesion among all the states exudes, to speak figuratively, from the soil of the West. To borrow a metaphor from my own profession, it is the interior of the sovereign body politic, embracing the vital organs, which distribute nourishment through the outer parts. Once more to change the figure, it is the part, where the cords of union are wound into a Gordian knot, which, cut assunder by the sword, would, under proper treatment, reunite, *by the first intention,* and not even leave a scar behind. Conventional regulations may be annulled, treaties abrogated and political confederacies dissolved, when they are not based on nature; but give them this foundation; rest the political and social upon the physical; and they will be preserved from all serious revolutions, but those which change the surface of the earth itself.

But to produce this effect on the union, the west must become and remain united with itself. Whatever retards or diminishes this subordinate but central and natural confederacy, weakens the general union; whatever strengthens it, invigorates the whole. The federal constitution cannot be overthrown while the Mississippi states remain in connection and harmony. To my own mind, this opinion is so conclusive and cheering, that I wish, most earnestly, to commend the grounds on which it rests, to the consideration of every intelligent patriot, who may apprehend our political dissolution.

Under these views, let us proceed to inquire into the duties of the people of the interior. They must weave among themselves a firm web of brotherhood, and become still more closely united in social feeling, literary institutions, and manners and customs; and then, no temporary or partial suffering, no conflicting interests, or state aspirations; no lawless ambition, no military power, nor reign of faction among their elder brethren of the sea-coast, can ever jostle from its place, a single column of the great temple of union. The objects thus presented to the people of the west are of the highest and noblest kind. In laboring to promote harmony among themselves, they are working for the harmony and happiness of the whole union. They have a holy task of patriotism to perform. The palladium of the Constitution is committed to them by nature, and they should faithfully preserve it. The destinies of brethren widely separated are confided to their care—let them not betray the trust.

But their own prospective interests should prompt them to action. Suppose they should neglect these labors, and the Union dissolve:—how deplorable then would be their own condition—geographically united—bound inseparably together like the Siamese twins—but attached by no civil ties—no pervading sentiments of kindness—no general plans of education—no common bonds of social harmony! Nature demanding union, but reciprocal prejudices —local animosities—contrarieties of education, and diversities of manners and customs, conspiring to array them against each other. It requires but few lessons from history, but a limited knowledge of our common dependence on the Mississippi to foresee, that should such a melancholy event ever happen, without social preparation on our part, the west would crumble with the rest into its political elements, and the immense valley, where brethren dwell together in peace, would become the Flanders of the new world in war, as it now is in corn. "Ploughs would be turned into swords, and pruning hooks into spears." The drum and trumpet would echo along our fertile valleys, and the midnight cry of hostile sentinels fall on the ear from the opposing banks of our beautiful rivers. The teeming steamboat would no longer depart from Pittsburgh or Cincinnati for New Orleans, but to a neighboring port within the limited jurisdiction to which it belonged. Never again would its decks present an epitome of the Union; a concourse of passengers from every state; greeting each other as brethren and sisters; originating plans of business, contracting new friendships, and forming the alliances of love, while the noble steamer held its way for a thousand miles, through peaceful and happy lands, which each might call his own. Fortifications would then frown from the magnificent cliffs on which the eye of the voyager now dwells with delight. The smoke of artillery would poison our evening mists, and contaminate the morning fogs, which rising from our plains, curl around the summits of the green hills. A sulphurous odour would blend itself with the aroma of our flowers. Armed steamboats would traverse all our rivers, and the glorious stripes and stars of our Union, be replaced by the hostile flags of every device. To the dangers of navigation would be added those of war; a brother's hand would apply the torch of battle, and a brother's blood mingle with our waters.

Would we, through all coming ages, avert these vast calamities, we must in due time, and at all times, labor to preserve ourselves in domestic harmony; make ourselves one brotherhood in our customs, affections, and feelings, however, distinct in political power: let us, in short, establish among the people of the Mississippi valley a literary and social communion, like that which New England presents, and then, should the old, the parent states; respectively, set up for themselves; should the demagogue undermine the foundations of the republic, or the reeking sword of the desperado cut assunder its bands, the West would go together; the largest mass of the ruin, the least mutilated in the fall, the most powerful, the most respected, the most prosperous! She might, as she would, mourn over the catastrophe; her daughters, like the damsels of Jerusalem weep fountains of tears, and her sons as those of Judea, clothe themselves in sackcloth and ashes; but they would still be safe and happy, compared with their brethren of the other states.

Thus, whether we seek to perpetuate the Union, or would prepare for its possible dissolution at some remote epoch, our duty is the same; to commune together from every part of the mighty West; to make acquaintanceship with each other; to correct each other's faults; sympathise in each other's joys and sorrows, and mould ourselves into one great social brotherhood as our flowing waters mingle and roll onward to the ocean. To these labors of love we are exhorted or commanded, by more considerations than ever prevailed among the people of any other land.

The millions who already flourish in the valleys of the Mississippi and the lakes, are chiefly emigrants. They have entered it on every side, and are derived, not only from all the original states of the Union, but from western and even central Europe. Bringing with them various national peculiarities, the common good requires that they should be speedily amalgamated into one social compound. On this will their stability and moral power depend. Every movement of the air or waters drifts about the loose sand; but consolidated into rock, it resists the action of the "winds and waves," and is fitted for permanent use. A community formed out of such elements would exert an attraction on the whole Union. But little emigration takes place between New England and Pennsylvania, New York and Virginia, Maryland or New Jersey and the Carolinas. There is no region east of the mountains, where natives of all the states congregate, and cherishing their early attachments, constitute a *Union Society*. It was reserved for the west to exhibit this interesting concourse. Elsewhere, general patriotism may be the offspring of policy or interest —in the west it must always be a sentiment of the heart.

In the character of the materials for Western society, there is much to encourage those who would labor to construct it. I believe them the best which ever came to the hands of the social architect. The old states were peopled by Europe, when she was far from the elevated grade of civilization she now exhibits, and which belongs equally to those states. Those who are emigrating to the west, have more knowledge and refinement than ever before belonged to any moving population. Even the pioneers of Kentucky, Tennessee, and

Ohio, were, in part, composed of men who would have been respected in any community. Our *Clarkes, Boons, Todds, Logans, Scotts, Marshalls, Shelbys, Putnams, Cutlers, Symmes, Ludlows, Benhams, Worthingtons, Lytles, Harrisons, St. Clairs, Robertsons, Seviers, Buchhanons, Jacksons, McNairys*, are but a small portion of the honored patriarchs of the three sister states; and those who accompanied them as associates or followers, possessed, like themselves, the sagacity, courage, and high aspiration which gave a good earnest in the infancy of the west of what we already see it in youth. The emigrants to the original states, were chiefly from Great Britain, and frequently in masses or streams which flowed and settled together. When the continent made contributions, it was done in colonies, which too often continued as such after they reached the shores of America. Thus the original materials of society in the Atlantic states were less diversified, and *therefore* inferior to those which past and present emigration has distributed over the region we inhabit. With such ingredients in the moral crucible, the resulting compound must, ultimately, have less alloy than is found in most communities, should those who watch over and direct the process of union, be diligent, harmonious, and persevering.

Other considerations press themselves on the mind; address themselves to the heart. Western Pennsylvania, Western Virginia, Kentucky, Tennessee, and Ohio, young as they are, compared with the seaboard, have long been emigrating states. From the unsettled feelings of a new community, their people have passed incessantly from one to the other, still, however, advancing into the wilderness. Thus, Western Pennsylvania has scattered its sons over Ohio, while the latter has peopled Indiana and Illinois with thousands, and Tennessee sent her children to Alabama, Mississippi, Missouri and Arkansas. But above all, Kentucky, the land of my earliest recollections, has spread herself over Ohio, Indiana, Illinois, and Missouri. Of all the new states, she has, indeed, been the longest an emigrating state. Even the plan and settlement of Cincinnati, which the orator of the West has pronounced the unrivalled queen of the West, were arranged in Lexington, at that time the infant metropolis of all the new settlements.

Thus, all the states and territories of the Great Valley contain the germs of a natural brotherhood. Every where we meet with men and women, whose feelings turn instinctively to some other spot of the interior, where in the may day of life, when filled with the love of nature, they joyfully collected the lilies of the untrodden valley, or rambled, without care, among its pawpaw groves. While the grandfather smokes his pipe in the wide hall of a Kentucky double cabin, his son follows the plough in Ohio, and his grandson opens a new farm in Indiana. Two brothers embark on the Ohio river; one will stop in Missouri, the other plant himself in Louisiana. Two sisters marry in the same week; one to be taken to Alabama, the other to Illinois. These unrestrained and apparently capricious migrations, so familiar to all the inhabitants of Ohio, Kentucky, and Tennessee, must inspire the people of those states with an imperishable interest in the entire West. They predispose to union. They invite the western patriot to action, and point out the delightful task, which love of country, and love of liberty, and love of offspring, alike call upon him to execute. Seizing upon these scattered elements of union, he should bind them together, nourish them with one blood, and harmonize them with a single nervous system. Thus they will come to work together, like the different organs of the living body endowed with the same sensibility, reciprocally sympathizing, and obedient to the same laws of morality, religion, and social order.

Other considerations still, arise to the mind while intent on this subject. The territory which is now divided into Kentucky, Ohio, Indiana, and Illinois, was once a part of Virginia. The recollection of this should inspire the inhabitants of the whole, with a feeling of affection for each other. At that time, all except Kentucky was an unpeopled wilderness. The history of their early settlement fills the heart with emotion. The region between the Ohio and the Lakes, *belonged* to the Ancient Dominion, but Kentucky was the child—a daughter settled in the wilderness, and exposed to every kind of peril and privation. Participating in the glory of the Cavaliers, her sons established their claims to such a heritage, by a chivalrous devotion to the younger sisterhood of the west, which posterity will never cease to admire. For years, Kentucky was a living barrier to Tennessee, against the tribes of the north; and when did Ohio, Indiana, Illinois, and even Michigan raise the cry of alarm, that the gallant state did not, by spontaneous impulse, send forth the choicest of her sons? Their battle cry has resounded through all the forests of the north—their blood has fertilized the plains of the Wabash, the Maumee, the Raisin, and the Thames; their bones still moulder among the rank weeds, from the banks of the Ohio to Lake Superior. Their fall in the defence of their younger brethren, has more than once clothed the mothers of their native state in mourning, and spread through the city where we are now assembled, a voice of lamentation more sorrowful than even pestilence could raise. Kentucky must even cherish towards those for whom she has thus fought and bled, the good will which kind offices create for those on whom they are bestowed; and the children of the people for whom she thus suffered, can never be unmindful of her gallantry. Here then we have another chain of friendship; one which applies itself to the heart, encloses its best affections, is alloyed with no selfishness, and may be brightened through all future ages, should our literature prove true to its charge, and our men of influence devote themselves to the great cause of civil union.

But let us leave the history and resume the physical and political geography of the West, for the purpose of considering the relations of its different regions —not to the *Atlantic States,* but to *each other.* In reviewing their boundaries and connections, we find much to excite reflection and inspire us with deep emotion. The geography of the interior, in truth, admonishes us to live in harmony, cherish uniform plans of education, and found similar institutions.

The relations between the upper and lower Mississippi States, established by the collective waters of the whole valley, must forever continue unchanged. What the towering oak is to our climbing winter grape, the "Father of waters" must ever be to the communities along its trunk and countless tributary streams—an imperishable support, an exhaustless power of union. What is the composition of its lower coasts and alluvial plains, but the soil of all the upper states and territories, transported, commingled, and deposited by its waters? Within her own limits, Louisiana has, indeed, the rich mould of ten sister states, which have thus contributed to the fertility of her plantations. It might almost be said, that for ages this region has sent thither a portion of its soil, where, in a milder climate, it might produce the cotton, oranges and sugar, which, through the same channel, we receive in exchange for the products of our corn fields, work shops, and mines. Facts which prepare the way, and invite to perpetual union between the West and South.

The state of Tennessee, separated from Alabama and Mississippi on the south and Kentucky on the north, by no natural barrier, has its southern fields overspread with floating cotton, wafted from the two first by every autumnal breeze; while the shade of its northern woods, lies for half the summer day on the borders of the last. The songs and uproar of a Kentucky *husking* are answered from Tennessee; and the midnight raccoon-hunt that follows, beginning in one state, is concluded in the other. The Cumberland on whose rocky banks the capital of Tennessee rises, in beauty, begins and terminates in Kentucky—thus bearing on its bosom at the same moment the products of the two states descending to a common market. Still further, the fine river Tennessee drains the eastern half of that state, dips into Alabama, recrosses the state in which it arose, and traverses Kentucky to reach the Ohio river; thus uniting the three into one natural and enduring commercial compact.

Further north, the cotton trees which fringe the borders of Missouri and Illinois, throw their images towards each other in the waters of the Mississippi —the toiling emigrant's axe, in the depths of the leafless woods, and the crash of the falling rail-tree on the frozen earth, resound equally among the hills of both states—the clouds of smoke from their burning prairies, mingle in the air above, and crimson the setting sun of Kentucky, Indiana and Ohio.

The Pecan tree sheds its fruit at the same moment among the people of Indiana and Illinois, and the boys of the two states paddle their canoes and fish together in the Wabash, or hail each other from opposite banks. Even villages belong equally to Indiana and Ohio, and the children of the two commonwealths trundle their hoops together in the same street.

But the Ohio river forms the most interesting boundary among the republics of the West. For a thousand miles its fertile bottoms are cultivated by farmers, who belong to the different states, while they visit each other as friends or neighbors. As the school boy trips or loiters along its shores, he greets his playmates across the stream, or they sport away an idle hour in its summer waters. These are to be among the future, perhaps the opposing statesmen of the different commonwealths. When, at low water, we examine the rocks of the channel, we find them the same on both sides. The plants which grow above, drop their seeds into the common current, which lodges them indiscriminately on either shore. Thus the very trees and flowers emigrate from one republic to another. When the bee sends out its swarms, they as often seek a habitation beyond the stream, as in their native woods. Throughout its whole extent, the hills of Western Virginia and Kentucky, cast their morning shadows on the plains of Ohio, Indiana, Illinois, and Missouri. The thunder cloud pours down its showers on different commonwealths; and the rainbow resting its extremities on two sister states, presents a beautiful arch, on which the spirits of peace may pass and re-pass in harmony and love.

Thus connected by nature in the great valley, we must live in the bonds of companionship, or imbrue our hands in each other's blood. We have no middle destiny. To secure the former to our posterity, we should begin while society is still tender and pliable. The saplings of the woods, if intertwined, will adapt themselves to each other and grow together; the little bird may hang its nest on the twigs of different trees, and the dew-drop fall successively on leaves which are nourished by distinct trunks. The tornado strikes harmless on such a bower, for the various parts sustain each other; but the grown tree; sturdy and set in its way; will not bend to its fellow, and when uprooted by the tempest, is dashed in violence against all within its reach.

Communities, like forests, grow rigid by time. To be properly trained they must be moulded while young. Our duty, then, is quite obvious. All who have moral power, should exert it in concert. The germs of harmony must be nourished, and the roots of present contrariety or future discord torn up and cast into the fire. Measures should be taken to mould an uniform system of manners and customs, out of the diversified elements which are scattered over the West. Literary meetings should be held in the different states; and occasional conventions in the central cities of the great valley, be made to bring into friendly consultation, our enlightened and zealous teachers, professors, lawyers, physicians, divines, and men of letters, from its remotest sections. In their deliberations the literary and moral wants of the various regions might be made known, and the means of supplying them devised. The whole should successively lend a helping hand to all the parts, on the great subject of education from the primary school to the University. Statistical facts, bearing on this absorbing interest, should be brought forward and collected; the systems of common school instruction should be compared, and the means of different school books, foreign and domestic, freely canvassed. Plans of education, adapted to the natural, commercial, and social condition of the interior should be invented; a correspondence instituted among all our higher seminaries of learning, and an interchange established of all local publications on the subject of education. In short, we should foster western genius, encourage western writers, patronize western publishers, augment the number of western readers, and create a western heart.

When these great objects shall come seriously to occupy our minds, the union will be secure, for its centre will be sound, and its attraction on the surrounding parts irresistible. Then will our state governments emulate each other in works for the common good; the people of remote places begin to feel as the members of one family; and our whole intelligent and virtuous population unite, heart and hand, in one long, concentrated, untiring effort, to raise still higher the social character, and perpetuate forever, the political harmony of the green and growing WEST.

Discourse on the History, Character, and Prospects of the West*

(1834)

Much of what Drake said to the young men of the Union Literary Society at Miami University concerning the diversity of the West and of its population, the rapid progress of social organization which seemed to drive pioneer days out of the contemporary consciousness even as it drove the larger mammals of the region to the uninhabited headwaters of the Arkansas and the Missouri, and the feasibility of establishing sophisticated cultural and educational institutions in so young a country, is reminiscent of his remarks before the Western Museum Society and the Medical College of Ohio in 1820. These appear with a new emphasis in this address of 1834, however, for Drake here is concerned less with the possibility of recreating the culture of the East in the West than with describing the West as a distinct region[1] and with assessing the consequences of this fact for the development of an indigenous American literature and an indigenous American civilization. Also noteworthy in this address is the manner in which Drake tailored his conventional Baconian advice on the desirability of systematic and objective observation of nature to fit the literary emphases of the Society by talking the language of romanticism rather than of empirical science. The implications, of course, remained the same.

Discourse on the History, Character, and Prospects of the West: Delivered to the Union Literary Society of Miami University, Oxford, Ohio, at their Ninth Anniversary, September 23, 1834. By Daniel Drake, M.D. Cincinnati: Truman and Smith. 1834. Pp. 5-17, 28-34, 40-45.

[1] In an appendix to the published address Drake provided a description of the western region in the form of extensive quotations from his *Remarks on the Importance of Promoting Literary and Social Concert in the Valley of the Mississippi* (*Discourse on the . . . West*, pp. 47-53).

IN APPEARING among classical scholars, within the walls of a university, as your orator on this academical occasion, I find myself in the situation of a Haw tree of the woods, left standing in the cleared ground, and planted about with foreign fruit trees. Being improved by grafting and the various labors of art, their products are savory, and by persons of good taste, are, of course preferred; but still the Haw is not useless, for it serves as a term of comparison, and shows the necessity and value of early cultivation.

In consenting, at a late period, to supply the place of the able civilian on whom you at first relied,* I felt all the embarrassment that could arise from the consciousness of my incapacity to discuss a theme of pure literature; but I have, finally, chosen a topic which commends itself to my own feelings, and will not, I hope, be unacceptable to yours—it is the character, history, and prospects of the WEST.

The ancient and venerable maxim, KNOW THYSELF, has been generally addressed to individuals, but is equally applicable to communities; who should be familiar with the natural resources of their country, and the genius and tendency of their social, literary, religious, and political institutions; or they cannot cherish the good, and successfully cast out the evil. This self-knowledge of nations, is especially necessary for one of recent origin, where everything is still green, and must be fashioned according to the skill of those who regulate its growth.

Society in these BACKWOODS, even in the most thickly settled parts, is but in its forming state; and we are, therefore, invited to scrutinize, with care, the principles which control its development; for otherwise its maturity may offer less of perfection, than is found in communities which sprang up at an earlier period, instead of displaying, in its own strength and beauty, the beneficial fruits of their experience and wisdom.

It may be asked, however, whether it is consistent with the peace and perpetuity of the UNION, to inculcate a devotion to one of its parts? I shall not give a general answer to this question, but reply, that a devotion to the WEST, is manifestly compatible with both, and indeed the most efficient means of promoting both. This results from the geographical relations between the Valley of the Mississippi and the Atlantic states; relations, which being founded on nature, cannot be dissolved by the hand of art, but are daily

---

* JUDGE LANE, of the Supreme Court of Ohio.

acquiring new strength, as the ligaments of the body bind its different organs more closely together in each succeeding year of its natural growth.

I do not propose, however, to go into the analysis of our young institutions; but in the spirit of the West, shall wander to and fro, expatiating on whatever may seem attractive, but still keeping within its ample bounds.

The first thing which strikes our attention, is the difference between the opportunities for intellectual and moral improvement, in old and new states of society, and their influence on the character of the people.

As the flavor of the grape depends greatly on the soil by which it is nourished, so the temperament of individuals is modified by the intellectual aliment on which their minds subsist in childhood and youth; and of course, in studying national character, it is of great service to know the different circumstances under which the people of different places have been educated.

Children who are born in old and compactly organized communities, are surrounded from infancy, with all the means of improvement which the inventive genius of civilization can create. Books adapted to every age and all varieties of taste—established institutions of learning, from the infant school to the ancient and venerable university—professional teachers of every grade of erudition—ingenious toys, which, in the very creaking of their wheels, speak instruction—full cabinets of the works of nature and art,—public lectures in lyceums—and laws of action, for the morning, noon and night of every day throughout the year, are but a part of the means of their education and discipline. They are thus made the objects of a sleepless superintendence; which not only supplies their minds with rich materials of thought, but lays down the rules by which their growth in intellect shall proceed. Educated under these advantages, they acquire a copious and varied learning, and exhibit, in manhood, a conformity more or less striking, to the standards of excellence which have been held up for their imitation.

Most of what gives them this excellence, is either imperfect or entirely wanting, in a new country; but are there no substitutes for these artificial advantages? I think there are several, and shall proceed to offer some of them to your consideration, leaving it with yourselves to assign the value of each.

Precious as may be the benefits which good establishments of learning afford, they are not the only means of intellectual improvement; for the pathless wilderness may be made a schoolbook, and nature is the institution, in which many of the ancients were chiefly educated, whose works of taste and genius, constitute an important part of your college course. It would be an error to say, that all children of the woods, are thus instructed; for all are not educated where the best institutions have been established; and many are incapable of being taught: but none, even for mere pastime, can roam over hill and dale, descend the precipice, and stray in the cavern that opens underneath, wade through the matted herbage, and part the tangled bushes, without acquiring knowledge at every step; as the bee which buzzes round him, loads its limbs with the *materiel* of its cells, while it flits from flower to flower to feast upon their honey. To derive substantial advantage from this intercourse with nature, the youth must give scope to his curiosity, and be fully aware that its gratification will bring a rich harvest of knowledge. He should, also, cultivate the faculty of observation; which, beyond every other, can be made to supply him with valuable information, in whatever situation he may be placed; and must be exercised early, or it will remain inactive and unproductive through life. An acute and vigilant observer finds improvement in the smallest object or humblest event, as well as in those impressive phenomena, which only can arouse the attention of the dull and heedless. He suffers nothing to pass without inspection; and from habit, connects all he sees, with the memory of something he has seen before. Even in his moments of deepest study, he glances on what surrounds him, and recognizes the new and curious; he unites contemplation with his observation, or passes from one to the other, with a facility that confounds those who cannot think, except they be secluded from every external influence. He supplies his mind with fresh materials of thought, instead of ruminating on the old; and nourishes it with food collected by himself, in place of what has passed through a hundred intellects, and been subjected to as many distinct concoctions; finally, he perceives new qualities, relations and functions, in the objects that lie along his path, and thus becomes original and inventive. Indeed, with a small number of exceptions, every branch of knowledge and all the duties of life, call for the active and accurate exercise of this faculty; and the world has had but few distinguished and useful men, in whom it was not cultivated and powerful. The WEST, as already intimated, presents an endless variety of new objects and operations, to stimulate and reward this faculty; and hence, our young men *may* attain strength of intellect, and treasures of useful knowledge, although comparatively destitute of the means of academical instruction. Here then have been, and still are, a number of sources of mental improvement, which may compensate, to a small extent, at least, for the want of those which abound in older nations.

The extended limits of the WEST, and the broad navigable rivers which traverse it in every direction, exert on the mind that expanding influence, which comes from the contemplation of vast natural objects; while the distant visits and long migrations, to which this condition invites, and the wide, reciprocal commerce, which it suggests and facilitates, perpetually call its inhabitants from place to place, opening new sources of observation, and establishing fresh and profitable modes of intellectual communion.

The want of those arts and inventions, by which the inhabitants of older countries accomplish their ends, renders it necessary for the people of a new state, to invent and substitute others, as emergencies may arise; whereby their faculties are strengthened, and a spirit of self-dependence is awakened, which comes, at length, to preside over all their actions.

The many opportunities for bold enterprize, compared with the population, which a new country presents, constitute a kindred source of improvement; for occasions call forth ingenuity, and where the mind is left free to execute its schemes according to its own suggestions, it becomes fertile in expedients, and even failure does not bring discouragement; while success inspires a taste for higher undertakings, and contributes to develop the power requisite to their achievement.

In old countries, the employments of men divide them into *castes,* and while each becomes distinguished in the business to which he is confined, and which he can seldom relinquish for any other, his mind is narrowed down to the limited circle of his employments, and like the rail-road car, he moves always on the one path. But in a country like the WEST, the same person is compelled to do many different things, and often tempted to change his pursuits. A high degree of perfection in any, is impracticable under this variety of objects; but the intellect, by such various training, expands in many directions, and the aggregate of its powers, is greater than when it is compelled to extend itself in one only.

In a new country, the restraints employed by an old social organization, do not exist—the government of fashion is democratic—and a thousand corporations,—literary, charitable, political, religious, and commercial, have not combined into an oligarchy, for the purpose of bringing up to one set of artificial and traditional standards, the feelings, opinions, and actions of the rising generation; and thus the mind of each individual is allowed, in a great degree, to form on its own constitutional principles; whence result those exhibitions of original character, of which the country has always been more prolific than the city, and which are oftener seen in new than old states of society.

When an individual from the depths of a compressing population, builds his cabin in the WEST, of the trees which grew on the spot selected for his future home, being speedily released from the requisitions of the society he left behind, he permits his children, like the bushes among which they ramble, to vegetate, almost unmoulded by the hand of art. Deep and enduring ignorance might be thought the lot of all who thus grow up in the forest; but observation has shown, that this condition of the mind is far more favorable to the reception of new truths, than that which prevails in the youth of older states of society. Hence, the WEST is pre-eminently the place where discoveries and new principles of every kind, are received with avidity, and promptly submitted to the test of experiment. The mental sensibility is alive to innovations, and the growth of intellect which they impart, has a corresponding activity.

It is the peculiar distinction of the institutions, and the public sentiment of the United States, that a youth of talents and virtue, may rise from the lowest to the highest walks of society, without being obstructed or frowned upon as he advances.—This is especially the case in the Western States, where the feelings of the people are in sympathy with young men of poor parentage; and the knowledge of this facility, arouses the emulation, strengthens the purpose, and enlarges the views of our native population.

For the first quarter of a century after the settlement of the WEST began, it had but few post roads, and its scattered inhabitants seldom saw a newspaper. In this comparative destitution of a political press, it became necessary for the candidates for office to visit the people, and address them, when assembled for that purpose in central situations. On these occasions, opposing aspirants often met each other in fierce or earnest debate; and departed from the arena, improved both in logic and the art of stirring up the passions; while the people themselves were instructed on subjects of legislation, and warmed in their political sensibilities. The practice has survived the necessity from which it was at first adopted, and may still be regarded as a valuable school of oratory and political knowledge.

The itinerant clergy are important teachers in a new country; for they present to the observation of the people, a perpetual succession of ministers, who lodge in their houses, converse with their families, and, from the pulpit, promulgate every variety of Christian doctrine, explained by the aid of as many different modes of illustration.

The emigration to the West is a perennial stream. The fertility and beauty of the Great Valley, have been proclaimed on both sides of the Atlantic, and the subjects of European despotism have started from their slumbers and felt new impulses to action. Captivated by the story of our social and political freedom, our native luxuries, and the amplitude of our unsettled territories, the mind of the peasant and the villager, has been raised above the venal condition of their forefathers, and fired with the desire of emigration; the cottage of three generations, and the overshadowing elm of a hundred years, have lost their spell, and the friendships of childhood their charm; brother has bid farewell to brother, the father has pronounced his blessing on the son, impatient to be gone, and the mother shed the tear of love and sorrow, on the daughter she was to see no more; compacts of emigration have been formed, and departing companies have thinned the population of the lordly estate, or left entire streets of the village unpeopled and deserted. Thus, day after day has brought into the WEST, the enterprising and ambitious from other realms; and each has been a schoolmaster to our native population—presenting them with strange manners and customs; arts, opinions, and prejudices, not seen before; and traits of individual and national character, as numerous as the kingdoms which have poured their little colonies into the bosom of our young society. Many of the advantages of foreign travel, are thus experienced by those who could never go abroad; the Atlantic states and the west of Europe have come to us; and without leaving our native woods we have seen specimens well fitted to enlarge our conceptions of character, and diminish the necessity of hazardous voyages, for the purpose of studying human nature, in its development under political institutions entirely different from our own.

The emigrants, themselves, generally the most enterprising members of the families to which they belonged, are improved by the change of place, for it affords new objects and associations; their curiosity is awakened, and their powers of observation are rendered more acute; their minds are thrown into fermentation and become heated; purer standards of excellence float before their eyes and lead them on, while brighter hopes illuminate the paths they are to tread—thus they aspire to a better rank in society, and the aspiration brings the means of its attainment.

The addition to the Union, of Louisiana, with its French and Spanish population, opened to the inhabitants of the Valley, a new source of intellectual improvement; for the trade between the Upper States and Lower Louisiana, has made thousands acquainted with the manners and customs and character, of a different people from ourselves, and thus augmented our knowledge of human nature. In the state of Missouri, the number of French inhabitants was very considerable, and even Indiana and Illinois had masses of the same population, whose intercourse with the Anglo-American emigrants contributed to the same effect.

The near neighborhood, the wars, and the monuments, insignificant as the last may be, of the Indians, have exerted a similar effect on the mental improvement of our young population, because they have been led, intently to observe and contemplate a peculiar variety of the human race, having a number of striking features, and far removed, in most of their qualities, from our own.

Additional means of intellectual improvement, which, like these, are in some degree peculiar to the WEST, may have been recognized by other observers; but a sufficient number have been enumerated to show, that new countries are not wholly deficient in substitutes for the academies and colleges of the old. It is true, that sound scholarship, in the present era of the world, is conferred only by institutions of learning, supplied with the requisite books, and confided to able professors; but much valuable knowledge, adapted to the immediate purposes of human life, may be amassed by observation alone, if the objects and wants which stimulate and satisfy that faculty are brought within its reach. In regard to the varieties of national character, that may spring from this diversity in modes of education, the estimate of a person who has not been familiar with both, may not, perhaps, be according to the fact; but I feel strong in the conviction, that with all its deficiencies in literature and science, the mind of the WEST is at least equal to that of the East and of Europe, in vigor of thought, variety of expedient, comprehensiveness of scope, and general efficiency of execution; while in perspicacity of observation, independence of thought, and energy of expression, it stands on ground unattainable by the more literary and disciplined population of older nations.

But it would be great injustice to the subject before us to stop here. We have considered some of the beneficial effects of new countries on the mind, but their influences are, perhaps, still more salutary on the heart. Without aiming at metaphysical accuracy, we may recognize in the human character, a love of nature for the enjoyment derived from contemplating her beauties, sublimities, and eccentricities—a feeling of romance and enthusiasm—a keen sensibility to whatever is touching or magnanimous in the human character—a taste, in short, for all which the natural and moral world can present, to stir the imagination, and warm and elevate the feelings. This susceptibility constitutes the true poetical temperament, although it may not often express itself in numbers. To do this it must be associated with an imagination, that is not merely effervescent but creative, and an understanding, that will enable that imagination to embody and put forth, in beauty and natural order, those images which, in common minds, play in a lively confusion among themselves, like fairies sporting amid the violets in the darkness of the night, but never moving in procession after the dawn of day. The influence of this temperament on the character of the individual is impressive, and, within proper limits, every way admirable. It is the animating power of the inquiring and reasoning faculties—the soul of the intellect—the vital fire of genius, and the fountain which encircles, with a halo of light, not a few of the noblest forms of human greatness. The influence of this temperament may be seen, *must* indeed manifest itself, in the opinions and actions of the individual, whatever may be his rank or pursuits; and when its intensity does not make him a visionary, it throws about his character an irresistible charm. Would you have examples of it, take the man of business, who stops in the street to admire a curious or beautiful object, or listen with delight to the story of a new act of generosity or self-devotion by one whom, perhaps, he never saw; and then, by a redoubled effort, overtakes the object from which his attention had been withdrawn; or take the young farmer, who turns away his scythe from a clump of sweet-williams, that may stand smiling in his meadow; or the student, who hastens on with his problem or his translation, that he may stray for an hour in the genial air, and register the forms of the passing clouds. The soul that was never warmed by this vivifying flame, like unbaked clay of the potter, is destitute of transparency, and will not vibrate to any stroke; and the greatest intellect in which it may have been quenched, resembles the half extinguished volcano, that obscures with volumes of murky smoke, the heavens which it once illuminated with sheets of fire.

Now it must be admitted, that new countries are more favorable than old, to the preservation and active influence of this temperament; and I cannot doubt, that their inhabitants have greater freshness of feeling, more lively impulses and deeper enthusiasm, than those who grow up and die, in the midst of a dense and struggling population.

Young Gentlemen: let me exhort you to cherish this temperament by every means within your power. Like the other dispositions of the mind, it may be nourished and exalted; or depressed, degraded, and even extinguished. By exercise it grows in strength, and by receiving a direction upon proper objects it acquires dignity. The means of its gratification and improvement are always at your command:—

Watch attentively the conduct of little children, for in them you see the workings of nature; be wide awake to the eccentric movements of those around you, for the human character is known by its extravagant flights, as the coruscations of the clouds reveal to us, that they are charged with electricity; treasure up the great and good actions that fall under your observation, for they will warm your own hearts, and fortify them against the mildew of a frigid selfishness; recall perpetually and dwell upon the memory of your young friendships; foster all your early local attachments, and cherish the wild and airy superstitions of your childhood. When opportunities offer plunge into the depths of the forest, alone, or with friends of kindred taste, and establish a familiar intercourse with nature—drink out of your hand at her gushing fountains, and wade in the pebbly brook below; bathe in the deeper stream, and give yourselves up to musing on the lonely banks of the majestic river; now cast your eyes through the green canopy of maples, and gaze at the vulture poised high in the regions above; then chase the humming-bird, as it glides among the flowers which dress out our prairies in the dyes of the rainbow, or watch the worm as it slowly penetrates the trunk of the fallen tree; seek a spot still more silent and retired, people it with the creations of your own heated imagination, and then hold converse with the spirits which you may fancy are dwelling in gaiety or gloom beneath its embowering trees; as the thunder-cloud rolls onward, emerge from the woods and contemplate the warring hosts of heaven; sympathize with the ancient and venerable oak when you see him scathed by the thunderbolt; take sides with the conflicting elements, and soothe your feelings with a view of the mild glories of the setting sun, when the west wind has swept away the angry and contending clouds.

Who is he that will sneer at this advice, and call it rhapsody; and guard you against its seductions; and tell you, "the soft grass waves smilingly, but the copperhead lurks beneath?"—Who is he that would subdue your admiration of nature, put out the fires of your enthusiasm, and plunge the ice bolt into your warm hearts? The man who forgets the divine command,—"Take no thought for your life, what ye shall eat, or what ye shall drink; nor yet for your body, what ye shall put on."—Who can *not* exclaim, with the inspired poet—"Praise ye the Lord. Praise ye the Lord from the heavens: Praise him in the heights: Praise ye him, sun and moon: Praise him, all ye stars of light. Praise the Lord from the earth, ye dragons and all deeps: fire and hail; snow and vapor; stormy wind fulfilling his word: mountains and hills: fruitful trees and all cedars: beasts and cattle: creeping things, and flying fowl: Let them praise the name of the Lord." Who is he that would dry up your fountains of sympathy, with all that is grand and lovely in man, or beautiful and inspiring, in the great field of external nature? It is he, whose feelings never rise above mean heat; whose idols lie on his work bench; and whose delight is in the music of the saw; who passes, heedless, by the tender leaves of the young ash, and looks with exstacy on those of his ledger; who counts his gold by day, and dreams upon it by night; plants in the morning, and hopes to reap at noon; talks only of profitable results; and would make the earth a great work shop, and convert the human family into a vast body of operatives—instigated by avarice and abandoned to deeds of rapacity: The self-styled utilitarian, whose scope of vision takes in but the lowest part of human nature; provides chiefly for the gratification of his animal wants, hoards up the excess of his earnings, and feels no pang in the hour of death, but that of separation from the stores which a life of toil and eagerness, had enabled him to gather into his vaults.

A cherished sensibility to all that is admirable in nature, is in no degree incompatible, with the acquisition of all that is necessary or useful in life. The sluggard, the glutton, and the drunkard, no less than the miser, do not, it is true, find time to indulge themselves in hours of fervent contemplation among the works of God; but all who are not delivered over to the tyranny of one, out of the many desires which belong to human nature, are enabled in the midst of business, to send forth their imaginations upon the world of matter and of man, and take into the warm embrace of their feelings, whatever is touching and noble in both.

He who fosters this sensibility, retains a youthfulness of taste, that keeps him in sympathy with the generations, which, like saplings that spring up around the aged and decaying tree, are at last to succeed him in society. This amiable condescension, spreads an irresistible charm over the character of age. Its maxims of wisdom become a law to the erring footsteps of youth; while the dark and dreary hours from which the most favored cannot escape, are lighted up by the flashes of gayety and innocent mirth, which beam from the eye in the springtime of life. On the contrary, the sullen old man lives only in the past, and dwelling alone in his dotage, goes down towards the grave, as the sun in winter descends through the mists and fogs of our western mountains, which extinguish his fires, while he still lingers on the verge of the horizon.

Dismissing, for the present, our inquiry into some of the intellectual and moral advantages, which our new country offers, as substitutes for the establishments of older states, let us proceed to speak of the duties and labors which it enjoins upon its sons.

In the first place, we should transmit to posterity a graphic description of the Great Valley, as it appeared in primitive loveliness to the eyes of the pioneers, as many of us remember to have seen it, and as it still smiles in spots unviolated by man. Civilization is a transforming power, and wherever its wand is raised, the surface of the earth assumes a new aspect. The native trees, cut down and consumed, are replaced by the apple and orange; the wild grape, which united their limbs, is succeeded by an exotic, resting on trestles; the rivers are constrained within narrower channels, or turned into canals; and the mossy rocks of their margins, are broken with the sledge or exploded with gunpowder; hills are leveled and valleys filled up; a macadamized road usurps the bed of the little brook, and the rumbling of the coach wheel falls upon the ear, instead of the soft music of its rippling waters; fields of wheat undulate, where the prairie grass waved before, and tobacco and cotton, are nourished on the wreck of the cane-brake, which formerly spread its green leaves over the snows of winter. Thus the teeming and beautiful landscape of nature, fades away like a dream of poetry, and is followed by the useful but awkward creations of art. Before this transformation is finished, a portrait should be taken, that our children may contemplate the primitive physiognomy of their native land, and feast their eyes on its virgin charms.

* * * * *

Young Gentlemen: The scenery, history, and biography, of the Valley of the Mississippi, constitute the very elements of our literature, and their retrospect naturally leads us to inquire into its resources, and the character it will probably assume. When the young planter, on the banks of the Yazoo or the Illinois, clears away the forest, and prepares his lands for tillage, his taste and judgment are displayed in the plan on which he marks out his fields, and the seeds with which he sows them. It will depend on himself, whether his farm be beautiful in its arrangement and varied in its products, or irregular, unsightly, and more prolific in weeds and briars, than the useful and elegant productions of agriculture. Thus must it be with the scholars of the Great Valley. They have a vast field to cultivate, but small portions of which are as yet laid off and planted, and its future beauty and abundance, will be according to their skill and industry.

As a part of the generation, to which are confided the rudiments of our infant literature, I would exhort you to study profoundly the elements you are to control, and labor to combine them according to the principles of taste and science. If the germs are deformed and sickly, the future plants must be shapeless, feeble, and unproductive of salutary fruit.

The materials placed at your command, and the age of the world in which you come up to the task, confer upon you many important advantages. When we contemplate the history, condition, and prospects, of the West, we cannot fail to perceive, that its literature will ultimately prove not only opulent in facts and principles, but peculiar in several of its qualities. Let us inquire into some of its present and prospective characteristics.

In the first place—The time is remote, when language in the West, will acquire a high degree of purity, in nomenclature and idiom. Many of our writers have received but little education, and are far more anxious about results, than the polish of the machinery by which they are to be effected. They write for a people, whose literary attainments are limited and imperfect; whose taste is for the strong rather than the elegant; and who are not disposed or prepared to criticize any mode of expression, that is striking or original, whatever may be the deformities in its drapery;—consequently, but little solicitude is felt by our authors, about classical propriety. Moreover, the emigration into the Valley being from every civilized country, new and strange forms of expression are continually thrown into the great reservoir of spoken language; whence they are often taken up by the pen, transferred to our literature, and widely disseminated. For many years to come, these causes will prevent the attainment either of regularity or elegance; but, gradually, the heterogeneous rudiments will conform to a common standard, and finally shoot into a compound of rich and varied elements; inferior in refinement, but superior in force, variety, and freshness, to the language of the mother country.

Second. Our literature, at present, is but slightly imbued with allusions and illustrations drawn from the classics; and although it may possess a portion of their temperament, they have not infused it; for they are cultivated by a small part of our scholars only, and seldom read, even in translation, by a majority of our educated people. I shall not prophecy on this subject, but nothing indicates, that the number of devotees to classical learning will be greater in proportion to our population, hereafter, than at the present time. I see as little to admire in this neglect, as in that preposterous idolatry to the ancients, which would substitute the study of their literature for that of modern times. A genuine scholar extends his researches as far as his opportunities will permit, and drawing from the literature of all nations—ancient and modern—whatever is good and beautiful in spirit, applies it to the embellishment and elevation of his own.

Third. Our literature will be tinctured with the thoughts and terms of business. The mechanic arts have become locomotive, both in temper and capacity—they travel abroad, and exhibit themselves in every department of society. To a certain degree, they modify the public mind; supply new topics for the tongue and pen; generate strange words and phrases, as if by machinery; suggest novel modes of illustration, and manufacture figures of speech by steam power. They afford canal transportation to the ponderous compiler of statistics; a turnpike to the historian; a tunnel to the metaphysician; a scale of definite proportions to the moral philosopher; a power loom and steam press to the novelist; fulminating powder to the orator; corrosive acids to the satirist; a scalpel to the reviewer; a siesta chair to the essayest; a kaleidescope to the dramatist; a balloon to the poet; a railroad to the enthusiast, and nitrous oxide to the dunce. While we devoutly indulge the hope, that our literature will not depend for its elevation on the lever of the arts, there can be no objection to a fellowship between them; nor any reason why it should not adopt, whatever they may offer, to diversify its objects and enrich its resources.

Fourth. The absence, in the Valley of the Mississippi, of those ancient and decaying edifices, which are scattered over Europe, and were once the seats of great political, military, or social events, must deprive our literature of an element of solemn and touching grandeur. It might be thought, that our own antiquities would supply the place of those; but we know nothing of the people by whom these were erected, and consequently, they inspire but little of that romantic and tender feeling, which results from associating the history of a people with the ivy-covered ruins of their former taste and industry.

Fifth. In the West there is no prevailing love or talent for music, the most delightful of all the liberal arts; and, of course, its softening and refining influences will not be exerted on our literature. To what extent a musical taste might, hereafter, be created by pressing the study of this science, as a branch of popular education, cannot be foreseen; but the interesting results that would flow from success, should animate us to a vigorous effort in the experiment. I have little doubt, that the musical temperament of Germany, is one reason why, on having her mind directed to the creation of a national literature, she so speedily and gracefully accomplished the object.

Sixth. A religious spirit animates the infancy of our literature, and must continue to glow in its maturity. The public taste calls for this quality, and would relish no work in which it might be supplanted by a principle of infidelity. Our best authors have written under the influence of Christian feeling; but had they been destitute of this sentiment, they would have found it necessary to accommodate themselves to the opinions of the people, and follow Christian precedents. The beneficent influence of religion on literature, is like that of our evening sun, when it awakens in the clouds those beautiful and burning tints, which clothe the firmament in gold and purple. It constitutes the heart of learning—the great source of its moral power. Religion addresses itself to the highest and holiest of our sentiments—benevolence and veneration; and their excitement stirs up the imagination, strengthens the understanding, and purifies the taste. Thus, both in the mind of the author and the reader, Christianity and literature act and react on each other, with the effect of elevating both, and carrying the human character to the highest perfection which it is destined to reach. Learning should be proud of this companionship, and exert all her wisdom to render it perpetual.

Seventh. The literature of the West is now, and will continue to be, ultra-republican. If we compare the constitutions of the new states with the old, we find that when republicans transfer themselves into the free and expanded solitudes of the wilderness, and proceed to organize new institutions, they display an increasing disposition to retain the political power in their own hands. It is possible to run into excesses in this respect, but that error is safer than the opposite; unless, indeed, they should carry their democratic principles so far, as to generate anarchy. Liberal political institutions favor the growth of literature; and, in turn, when its powerful energies are exerted in the great cause of personal freedom, the liberties of a reading people are placed beyond the grasp of tyranny.

Eighth. The literature of a young and a free people, will of course be declamatory, and such, so far as it is yet developed, is the character of our own. Deeper learning will, no doubt, abate its verbosity and intumescence; but our natural scenery, and our liberal political and social institutions, must long continue to maintain its character of floridness. And what is there in this that should excite regret in ourselves, or raise derision in others? Ought not the literature of a free people to be declamatory? Should it not exhort and animate? If cold, literal, and passionless, how could it act as the handmaid of improvement? In absolute governments all the political, social, and literary institutions, are supported by the monarch—here they are originated and sustained by public sentiment. In despotisms, it is of little use to awaken the feelings or warm the imagination of the people—here an excited state of both, is indispensable to those popular movements, by which society is to be advanced. Would you rouse men to voluntary action, on great public objects, you must make their fancy and feelings glow under your presentations; you must not merely carry forward their reason, but their desires and their will; the utility and loveliness of every object must be displayed to their admiration; the temperature of the heart must be raised, and its cold selfishness melted away, as the snows which buried up the fields when acted on by an April sun; then —like the budding herb which shoots up from the soil—good and great acts of patriotism will appear. Whenever the literature of a new country loses its metaphorical and declamatory character, the institutions which depend on public sentiment will languish and decline; as the struggling boat is carried back, by the impetuous waves of the Mississippi, as soon as the propelling power relaxes. In this region, low pressure engines are found not to answer— high steam succeeds much better; and, although an orator may now and then explode and go off in vapor, the majority make more productive voyages, than could be performed under the influence of a temperate heat.

Ninth. For a long time the oration, in various forms, will constitute a large portion of our literature. A people who have fresh and lively feelings, will always relish oratory; and a demand for it will of course bring a supply. Thus auditors create orators, and they, in turn, increase the number of hearers. In a state of society where an indefinite number of new associations, political, religious, literary, and social, are to be organized, it is far more effective to assemble men together and address them, personally, than through the medium of the press. If an excitement can be raised in a few, it spreads sympathetically among the many; and is often followed by immediate results of greater magnitude, than the pen could produce in years. Hence, I regard the study of oratory, as among the most important objects of an academical and collegiate course; and would earnestly commend it to your consideration. None of you should assume, that he will never be called upon to speak in public, and may, therefore, omit the cultivation of eloquence. In this country, occasions for doing good by public speaking come up when little expected; and are not confined to the learned professions of theology and law. The opportunities and calls are numerous beyond computation; and the variety of objects so great, as to extend to every intelligent man in society. Even the merchant, the mechanic, and the agriculturist, are often placed in situations where an expression of their opinions, before assemblies of their own brethren, may be followed by beneficial effects to themselves, as well as to those whom they may address. I am so far from wishing to discourage this practice, that I would promote it by every argument, as an instrument of social advancement, a method of popular instruction on specific subjects, and a means of preserving our free institutions.

Tenth. The early history, biography, and scenery of the Valley of the Mississippi, will confer on our literature a variety of important benefits. They furnish new and stirring themes for the historian, the poet, the novelist, the dramatist, and the orator. They are equally rich in events and objects for the historical painter. As a great number of those who first threaded the lonely and silent labyrinths of our primitive woods, were men of intelligence, the story of their perils and exploits, has a dignity which does not belong to the early history of other nations. We should delight to follow their footsteps and stand upon the spot where, at night, they lighted up the fire of hickory bark to frighten off the wolf; where the rattlesnake infused his deadly poison into the foot of the rash intruders on his ancient domain; where, in the deep grass, they laid prostrate and breathless, while the enemy, in Indian file, passed unconsciously on his march. We should plant willows over the spots once fertilized with their blood; and the laurel tree where they met the unequal war of death, and remained conquerors of the little field.

* * * * *

Eleventh. Our literature cannot fail to be patriotic, and its patriotism will be American—composed of a love of country, mingled with an admiration for our political institutions. The slave, whose very mind has passed under the yoke, and the senseless ox, whom he goads onward in the furrow, are attached to the spot of their animal companionship, and may even fight for the cabin and the field where they came into existence; but this affection, considered as an ingredient of patriotism, although the most universal, is the lowest; and to rise into a virtue it must be discriminating and comprehensive, involving a varied association of ideas, and embracing the beautiful of the natural and moral world, as they appear around us. To feel in his heart, and infuse into his writings, the inspiration of such a patriotism, the scholar must feast his taste on the delicacies of our scenery, and dwell with enthusiasm on the genius of our constitution and laws. Thus sanctified in its character, this sentiment becomes a principle of moral and intellectual dignity—an element of fire, purifying and subliming the mass in which it glows. As a guiding star to the will, its light is inferior only to that of Christianity. Heroic in its philanthropy, untiring in its enterprises, and sublime in the martyrdoms it willingly suffers, it justly occupies a high place among the virtues which ennoble the human character. A literature, animated with this patriotism, is a national blessing, and such must be the literature of the West. That of all parts of the Union must be richly endowed with this spirit; but a double portion will be the lot of the interior, because the foreign influences, which dilute and vitiate this virtue in the extremities, cannot reach the heart of the continent, where all that lives and moves is American. Hence a native of the West may be confided in as his country's hope. Compare him with the native of a great maritime city, on the verge of the nation,—his birthplace the fourth story of a house, strangulated by the surrounding edifices, his play-ground a pavement, the scene of his juvenile rambles an arcade of shops, his young eyes feasted on the flags of a hundred alien governments, the streets in which he wanders crowded with foreigners, and the ocean, common to all nations, forever expanding to his view: estimate his love of country, as far as it depends on local and early attachments, and then contrast him with the young backwoodsman, born and reared amidst objects, scenes, and events, which you can all bring to mind;— the jutting rocks in the great road, half alive with organic remains, or sparkling with crystals; the quiet old walnut tree, dropping its nuts upon the yellow leaves, as the morning sun melts the October frost; the grape vine swing; the chase after the cowardly black snake, till it creeps under the rotten log; the sitting down to rest upon the crumbling trunk, and an idle examination of the mushrooms and mosses which grow from its ruins; then the wading in the shallow stream, and upturning of the flat stones, to find bait with which to fish in the deeper waters; next, the plunder of a bird's nest, to make necklaces of the speckled eggs, for her who has plundered him of his young heart; then the beech tree with its smooth body, on which he cuts the initials of her name interlocked with his own; finally, the great hollow stump, by the path that leads up the valley to the log school-house, its dry bark peeled off, and the stately polk-weed growing from its centre, and bending with crimson berries; which invite him to sit down and write upon its polished wood, how much pleasanter it is to extract ground squirrels from underneath its roots, than to extract the square root, under that labor-saving machine, the ferule of a pedagogue! The affections of one who is blest with such reminiscences, like the branches of our beautiful trumpet flower, strike their roots into every surrounding object, and derive support from all which stand within their reach. The love of country is with him a constitutional and governing principle. If he be a mechanic, the wood and iron which he moulds into form, are dear to his heart, because they remind him of his own hills and forests; if a husbandman, he holds companionship with the growing corn, as the offspring of his native soil; if a legislator, his dreams are filled with sights of national prosperity, to flow from his beneficent enactments; if a scholar, devoted to the interests of literature, in his lone and excited hours of midnight study, while the winds are hushed and all animated nature sleeps, when the silence is so profound, that the stroke of his own pen grates, loud and harsh, upon his ear, and fancy, from the great deep of his luminous intellect, draws up new forms of smiling beauty and solemn grandeur; the genius of his country hovers nigh, and sheds over his pages an essence of patriotism, as sweet as the honey-dew which the summer night distills upon the leaves of our forest trees.

Young Gentlemen: I have directed your attention to some of the circumstances that will exert an influence on the character of our literature. It is for you and your cotemporaries to recognize others, and so control and animate the action of the whole, as to bring out results in harmony with the nature that surrounds you. To do this, successfully, you must study that nature and comprehend its temperament. With the elements of learning and science, conferred by your honored *alma mater,* you should go forth, and make acquaintance with the aspects, productions, and people of your native land. Few of you can travel in foreign countries, but all may explore their own; and I do not hesitate to say, that the latter confers greater benefits than the former; though both should be enjoyed by those who possess the means. But to render traveling beneficial, it must not be performed in steam boats and railroad cars, darting with the flight of the wild pigeon before the north wind, and cutting through whole states in the darkness of a single night. Thus borne impetuously onward, you see only the great commercial points, which, from their constant intercourse, become so assimilated, as to afford but little variety. The *diversities* in aspect and productions; in natural curiosities; in works of art, both elegant and useful; in public improvements and resources; in political, literary, social, and religious establishments, and in personal and national character, the study of which should be the chief end of travel, are found in places remote from the commercial highways of the nation, not less than in those which lie upon them; and can only be seen and studied, by him who departs from the beaten track, and views every spot with the eye of a curious and disciplined observer. The copious stores of knowledge, and the vigor of intellect, which may thus be acquired, are not the only advantages which traveling in your own country can yield; for it will confirm your native tastes and feelings, preserve your love of home, and strengthen your nationality—so often impaired by premature or protracted residence abroad. Hence you will become better qualified as writers; and, when time shall ripen your judgments into perfect maturity, you will be able to lend important aid to your countrymen, in the formation of an American literature; that shall be rich in illustrations drawn from your native land, glowing in its patriotism, attractive by its freshness, and intense in its strength and fervor.

My Young Friends: When you return home as men, you will find that other duties await you, than those which relate to our literature. Your fathers have done little more than clear the ground, and scatter the first seeds of society; and you must not only weed and water the young plants, but enrich the soil with others, to which their limited means could not extend. Thus you, and even the next generation, will be pioneers, like the last; but your pioneering will be less difficult and arduous. I cannot indicate all the labors and enterprises which lie before you; but as specimens may say, that new political constitutions are to be formed, and the older remodeled, as experience may dictate; laws adapted to the character and genius of a varying population, and to the wants and productions of different parts of the Valley, are to be devised; a machinery of civil and municipal government, and systems of jurisprudence, in unison with the taste and temper of our rising communities, are to be instituted; inventions and manufactures, appropriate to our various situations, are to be naturalized, or brought forth on the spots where required, and put into operation; our plans of internal improvement must be extended, and made to unite with each other, in such manner as to spread over and connect all parts of the Valley; institutions of learning, from common schools up to universities, must be organized where they do not exist, and re-organized and improved where they do; public hospitals on all our great rivers should be erected, for the relief of our trading population; new associations, for purifying the morals of the great mass of the people, should be formed; and religious societies constituted, wherever they are rendered necessary, by the extension of our settlements.

Thus, you will be called to participate in grand and noble objects, and enjoy the high prerogative of creating—of giving the first impulse—of prescribing the direction, and laying down the rule of action. In performing these momentous functions, you will fix the course of future events, as far as human agency can regulate them. A great responsibility rests upon you—the destinies of millions will be lodged in the hands of your fellow laborers and yourselves. Keep those hands free from stain; look into your own hearts, and cast out all unholy selfishness; chasten your ambition; cherish your benevolence, till it shall expand over every object of philanthropy; cultivate your religious feelings; preserve your simplicity of manners; rebel against the tyranny of fashion; study profoundly the character of your countrymen, that you may know how to supply their intellectual and moral wants; enrich your minds with the maxims of wisdom furnished by other ages, and modify them to suit your own; learn to concentrate your thoughts, successively, on every scheme of public utility; mould yourselves into practical patriots; declare a war of extermination against the whole class of demagogues; finally, school all your faculties and affections, till you can come to feel powerful in your *country's* strength, exalted in *her* greatness, and bright in *her* glory.

With this preparation of mind, and willing devotion of heart, you will labor, in harmony, till the monuments of your skill and industry shall cover the land, from Michigan to Louisiana—from the mountain rivulets of our own unrivaled Ohio, to the grassy fountains of the savage Arkansas. You will contribute to raise up a mighty people, a new world of man, in the depths of the new world of history, and the friends of liberty, literature, and religion, in all nations, will look upon it with love and admiration: composed of the descendants of emigrants from every country, its elements will be as various as the trees which now attire our hills; but its beauties as resplendent as the hues of their autumn foliage.

Then, in the hour of death, when your hearts shall pour out the parting benediction, and your eyes are soon to close, eternally, on the scene of your labors, you will enjoy the conscious satisfaction, of having contributed to rear in your native Valley, a lovely sisterhood of states, varying from each other, as the flowers of its numerous climates differ in beauty and fragrance; but animated with the same spirit of patriotism, instinct with one sentiment of rising glory, and forever united by our Great River, as the Milky-way, whose image dances on its rippling waters, combines the stars of the sky into one broad and sparkling firmament.

*Discourse on the Philosophy of Discipline**

(1834)

After the late 1820s Drake's introductory lectures to his medical classes in Lexington, Cincinnati, and Louisville and his hortatory addresses to young people and students regularly stressed the need for personal discipline in their studies, their personal behavior, and in the conduct of their lives. His *Discourse on the Philosophy of Discipline, in Families, Schools, and Colleges,* was an attempt to establish a scientific basis for the necessity of personal discipline and to examine the possibilities of an external imposition of that order which should more properly come from within. It stands as a characteristic document of his increasing concern during the 1830s with the need for artificial institutions (albeit based on nature) to direct and assist individuals in becoming mature men and women and responsible citizens, and of his increasing fascination with the physiological basis of social behavior. Delivered at the annual meeting of the Western Literary Institute and College of Professional Teachers in Cincinnati in 1834, it appeared in their transactions but was "bound up separately, for gratuitous distribution, in the Valley of the Mississippi, by the author,"[1] and may be seen as an additional example of Drake's continuing efforts to instruct his fellow citizens in the means for achieving a higher and more moral civilization in the West.

* *A Discourse on the Philosophy of Discipline, in Families, Schools, and Colleges; delivered before the Western Literary Institute, and College of Professional Teachers, In Cincinnati, on the 6th of October, 1834, By Daniel Drake, M.D.* Cincinnati: Published by U. P. James. 1834.

[1] *Discourse on the Philosophy of Discipline,* p. 2.

THE UNIVERSE IS an empire, and God is its sovereign. It consists of masses of matter suspended in space; one of which is our earth. Of the others, we know very little from observation; but, relying on several ascertained analogies, presume, that in their intimate structure, they may not be unlike our own. In it we observe two great divisions, the mineral, and the organized or living kingdoms. Passing by the former, we find the latter divisible into two classes, vegetable and animal; the last of which, may be subdivided into two orders, the inferior animals and the human race. Thus we know, that our globe comprehends and sustains an innumerable variety of bodies.

The different objects which compose the universe, are not at rest, nor do they remain in the same relation. Motion is the condition in which most of those on the earth's surface exist; the mass itself is in motion, and even the sun turns on its axis; the other planets of the solar system, have the same movements with ours. It is even probable, that the constellation to which our sun belongs, has a progressive motion in the heavens; and, if this is the fact, we may suppose that the whole—the entire universe, is in action. Such being the probability, and in reference to our earth and its productions, the actual fact, it follows, that a state of chaos would sooner or later arise, unless these complicated movements were made on some kind of system. But the experience of the human race in past times, and every day's observation, convince us, that disorder is *not* the consequence of this action, and, of course, there must be laws of motion; and we believe that God, who made the worlds and all who inhabit them, is the great law-giver. To regulate the revolutions of the planets, he has enacted laws; to guide the actions of atoms of matter on other atoms, he has made other laws; to direct the arrangement of those atoms in organized bodies, he has established other laws; and, lastly, to govern man, he has made others, which refer both to his mind and body. Thus, every movement, from that of a satellite round the earth, to the revolution of the sun on his axis; from the rise and fall of a particle of dust, or the growth of a blade of grass, to the voluntary actions of man himself, is regulated by laws, which God only can modify or repeal. The government, then, of the entire universe, is a government of laws, and without them, it would stand still, or speedily run into confusion.

If such be the fact, and who can deny it? we come directly to the conclusion, that a violation of any of the laws of nature is eventually followed by disorder; and this disorder, involving as it does or should do, the agent which commits it, constitutes the punishment or penalty. Thus, on the plan of nature, every violation is punished; for a law without a penalty is a dead letter. Let us apply this reasoning to the human race.

Was there but one man, it would be necessary to his welfare that he should not violate the laws, which regulate the relations between him and the surrounding elements; for if he did, he would suffer bodily pain, and perhaps perish. Thus, if he exposed himself, unprotected, to the north wind, at midnight in winter, he would be frozen; or, if he walked into the fire, he would be burnt—in both cases, receiving the penalty imposed on the violation; while on the other hand, if he scrupulously observed the laws which regulate the relations between his system and heat and cold, his feelings would be pleasant, and, in that pleasure, he would find the reward of his fidelity to the requirements of his physical nature.

Again, if we contemplate him associated with others in society, and suppose him to violate the laws which are necessary to its government and well being, we see him doomed to suffer a penalty; while, on the contrary, a strict observance of all the regulations of the social compact, never fails to preserve his peace, and procure for him the reward of conscious rectitude, and the approbation and confidence of his fellow. Thus, both in the world of matter and the world of mind, we find punishment the consequence of violation, and reward the beneficial effect of obedience.

When we come to inquire into the reason of this relation between the act and its consequences, we at once perceive, that without it, no law would be respected; and that, in the economy of the world, rewards and punishments are the appointed means of securing obedience, and maintaining the supremacy of those enactments, domestic, social, political, and moral, without which men could not live in each other's society.

Hence, from a survey of the physical and social world, we derive a warrant for rewards and punishments, and acquire a conviction of their justice and necessity.

* * * * *

Having then the lights, both of reason and revelation, to guide us, we possess the highest assurance which the human mind can attain, that both rewards and punishments are not only right, but indispensably necessary; and that in all cases, where, as individuals, it is our right to govern, it is our duty, in imitation of Him who ruleth all things in wisdom, to punish offences and reward virtuous obedience. And what is the philosophy of this system? One easily understood; one that he who runs may read. It is to associate pain with the transgression, and pleasure with the observance of the law. By pain and pleasure God governs the whole animal world. In the lower orders they are limited to the body—in man, they extend also to the soul. God has not required of us the observance of any law, without making that obedience a source of pleasure, corporeal or mental; nor permitted the violation of any, without annexing the penalty of pain, either present or prospective. The object and effect of all punishment should be, to establish this association of ideas, that, when the temptation comes, the fear of the punishment we have felt, may come also, and deter us from the act—and the end of every reward should be, to make the resistance of temptation an immediate source of pleasure. As far as we can fathom this matter, the moral government of the world could not be maintained by any other system; neither punishments nor rewards alone, could accomplish the object.

Has, then, a parent the right to govern his child? If he have, it is his duty to reward and punish it, according to the manner in which it acts, under the just and necessary rules which he lays down for its government. That he has such a right, cannot be doubted by any, who reflects on the relations of parent and child. It results from the dependance of the latter upon the former; a dependance as great as that of the young scion on its parent root. Here, moreover, as in the other branch of our argument, we are not left to the lights of our own understanding, for revelation throughout, recognises this dependance, and *commands,* what reason and instinct had already made manifest.

* * * * *

But can this right with propriety be delegated to another? It certainly can. The object is not to gratify the parent by the exercise of power, but to preserve the child from danger, qualify it for usefulness in life, and prepare it for happiness after death. But if both the parents should die, this must be done by friends or strangers; and when its education and discipline require it to be separated from them, the punishment must be inflicted by those who have it in charge, or else the duty which God enjoins and nature requires, will not be performed.

It follows, therefore, from these premises, that children require government; that this government must be by laws, for where there is no rule of action, there can be no offence; that rewards and punishments are the appointed means of securing obedience to the system, and that these cannot be dispensed with, either by the parent or the teacher.

Let us now inquire, what these rewards and punishments should *be.* To prosecute this investigation in a proper manner, a thorough knowledge of the constitution of human nature, as it exists in childhood and youth, is indispensable.

Man being a compound of mind and body, can only be understood by observing and studying both, for they act and re-act upon each other. In the successive periods of life, in different individuals, and in the various grades of civilization, the relative power of the mind upon the body, and the body upon the mind, is different. Thus, in the civilized and intellectual state, the mind exercises greater power over the body, than in the savage state; and the mind of a philosopher, or a christian, governs the desires of his body more effectually, than the mind of an ignorant or wicked person controls his appetites; and, finally, the mind of an adult rules over his bodily wants, with greater success than the mind of a child. In the tender stages of infancy, the reasoning powers and the moral sentiments, are but little developed, and the corporeal appetites and desires are strong. The reason is obvious. The body must be built up, and hence the appetite for food, and the pleasures of indulgence, are great, sometimes almost insatiable. The impatience of labor is quick, because its industry can seldom be turned to good account, and its limbs are soon fatigued, while they are growing; its natural repugnance to close or long continued confinement, is equally strong, for fresh air and unrestrained exercise, are requisite to the proper maintenance of health; its curiosity for wandering among new objects is intense, because, observation is the food of the young intellect, and indispensable to its growth; finally, its love of play and of pleasure is almost indomitable; because on the plan of nature, no responsibility in regard to the future rests upon it; and if it had not a desire for play, it would not take the necessary exercise, nor acquire the proper use and discipline of its limbs. Thus, almost all the pains and pleasures of infancy and youth, connect themselves with the body. The gratification of the physical or material part is the great object; that which answers to the wants and desires of the body affords the chief pleasure. Like the lower animals, it lives for the body, and for the present moment. Its enjoyments are physical—its sufferings are physical; and, when they extend to the mind, it is because something which administered to the pleasures of sense has been withheld, or applied in such manner as to mortify the few feelings and sentiments of the soul, which, at that early period, are in a state of susceptibility.

What is the deduction from these views? Undoubtedly, that there is in the constitution of childhood, a foundation for physical correction; and that punishment of the body is the most efficient mode of reaching and affecting the mind. Such are the conclusions of reason, applied to this subject. And what are the results of experience? Let the practice of the whole world return the answer. In every age, and in all nations, we find the hand of the parent uplifted in physical correction, or some other mode adopted, of punishing the body through its desires and sensibilities. It is, indeed, an instinct on the part of the parent, and, by an instinct equally intuitive, unerring, and universal, is acquiesced in by the child. Nature, in fact, is at the bottom of the matter, and prompts, if she does not regulate, the whole discipline.

But does God in his revealed will, bear us out in these conclusions? The Bible shall give the reply. *"He that spareth his rod, hateth his son; but he that loveth him, chasteneth him betimes." "Foolishness is bound in the heart of a child; but the rod of correction shall drive it far away." "Withhold not correction from the child, for if thou beatest him with the rod, he shall not die. Thou shalt beat him with the rod, and shalt deliver his soul from hell."*

Thus we find punishment of the body, even with the rod, expressly enjoined by Heaven, as a parental duty; and declared to be powerful, not only in driving away foolishness, and qualifying the child for the duties of this life, but in preparing it for the enjoyments of eternity; and we are thus supplied with new evidence of the conformity of the law of the Bible, to the laws which govern the constitution of man.

Corporeal punishments are of two kinds, those which act upon the body in a positive manner, and give pain, as the hand, the ferule, and the rod; and those which act negatively, and give pain to the unindulged appetites, as withholding luxurious articles of food and drink, and confinement to the house, or to a certain position. The latter, at first view, might seem preferable; but they are not always practicable with the great mass of parents, who are poor, and are obliged to work, and for whom all *general* rules should be formed; and they cannot always be conveniently resorted to by teachers. There is, moreover, an objection of a different kind, which detracts something from their character. If the child be not hungry, or its appetite be destroyed by its emotion of mind, the denial of good things will inflict no punishment; and confinement will give no bodily pain if there should, at the moment, be no disposition to go abroad. Still further, there are moral objections to restraints upon the appetites, which deserve deep consideration. The child is taught, by the estimate which it perceives the parent to place on the enjoyments of sense, when he withholds them as a punishment, to regard them as of paramount value, and is thus rendered more sensual; when, perhaps, the very offence for which he was punished, was an act of improper indulgence, or of depredation for the gratification of his appetite. Finally, if the hunger of children be not satisfied, they are tempted, secretly, to acquire the means of gratifying it; and are thus led into habits of concealment, deceit, and theft, which practised towards the parent for a time, may at last be exercised on society.

On the other hand, it has been said, that the use of the rod degrades the child in its own estimation; debases it in the view of other children; exasperates it towards its parents; is liable to be excessive; and contributes to maintain on the earth, the system of violence and war, which must be abolished, before the world can be christianized. These are serious objections, and it is our duty to consider them separately.

I begin by appealing to every judicious and observing parent and preceptor, to say, whether they have witnessed, under the application of the rod, any evidence of improper self-abasement in the child; and would ask all who have felt it, to recollect, whether its *merited* and *proper* infliction, sunk them in their own estimation, below the point of that humility which children ought to feel, under the deserved chastisements of their parents or teachers? From my own observation and experience, I should answer these questions in the negative; and, believing, as I have already said, that the use of this instrument of correction, is a kind of instinct on the part of the parent, acquiesced in by the feelings of nature in the child, I cannot suppose that its employment, under proper regulations, can debase the feelings, or break down the manly spirit, but rather contribute to purify and elevate both.

That it necessarily lowers the child in the estimation of others, there is as little reason to believe. If it be a *natural* punishment, such an effect *cannot* flow from it; and that it does not, is a matter of observation; for we generally see the surrounding children, if relatives or friends, disposed to pity the one which has been chastised, and often find them, subsequently, engaged in offering it their little consolations. That children who are frequently whipped, sometimes become objections of derision with their playmates, is certain; but, as a general rule, such children are great *offenders,* and among children as in society, those who continue to offend in the midst of correction, will, at length, fall into contempt.

That the rod *may* exasperate the child towards its parent, there is no doubt, if it be used when the child is innocent, or applied to a degree disproportionate to the offence, or with partiality, in reference to other children; and under such circumstances, it *ought* to feel indignant. But where is the individual, who can say, that he ever loved a parent the less, for inflicting personal chastisement in a proper degree, when he had a consciousness of having done wrong? So far from producing the alledged effect, it generates the opposite; and children never love their parents *more,* than in the hour of repentance and returning joy, which follows this kind of punishment, inflicted in a suitable manner and to a merited extent.

That the rod is liable to be handled to excess, is an evidence of its power, but no objection to its regulated use. Any other mode of punishment may be abused; and he who has not sense and self command enough, to use the rod discreetly, might be expected to err in any other means of correction. The objection, that being at hand, it is employed while the parent is still in anger, we shall consider hereafter.

The last objection, that it keeps alive a spirit of force and violence, and contributes to maintain war in the world, we may meet, as we might, indeed, have met the others, with the remark, that its use is of Divine appointment in the Old, and no where forbidden in the New Testament; and that it cannot, therefore, remotely promote the effects ascribed to it, for God is not the author of any commandment that leads to violence and war; nor would he have failed to prohibit every thing which interferes with the spread of his moral dominion on the earth. It is not, moreover, by abolishing war, that the world will be christianized, but, becoming christianized, war and violence will cease.

Although the advocate of corporeal punishments, I am far from intending to favor a system of cruel discipline; and should, moreover, think little of the head and heart, or rather, think much that was bad of the parent or teacher, who might overlook the circumstances under which they should be inflicted. Let us inquire into a few of these conditions.

Corporeal punishments influence the actions, but carry no instruction to the understanding. They should then, in all cases, from the cradle upwards, be preceded by a statement to the child, of the offence and the reason for the punishment; that is, it must be made to know and remember, that the act was wrong, and that its repetition will bring a return of the pain of correction. It should also be instructed in the *nature* of the duty it has violated, and made to see that it has trampled some *law* under foot, the *penalty* of which it is about to suffer, under a warrant of execution, derived both from nature and God. It will thus get considerations of duty, and a cultivation of its young moral sentiments, associated with the punishment, and the whole will be the better understood and recollected, from that painful associations of ideas. It should likewise, when practicable, be corrected in secret; for secret correction is most efficient, and it is less likely to lose its standing with its fellows, if they remain ignorant of its vices: Finally, in the midst of his anger or his regret, the parent or teacher should manifest affection, and by all his eloquence, arouse that of the little offender into activity.

Thus regulated in its use, the rod will be found, not merely an instrument of fear, but of penitence and respect, and such has been the experience of the world.

We come, now, to physical *rewards,* the opposite of physical *punishments.* These act by giving bodily pleasure, and, of course, address themselves to the senses. Let us consider them in succession, beginning with the sense of taste. This is the earliest on which we can act, because it is the first that requires to be indulged. There can be no objection to granting a child the means of this indulgence as a reward for good conduct; but as it generates a taste for luxury, it should not be continued after the other senses are so far developed, that we can act upon them with effect, which happens in different children, at various ages.

The sense of smell is next developed, but the means of gratifying it are not so convenient as those of the sense of taste. Its gratification, however, is less dangerous to the future, than that of taste, and need not be abandoned, as long as its special enjoyments can be made a means of reward.

Hearing is a sense, developed at an early period, as all who have observed the effect of music on young children are aware. Through this sense they may be pleasurably and powerfully affected; but the frequent resort of mothers and nurses to its soothing influence, prevents, in some measure, its use as an occasional reward. Whenever it can be employed, however, it should not be omitted; and as the indulgence of this desire does not contribute to debauch the mind, but to soften and elevate it, the reward may be given, as long as discipline is required, or the child continues to regard it as a favor.

The sense of feeling includes the sensibility of the skin to heat and cold and fresh air, that of the lungs for the last, and also, a want or desire seated in the muscles, for active exercise. These desires are all gratified, by excursions in the open air; and, while confinement is a corporeal punishment, going abroad for play, is, to children who are not permitted to run at large habitually, a real and most admirable reward. Its use, in no manner or degree, contributes to impair the intellect, pervert the moral sentiments, or excite the animal propensities; but to elevate the two former and promote health and symmetry of body, with buoyancy of animal spirits.

The last of the senses to which I refer, is that of sight. At a very early period, infants, as all mothers know, are attracted by light. The young child, as instinctively and steadily turns its eye to the candle at night, as the plant in a dark cellar directs its branches towards an opening in the wall. As it grows, the desire for this gratification also increases, and, finally, exceeds in energy, that of smell, touch, and hearing. Hence, the confinement of a child in a dark room, even where it is not afraid, is a bodily punishment; while the gratification of its vision with masses of light and shade, and variety and brilliancy of colours, may be made a most cherished reward. Vision has, with much propriety, been called the *intellectual* sense, for, of the whole, its indulgence approaches nearest to the indulgences of the mind. It involves nothing sensual, in the bad acceptation of the word, and may, therefore, be employed as a reward, till they shall cease to be necessary, whatever may be the age of the child.

In resorting to the pleasures of sense, as a reward, we may press several, or the whole of them, into our service at the same time; and, when skilfully used, their united influences are of the happiest kind. Children are great lovers of nature. A flower, a little bird, a branch of missleto with its pearl colored berries in winter, a babbling brook, which they can dam up in an hour, a fall of snow which lodges on the limbs of the shade tree in front of the door, or half buries up the grass in the yard, a butterfly, or a lightning-bug, the taste of a new fruit, the smell of a new flower, a whiter pebble stone, or a more retired play-ground surrounded by fresher natural objects, acts pleasantly on their senses, and may be made an indulgence and a reward. But when the sensible and benevolent parent, or teacher, combines a visit among the various objects of the natural world, as the reward he would bestow for obedience, or great effort at labour or study, he presents the highest sensual gratification, which God has placed at his disposal.

Diligence and propriety have characterized the deportment of the children or pupils, and he who has the care of them announces as the reward of those virtues, a ramble of all who have thus carried themselves, he being the leader and mentor, but not the *master* of the little company. What joy instantly beams from every countenance! and how strikingly must each contrast *his* happy lot with that of the *offender* who is left behind in confinement! how directly must he associate the reward with the observance of duty which procured it! What bustle of preparation then ensues, what contempt of bad weather, and bad roads, what feelings of young enterprise and impatience to be gone, start up in every palpitating heart! Spring is unfolding her beauties —the air is genial—the light is now and then interrupted by a passing cloud, raised high in the heavens, and threatening no shower to damp their ardor— the meadow lark, perched on the crag of a decaying stump, and the cat-bird in the thicket, raise their notes, and the urchins hasten to the spot and put the songsters to flight—the squirrel is then *treed,* and lies flat and quiet on the limb, while club after club, passes harmless by; one boy, more aspiring than the rest, attempts to climb the trunk, becomes dizzy, and slides sheepishly down over its rough bark, ashamed to catch the eye of her, whose admiration he sought to win, and half provoked at the shouts of merriment which his failure called forth, to die away the next moment, when some straggler announces a new violet, raising its timid head through the faded leaves of the preceding autumn! Then the steep hill, and the race of boys and girls to its top, the descent to the new and shaded hollow beyond, the jumping of the little brook, with the young gallantries it brings forth; the lying down to drink, by some thirsty boy, and another, filled with mischief, pushing his face into the water from behind; the discovery of a petrifaction and the gathering together, to wonder at its form, and struggle for its possession! Now, the admiration of the half expanded buds, and a transient comparison of those of different bushes! Then, the union of all the boys, under some leader, designated as it were by instinct, to roll over the rotten log, and the discovery of a harmless little snake; the instinctive impulse to kill, the haste and uproar of the execution, and the terror of the girls, who, afterwards, see a snake in every stick they are about to tread upon! The continuance of the ramble, till it reaches the dogwood, the red-bud, and the buckeye, with their blooming limbs, the climbing, the breaking, the throwing down, and the scrambling below, till all are loaded to their hearts content, and by some new route they return home, fatigued and hungry, to tell of great discoveries, and boast of great deeds. And where has been the parent or teacher throughout this scene of pleasure? If at the post of duty, in the midst of every pastime, and attentive to every opportunity of doing good; explaining each object, pointing out every relation, disclosing the properties and qualities of each attractive plant, separating the different parts of its flower, and teaching their names and connexions, lecturing on the woods, commenting on the thunderbolt which destroyed the ash, but passed instinctive and harmless over the beech tree, by its side; calling attention to the backwardness of vegetation on the north side of the hill compared with the south, and teaching that it is the effect of differences in heat; thus inspiring a love of knowledge in the young mind, when excited by the pleasures of the body, disclosing to it some of the most beautiful laws of nature, and directing the young heart up to her great and benevolent Author.

Such are the fruits of an excursion made in such manner as to gratify the senses of childhood, and none can fail to see in them, a reward that may be pressed into the service of school and family government with the happiest immediate results, and the most admirable effects upon the future character of the objects of our affection.

We come now by a natural and easy transition, to rewards and punishments which belong primarily to the mind. These connect themselves with the desires and motives of the soul, as those we have just travelled through, are connected with the appetites and sensibilities of the body. To view them accurately, we need not change our ground, but merely extend our vision a little deeper into the constitution of man. We have already seen, that he is a compound of body and soul—of flesh and spirit, and that each half, has its peculiar appetences and wants. It is the *improper indulgence* of these, that leads to transgression, and it is by acting on these, that he is both rewarded and punished. We have disposed of what relates to the body, let us now ascend to the sentiments and propensities of the mind, considering them as nearly as practicable in the order of their development with the growth of the child.

The first affection developed, is the love of mother; to which succeeds, in due time, that for the father, and at length, (the conduct and character of both parents being alike) the affection for both seems, in general to be equal. Now, at the earliest dawn of intellect, the child may be rewarded and punished through this affection. When the mother frowns upon it or turns away her face, the sun of its happiness is dimmed—it is distressed and punished, through the medium of its filial affection. On the other hand, when the soft music of her voice falls upon its ear, and her countenance beams with love and praise, it rejoices, as the chilled and tender lily of spring expands, when the clouds are chased away, and the fountains of light and heat are opened afresh.

Here then is the first, and, let me add, the greatest of the means of moral government, which God has given us; and no mother honors the name, or deserves to be blessed with children, who neglects its use. Early and skilfully exercised, it fixes over the child a dominion, that, like the permanent colors which the light of the sun stamps upon the opening rose, must be felt, till the individual is gathered with that mother in the grave. To *maintain* this influence, the parents, however, must attend to all that is necessary. They should view the child as having a rational soul, capable, as it grows in years, of observing and reasoning, and having other desires and wants, than those which, through infancy, make it cleave to its mother's bosom as the source of all its enjoyments, and its place of refuge in every danger. They should know, that to preserve an influence founded on filial affection, they must, as the child increases in age and knowledge, keep themselves in its respect and veneration. To do this, they should administer the reward of their approbation, and inflict the punishment of their displeasure on such occasions only, as demand them, and apportion them to the acts that are to be rewarded or punished. They must, in the very midst of their chastisements, convince the child of their affection, and that they are but discharging a duty of love. They should again and again recite the law of duty it has violated, and instruct it anew as far as practicable, on the reasons for the law; thus making it conscious that the punishment was merited, and will, finally, be for its own happiness. In this way, they will associate mental instruction with mental pain, and, at the same time, appear as benefactors instead of tyrants. They will excite repentance, which never comes from punishment unaccompanied with the conviction of error, and instead of anger inspire a sentiment of reverence, when the parental government is placed on a foundation that cannot be shaken.

To accomplish this great object, however, it is indispensable that parents should look to their own conduct. In their lives they must evince, that they are governed by moral laws, which are but a stretching out to greater objects and duties, of the laws they lay down for the government of the child. They should come into the family tribunal with clean hands, and engrave on the rod of correction, *"Let him that is without sin, cast the first stone."*

How is it possible that parents who give themselves up to passion and caprice, to deception and petty falsehoods, to instability of principle and fickleness of pursuit, to backbitings, to gluttony and drunkenness, to profanity, grossness and impiety, can by any rewards or punishments, make themselves objects of veneration, or acquire over their offspring a moral power? To do this, they must practise what they enjoin, show obedience to the laws of society and God, and present themselves as examples of whatever purity human nature can acquire.

If I dwell on this subject, it is because it must be regarded as the root of all moral government, and viewing it thus, it is proper to say still more, addressed especially to mothers. By the plan of creation, and the providence of God, it is the peculiar duty of the mother, to watch over her child for many of the first years of its life; and on her more than the father rests the parental responsibility.

It has been said, that most great men, have had talented mothers. How much of their superiority might have been a birth-right, we need not stop to inquire, but there is little doubt, that much of it, as far as the mother was concerned, arose from her instruction and discipline—training the faculties and affections by times, insisting on their supremacy over the appetites, and directing, even the tottering footsteps of INFANCY into paths that finally, led up to the temple of fame; a height that is never reached, by those who loiter on the way to eat and drink beyond the comforts of nature, or join in wild revelries, or prosecute schemes of vanity, avarice, or revenge.

Much has been said and written, on the influence of woman. This influence depends on two of our affections, conjugal and maternal love. But all the power she can exert on the *man,* sinks into insignificance, compared with that upon the *child.*—It is in shaping the character of the child, that her influence on society and its destinies, is distinctly perceptible. If she neglect to exert this power, or exert it in favor of wrong objects, no labors of the teacher or the moralist, can correct the bad effects of her errors. She *may* carry with her a mighty power on the earth, but must rely chiefly on those means which act on her offspring. Using these with talent and skill she will, indeed, direct, if she does not govern the world. But how few mothers, of all whom I now have the honor to address, can lay their hands on their hearts, those hearts which burn perhaps with the purest flame of affection, and say that they are conscious of having discharged their duty in this respect! How many are negligent and irresolute! How many overlook offences which do not happen to annoy themselves! How many from their necessary engagements, or from indolence, omit to find out, with certainty, that the crime was not committed by another! How many reward, when they should punish,—thus bribing the child to do its duty, so far as to save themselves the pain of inflicting salutary correction! How many sink themselves in the respect of their children, by appealing on all occasions to the father, and suffer themselves to be trampled upon, till he shall return to interpose! In this way mothers lay up for themselves *"wrath against the day of wrath."* The father at length dies—the governor is gone, and the rod of correction is buried with him in the grave! For a time the sorrow of the family may keep the house in order, but the elements of disobedience, discord and vice, are only smothered; the devouring flames at length burst forth, and the happiness and dignity of the household, are consumed like the withered grass of our fields. In the midst of this beginning desolation, she may have great amiability of heart and undying love, but the hearts around her do not respond, to her affections, and let loose from all salutary restraints, indulge themselves in every evil propensity, regardless of duty, and cold to the sufferings they raise, in the bosom which cherished them in the hours of their infancy. She exhorts in vain, and, for the first time, undertakes reproof and correction; but her hand is inexperienced and powerless; they do not fear and reverence her; they absent themselves, for scenes of idleness and vice; they come home altered in conduct and character, till they begin to seem to her like the children of strangers; they grieve her spirit by day, and fill her nights with dreams of anguish and terror; they eat out her substance, her spirits droop, she resigns herself to despair, her health consumes away, and like our beautiful locust, when the worm eats to its heart, she sinks into an untimely grave,— from the verge of which she looks back on the floating wreck of her once innocent and playful family, and then turns her eyes for ever to her husband and her God.

The next propensity in children of which I shall speak, is the love of ornament. This is a universal principle, for we find it as deeply infixed in the children of the Indian, as in our own. It is stronger in female than male children, because they are designed to be more ornamented. The indulgence of this taste is a high gratification in early life, and withholding its objects of desire, is of course a punishment. Much, then, may be done, at a small expense, to reward, and much may be omitted to punish on this principle. The objection to it is, that the natural love of ornament is increased, and to this due regard should be had; but, on the other hand, it cultivates the taste of the child, especially the daughter, and prepares her for appearing in society, in a better style of personal appearance than might otherwise be attainable, an object which deserves attention. I believe, that not a little may be effected, both of reward and punishment, through this principle, without vitiating the character, but "let every one be fully persuaded in his own mind."

Love of play has been already mentioned, in reference to its bodily effects; but it deserves a place among the moral influences; for when children play, they exercise their minds, call into action their ingenuity, give activity to their enterprise, and set various feelings into operation. To this gratification there can be no possible objection, but that founded on its too frequent recurrence; and as it promotes health of body, it may with great propriety be granted as a reward and denied occasionally as a punishment.

Love of property is an inherent and powerful passion. In childhood it is feeble, but increases with years, as other desires fade away, and in age too often swallows up every nobler propensity, leading the individual to horde up, and give nothing out but what is extorted; as an old pond in the field, swallows up all the muddy waters that flow towards it, and gives back only to the power of the sun and winds, which carry off its surface. Children desire merely that kind of property which they can use, for their object is not prospective but immediate gratification. Within this limit, however, the desire is importunate; and hence we may act strongly upon them, by giving or withholding such toys and playthings, as are adapted to the taste of different ages. In selecting these, a judicious parent or teacher, will constantly prefer those which improve the taste and enlarge the knowledge of the child; for in this way much useful information may be conveyed, on the mechanism and movements of the works of art; or some of the first rudiments of natural history inculcated, by choosing the productions of nature. The first lessons of economy may, also, be given, for the child will listen attentively, to the injunction, not to destroy that which it prizes as a reward, or values as an acquisition.

Curiosity and wonder are strong passions in childhood, and may be turned to good account in our systems of discipline as well as instruction. All activity and acuteness of observation, depend on our curiosity to see new objects, and find out new properties and relations; and upon our natural capability of feeling the emotion of wonder or admiration, at what is novel, or intricate, or beautiful or sublime, either in nature or art. The indulgence of these desires is not only another means of reward, but an actual duty towards the child, as contributing to the growth of its intellect; and the denial, is a punishment, which may be occasionally administered, with the effect of increasing these laudable desires, by refusing to it for a time the means of their gratification; as the appetite is whetted by withholding food. They are, indeed, designed to procure aliment for the mind; and may be played upon without any possible injury, either physical or moral.

The love of knowledge generally, is but an extension of the principle just considered. I speak now of every kind of learning and all the branches of science, which man has need of knowing. It *seldom* happens, that we meet with a child or youth, in whom it is necessary to *moderate* this desire, or who might be injured by offering new and special facilities for study, as a reward. Such, however, there are, and the destruction of health or intellect, is occasionally the touching result of too much indulgence of this desire. Parents and teachers should be on their guard in respect to such uncommon pupils, and moderate them in their application, so as to ward off its future consequences. These are but exceptions to our rules, which should always be adapted to the character of the many. I would say, then, that in *them,* the natural love of sound learning and useful knowledge, is adapted to the wants and duties of man in a state of nature, rather than civilization; and, that care and address, are necessary to raise it to the proper degree. *Hic labor hoc opus est;* but when the work *is* accomplished, the teacher has little left to do, for as the steam-boat when in rapid motion is easily directed, so the pupil that is bent on study, is governed with facility, and, indeed, seldom falls into transgression. Moreover, the chief object of all rewards and punishments in our schools and colleges, is to exact a compliance with those laws which require regular and accurate recitations; and he who, from love of knowledge, complies with this part of the system, *can* violate but few other rules of our institutions. The love of knowledge is not a desire, which we can press into our catalogue of principles to which we address our rewards and punishments, but goes very far to render them unnecessary; and may be placed high in the list of the preventive means of offences. It is then a great auxiliary to the teacher, but how is it to be inspired? As it is a duty to study, all the means enumerated as far as they can be used, may be employed, in turn to reward and punish him who is idle; but still the assigned lessons *may* be studied through *fear of punishment* and not *con amore;* and when the pupil leaves the institution, he may loathe the acquisition of further knowledge, even the more for having been punished into what he has acquired. Nevertheless, that which, to speak figuratively, has been whipped into the mind, is not without its use, as it has often happened, that he who at first studied only from fear, comes at length to study from love. Severity of punishment in these cases should, however, be the *ultima ratio præceptoris,* and always connected with other means, calculated to awaken the dormant passion. The plan of this discourse does not carry us into the consideration of this subject, and I should be little qualified to illustrate it before a body of enlightened practical teachers; but I will throw out a few hints, although foreign in some degree to our immediate object. But are they in fact foreign? Will not the scholar study it if he derives pleasure from it. He undoubtedly will, and this pleasure will reward him, and incite him to renewed efforts. Let the teacher then secure to him this pleasure, and it will generate the love of knowledge. But how, in many minds, can this be done? In some it *cannot* be done, for all intellects are not equal; and some were never designed to comprehend the properties and relations of things. But omitting a reference to these, I would say, First. That the philosophical maxim—*pass from the known to the unknown,* should be observed, and that its violation has prevented many a scholar from acquiring a love of study; because he was put, carelessly or unskilfully, on such plans as rendered the acquisition of knowledge difficult or impossible. Secondly. Different minds are differently constituted, as to the balance among their faculties and tastes. One will have a strong talent for languages; another for collecting and treasuring up historical facts; another for the relations among natural bodies, and another for the idealities of the imagination. But our plans of school classification do not recognize this important fact; and it must happen, that many are repulsed from study, and go through school or go from it, without acquiring a love of knowledge, simply by the influence of some branch, for which they had no capacity; who, if they had been tried separately, or in succession, on all the branches, might at length, have met with one, which was to their taste because adapted to their mental capability, and making progress in this, they would have passed by an easy transition to others, and finally acquired a love for the whole. Third. Something, I think may be done, by substituting the didactic conversation of the teacher, for the authors that are usually provided; as many things are rendered clear and attractive in coloquial intercourse, that seem obscure and incomprehensible in the formality of the books. Fourth. It may be possible to arouse the dormant attention, by showing the useful applications of knowledge of various kinds, in visits to works of art, where that knowledge manifests its utility and power. Fifth. Going into the great domain of nature, where *every* young heart palpitates more actively, and directing the attention of the pupil first, to curious or beautiful productions, as mere objects of sense; and then calling his awakened attention, to their structure, properties and relations, so far as to excite his curiosity, and put his faculties of knowledge into action; and, finally, referring him for a full account to the books, which he may then be induced to read.

By means like these, I have seen a love of knowledge aroused in the minds of students of medicine; and, therefore, speak from some experience, while I say that which seems to me to be in accordance with the laws of the human mind.

We come now to other principles of action, and I ask your attention to self-esteem, the foundation of pride. This sentiment exists in very different degrees in different children, and, consequently, the control which may be exercised by its instrumentality is various. We generally entertain and express, but a poor opinion of the child which has no pride, but on the other hand, we consider its inordinate manifestation a crime. Its offensiveness to man, and criminality to God, depend entirely on its intensity, and the objects on which it sustains itself. To feel proud of the character of a father or mother, one's friends, or a good reputation is noble, and he who does not, we look upon as degraded; but we despise him who is proud of dress, of wealth, of personal appearance, of slender attainments, or of his own opinions. Self-esteem may, with propriety, under proper restrictions, be pressed into the service of family and academical government. The child should be taught to esteem himself in proportion as he discharges his various duties. When he has done well, the gratification of self love may be extended to him in moderation, by an acknowledgment of the fact, and when he trespasses at home, in the primary school, or at the university, he may, if possible, be mortified, in his own estimation, as a punishment.

Nearly connected with self-esteem in many of its external manifestations, but distinct in much of its internal constitution, is the love of approbation, at once the fountain of vanity, ambition, and emulation. All the world condemn the paltry and ridiculous displays of this sentiment, which they have branded with the epithet of vanity; but a different, though not unanimous estimate, is made of those higher manifestations, which have received the names ambition and emulation.

Ambition, in the ordinary acceptation of the term, involves a thirst for power and conquest over others, and in this view must be condemned; but when confined to objects of public utility, and studies that lead to knowledge and wisdom, although for the *purpose* of acquiring distinction, it assumes a different character. The principle which sustains it is too deeply rooted in the human mind to be *eradicated* by *any* discipline; and, we should rather seek to direct it on proper objects, and limit its invasion of the rights of others, by hemming it in with the principles of justice, benevolence, and piety, than to aim at its abolition; confining ourselves to that love of relative distinction which is known under the name of emulation, let us inquire whether it should be employed in families, schools, and universities, as a means of reward and punishment, supposing it always to be directed upon admissible objects.

That the love of relative distinction, connected with the approbation and applause of those we respect, may be made a powerful means of restraining children from bad conduct, and animating them to good, including study, is generally acknowledged; but several evil consequences are thought to flow from the operation of this principle.

First.—It is said to stimulate some minds to excess. This is true, but we must oppose to these cases, the far greater number, in which it excites the sluggard, the one who has no innate love of knowledge, and him who is prone to vicious habits, by rendering them unwilling to forfeit the good opinion of those whom they are taught to reverence, and mortifying them by being placed below their fellows.

Second.—It is charged with generating unkind feelings among brothers and sisters and class-mates, which often amount to envy and strife, and sometimes involve both parents and teachers in the charge of partiality and injustice. That all these, and other bad consequences may, and do, in fact, very often come from it, must be admitted; but most of them are feelings that soon die away, the estimate of things made by the unsuccessful in the hour of disappointment being reversed, perhaps, the next day; and the friendships that were severed for a moment, in most cases, becoming speedily restored.

Third.—It is said to be the substitution of an inferior motive of action—the applause of man, for the approbation of God. We act incessantly, however, from motives inferior to that of direct duty to God, and such is the economy of nature and of providence. According to the Bible, all men, except christians, act from motives inferior to that, in every thing. Till religion takes possession of the soul, and transforms our principles, we neither practise virtue and refrain from vice, nor acquire knowledge, nor prosecute any object whatever, because God has commanded us; but because he has implanted in us desires both of mind and body, and connected with their exercise, the sensations of pleasure and pain; the former to animate us to action, according to the physical and moral laws that govern our systems; and the latter, also, to incite us to action, in certain cases, as when we take food to relieve the pain of hunger, or make great exertion to rescue a suffering child from danger, and thereby relieve ourselves from the pain of agitated parental love. Thus, we may, indeed we must, (before the influence of religion changes our motives,) act from other considerations than immediate obedience to God; and all that He requires, is, that our actions should in sincerity be the offspring of our natural desires, and such as his revealed will does not forbid.

But does not the christian as well as all others, necessarily act from motives, that have their foundation, in his natural desires; for if *they* were extinguished, what would prompt him to action? If, for example, the desire for fresh air were destroyed, no one would always breathe, because it was commanded of God. The moment his attention was directed on another object, he would, of course, forget that he was commanded to carry on that function. If the love of offspring were abolished, what would recall the minds of parents, from other pursuits, to the duty of looking after their infant children? And if the desire for knowledge and for property were expunged, who would recollect to leave off the search after the former, to acquire the necessary amount of the latter for present and future support; or think, when his mind was imbued with schemes of business, of the command to cultivate knowledge? We *must,* then, act from the subordinate motives established by God for impelling us on, as certainly as the particles of inanimate matter in the physical world, must move at the bidding of attraction and repulsion, according to laws which, like those of human nature, are a part of the system that governs the universe.

Revelation, in fact, was not given to instruct us in our duties to each other; but to enforce their observance, by presenting a system of rewards and punishments beyond the present life. It does not abolish our inherent desires, but teaches us to curb the unruly, repress the inordinate, and preserve such a balance among the whole, as that none shall gratify themselves at the expense of the rest; or of the rights and happiness of others. He who does this, *because* God has commanded it, lives in *duty to God,* though every action of his life may be the immediate offspring of the fundamental principles of his mind and body. The principle of emulation, then, is subordinate to the principle of duty to God and not at variance with it, except when improperly directed or excessively exercised.

Under this view, I regard the workings of emulation as not *necessarily* immoral, and, in reference to its influence in schools, the inquiry should be, how to obtain its valuable exciting influences without its disadvantages. This must be left to judicious practical teachers, who should always recollect, that of all the motives to action, emulation stands least in need of being stimulated; that in many minds it requires to be moderated; and that it should be kept under the supremacy of the nobler motives of benevolence, conscientiousness, and veneration. It is undeniable, that it has not always been thus regulated; and that its abuses have brought it into discredit. Teachers have found it a principle easily acted upon; and through indolence or an ignorance of consequences, or indifference to the moral character of their pupils, have too often made it the sole means of animating them to study and regular conduct, instead of restraining it within limits not incompatible with other principles.

Benevolence, or an interest in the welfare of others, is an innate sentiment, against which, as a means of discipline, moral and intellectual, there can be no possible objection, but its influence is rather preventive than corrective. The cultivation of the benevolent feelings of children, modifies and controls the operation of their lower passions and propensities, purifies their desires, and, on the whole, predisposes them to other acts of duty, than those of beneficence. This cultivation should, therefore, be carefully made, both by parents and teachers, and their labor will be bountifully rewarded, by a diminution in the number of their transgressions. One mode of training this sentiment, and pressing it into the cause of education, is to direct the attention of your children to objects of charity, whenever we reward them with money for obedience. We are thus enabled to incite them to study, or good conduct, without administering to a sordid love of property, and at the same time augment their benevolence, by affording them the means of purchasing the laudable pleasure, which comes from its practice.

The last principle of action to which I shall direct your inquiries, is veneration for God. This, like the others, is innate, and the highest of all the moral sentiments. I have already spoken of its influence, when the parent is its object.

Veneration, in its perfect degree, involves gratitude, love, and respect; but the two former are not indispensable, for we often cherish the latter alone. Indeed, respect is but a lower degree of veneration, and this is what we feel, for a great and good name of antiquity, or for an ancient and beneficial custom. Reverence is the same feeling, cherished for things that are divine, or for persons who seem to stand as representatives of the divinity, such as pious and aged parents, or exemplary and hoary-headed teachers, or ministers of the gospel.

The veneration or reverence of children for their parents, and preceptors, should comprehend love and gratitude with respect, and be ennobled with a looking up to God, as the fountain of whatever is lovely and reverential in them. Thus formed and directed, this sentiment gives to the parent and teacher, a control over the will and actions of the child, beyond every other. Of the means of forming it, nothing need be added to what was said, in speaking of the relations of parent and child. When this feeling exists, the fear of incurring the displeasure of the parent or preceptor, is constantly present, and constitutes a powerful means of prevention; while it keeps down anger and resentment under correction, if that should be necessary. The setting up of the authority of this sentiment of adoration to God and reverence for the parent, in the heart of the child, is the great *desideratum* in discipline, from the cradle to the theatre of life—from the primary school, to the university. It is an ægis of brass against immorality, and the palladium of liberty, in every land where freedom is sustained by a constitutional government. The power of this principle, in a national point of view, is disclosed, by the hesitation with which the subjects of a throne, held venerable by tradition and early impressions, come up to its overthrow; although it may have sent forth none but the edicts of despotism. The heroes of the revolution, and the authors of our federal constitution and the union it establishes, should be held up to our children, as patriots whom they ought to reverence—the works themselves as political institutions which deserve the deepest veneration. This should be a part of their education, at home, in society, in the primary school, the academy, and the university; for a great object of education in this country, is, to make good citizens, and devoted friends of the liberty we now enjoy. The spread of this feeling of reverence throughout the whole republic, would in no degree interfere with all necessary amendments to the constitution, but rather contribute to promote them, while it would afford the greatest of all possible guarantees against its abolition, by combinations of wicked men, in whom the sentiment of reverence for what is good, never finds a place.

I am sorry to say, that in the United States, especially in the valley of the Mississippi, the sentiment of veneration is not as carefully cherished in our children, as it is in some other countries, where its power is pressed into the service of tyranny; while here there is nothing which it could operate to sustain, that ought to be destroyed. The neglect arises, perhaps, from the very nature of our free institutions, which give to all, even in youth, a very great amount of liberty of speech and action; but we should take care that the altars of liberty are not profaned and demolished, by a licentiousness of feeling, the offspring of that very freedom. Children who are taught to venerate their parents and teachers; the fathers of the land who have labored for its prosperity; our aged and virtuous matrons; our benevolent, literary, and religious institutions, and those who conduct them on correct principles—finally, Heaven itself, for which they all labor, become a law unto themselves, and conform, in manhood, to what they had venerated in youth.

Reverence for God, as a first and great unseen, governing power, is a universal principle of human nature, which in different ages and nations, has made itself manifest in various ways, according to the lights of the understanding. Thus among the ancients, while the Egyptians bowed down, in blind and stupid adoration, to the filthiest reptiles, the Greeks paid homage to the creations of a bright but licentious imagination; and in one of the kingdoms of modern Europe, when delivered over to a civil war and drenched with innocent blood, though philosophy raised her voice above the din of anarchy, and proclaimed *there is no God,* the people erected altars to the worship of nature! The sentiment of devotion may be sunk, obscured and perverted, but cannot be abolished. Among ignorant and savage tribes, it is merely a passion of terror, and in this debased condition we observe it, in such of our own countrymen, as have, from their ignorance, vice and superstition, but few claims to the character of civilized men, beyond that of being blended with them. But they who are instructed in the Bible, view the Creator as the author of rewards as well as punishments, and *love* him with *gratitude* while they *fear* him in *humility*. They know his attributes and decrees, and humble themselves before him as a being of infinite wisdom and goodness—worthy of all veneration—whose revealed will commands every moral duty—whose law is a law of universal kindness—who enjoins justice and generosity—and whose all seeing and sleepless eye, watches over every object, from the sun glowing in the purple east, to the little child, that sports in his morning beams.

When this fear of God is once established in the child, it becomes docile and dutiful, not prone to vice, easy to be admonished, and given to repentance under correction. On this fear depends the influence of the morality of the Bible. We cannot dispense with this morality, but it would be powerless if separated from the theology of the Bible. Should the latter be despised and rejected by parents and teachers, the former would follow its fate in the estimation of the child. And this for the plainest of all reasons, that the morality is every where presented as the command of God—an expression of his will—a law enacted by himself and promulgated on the earth for his own pleasure. If then the child should reject the author, according to the established laws of the human mind it will neglect his decrees. Let every teacher ponder deeply on this matter. He would not hope to see the rules of his institution obeyed, after he had fallen into contempt with his pupils, and why should he expect to see them obey the moral law if they do not reverence its giver? Such logic would afford but a barren sign of talent, and he who might display it, should be advised to adopt some other profession. He is not, either in head or heart, intended for the instruction of youth in any country; much less, in our own, where christianity is, in fact, the sustaining principle of all our valuable institutions.

Although I have detained you long with a survey of the principles upon which our discipline of children should rest, I cannot close without recapitulating a few points, which must be thoroughly understood and conscientiously practised, or no system of rewards and punishments can be successful.

First. Children, like grown persons, act from motives: and when they transgress they have an object in view, which at the moment is dear to them. They should then be carefully and patiently instructed in their duties, and have the reasons for the laws, by which you govern them, as fully explained as possible.

Second. As there is among them a great variety in bodily and mental temperament, the characters of each should be studied, and the appropriate means of rewarding and punishing, selected accordingly.

Third. Children as well as adults, have their periods of undefinable indisposition, and consequent irritability of the nervous system and feelings, when of course they are froward, peevish and disobedient. Those who govern them should look into this matter; and in meting out their punishments, have respect to its influence, or, while the disease, not known perhaps by the child, shall continue, omit them altogether.

Fourth. The excitation of fear is a legitimate means of correction, for all correction operates indeed by exciting it, but children should not be frightened by goblins, or threatenings connected with supernatural appearances, for an association of ideas may make them superstitious and timid throughout life.

Fifth. Both rewards and punishments should be proportioned to offences. They should be dealt out with all the impartiality a man requires from a court of justice. Those which are promised and deferred should never be forgotten, and those which are inflicted as soon as the offence is committed, should not be greater, than if the parent or teacher had no excitement of feeling. It is best to punish and reward upon the spot, that both may become associated with the offence in the memory of the child; but he who cannot apportion them in the right degree, while his passions are up; should wait for them to become tranquil. His manifestation of anger is not objectionable, for children have the laws which are to govern them, so much identified with the will of the governor, as to think it a matter of course, that he should feel indignant or angry; and if punished, when he is in that state of feeling, they are less likely to be resentful or to regard him as cruel, than if it be done in his cooler moments.

Sixth. It has been said of rewards and punishments, that they do not change or purify our motives, but leave the desire to do wrong uncorrected, while they deter us from the act. The Bible says, however, *"train up a child in the way he should go, and when he is old he will not depart from it;"* and who has not seen and felt, that if we habitually make our actions right, our motives will gradually improve. It is of great importance, then to compel children into regular conduct; for if their bad desires are not gratified, they are starved out and at length cease to grow, while the good motives from being exercised on their proper objects are established in power; in which respect the mind and body are under the same laws of habit.

We have thus traced the outline of a system of discipline for children and young persons, embracing, both rewards and punishments, and founded equally on the constitution of the body and mind. We, affirm of nothing set forth, that it is absolutely the best which could be suggested, and claim nothing as original. Principles have been embodied which are afloat in society, for the purpose of presenting them in order, to those who are competent judges. In doing this, no book has been consulted but the Bible; and that for the purpose of discovering how far its wisdom is in accordance with the opinions of philosophy, when directed to the study of man in his physical and moral constitution, and on all points we have found them in perfect harmony. Throughout the inquiry, we have plead the cause of both parent and child; but above all, that of the conscientious and benevolent teacher, who can do nothing without the previous labors and continued aid of the *natural* master.

We have catered for home consumption—for our own adopted and native West—for a western college and a western audience—for a new people who must devise their own plans of education—establish their own systems of discipline—and teach their own children, like their elder brethren of the East; from which the West is in fact but a scion, transplanted, and struggling for air and light, in the depths of the wilderness. Its tender leaves are as yet scarce unfolded; but their form bespeaks, the sturdy and giant oak, that shall live on through a thousand years unless blasted with the lightnings of an angry Heaven.

The West will not go backward in numbers—no, not till her great river shall turn from the sea, and seek its icy cataracts, among our distant hills. Forward will be her march—and day after day must add to her *physical* strength;—but she should not rejoice in *this* power, and become the Mammoth of the Union, or the bones of her prosperity will, at last, lie unburied in the valleys, and mingle with those of her lost archetype.

Let all then who love its name—who, beholding it in the dim and distant future, can now take delight in the strength and beauty, which should mark its perfect growth, or mourn,—while the day is yet afar off—at the vice and anarchy which may overwhelm it, as the angry snows of the mountain, dissolve and swell with troubled waters, the peaceful Ohio, till they deluge our pleasant places and rush in desolation along our streets: Let all who feel proud that the voice of its infancy has called the enterprising stranger from lands beyond the sea—from the isles of Britain—from the banks of the Danube and the valleys of the Alps—from the frozen coasts of the Baltic and the classic shores of the Mediterranean—from the olive and the vine,—to build his cabin beneath our embowering sycamores: Let all who would rejoice to see it, not only the asylum of the exile, from the uttermost parts of an oppressed world, but the chosen and permanent abiding place of knowledge, religion and liberty,—stand forth, while it is yet in the morning of its days, and will bow its head to the rod of discipline, to lend a helping hand, in training its young footsteps, and giving them an impulse on the paths of loveliness and peace.

Rail-road from the Banks of the Ohio River to the Tide Waters of the Carolinas and Georgia*

(1835-1836)

On August 10, 1835, Drake attended a public meeting in Cincinnati to promote the construction of a railroad from Newport or Covington, across the river from Cincinnati, to Paris, Kentucky, northeast of Lexington. At the conclusion of the meeting Drake proposed that a committee be appointed to investigate the "practicality and advantages of an extension of the proposed Rail-Road, from Paris into the State of South Carolina." Drake was appointed to chair such a committee and made his report at a second meeting held on August 15. At this time Drake proposed the establishment of a permanent Committee of Inquiry and Correspondence to pursue the matter further. This committee met on August 21 and designated Drake and Edward D. Mansfield to prepare a public letter on the subject of the proposed railroad. Drake's "Address to the People of South Carolina, Georgia, East Florida, North Carolina, East Tennessee, Kentucky, Western Virginia and Pennsylvania, Ohio, Michigan, Indiana, Illinois, and Missouri" was reported back on August 26, 1835.[1]

During the following spring Drake was accused of presenting as his own an idea which had long been in the minds of others. His reply was characteristically imperious, yet humble: "I deny having received it from anyone, or from any source, but that observation and study of the physical geography of our country, which may have suggested it to others, as early perhaps as to myself."[2]

* *Rail-road from the Banks of the Ohio River to the Tide Waters of the Carolinas and Georgia.* Cincinnati. Printed by James and Gazlay. 1835. Pp. 4–12, 27–30.
[1] *Rail-road from the Banks of the Ohio River* (1835), pp. 3, 13, 27.
[2] *Rail Road to Carolina from the Ohio River,* n.p., n.d. [Cincinnati, 1836], p. 5.

THE STATES which border on the Ohio, or are watered by its great tributary streams, are western or tramontane Pennsylvania and Virginia, Ohio, Indiana, Illinois, Kentucky, and Tennessee; nearly through the centre of which that river flows, almost parallel with the sea coast of the old southern states. From the seven states above mentioned, there are highways of communication with the ocean in but two directions—north-east, and south-west. The former, consisting of several distinct lines of river, canal, macadamized and railroad communication, reaches the Atlantic ocean between the west end of Long-Island Sound and the mouth of the Chesapeake Bay—from New York to Norfolk—a distance, on a straight line, of 300 miles: The latter communicates with the Gulf of Mexico by the delta of the Mississippi. Between these two points of marine connection with the interior, is a coast nearly 3000 miles in extent, constituting the sea-board of southern Virginia, North and South Carolina, Georgia, Florida, Alabama, and Mississippi, with which the states in the Valley of the Ohio have no direct communication, even by means of a good post-road, so that the mail to the northern frontier of Georgia and the Carolinas, not three hundred miles distant from the banks of the Ohio, in a straight line, is actually sent by Washington City, on a route nearly four times as long. With that part of the southern coast which lies west of the peninsula of Florida, the Ohio states have ready intercourse, by the Mississippi river; but with the region east of that peninsula, they are destitute of all adequate means of commercial and social connection. Here then is a great *desideratum,* which can be supplied in no other manner than by the contemplated RAIL-ROAD.

Starting, perhaps from more than one point on the Ohio river, in the state of Kentucky, this road should stretch nearly south; and branching, when it enters the Carolinas and Georgia, to reach their tide-waters at several different places. Taking Cincinnati as a city intermediate between Maysville and Louisville, and Charleston as intermediate between Wilmington, in North Carolina, and Augusta, in Georgia, the road might be said, more especially, to connect Cincinnati and Charleston, and may for convenience in this report, take its length and designation from those two cities. Starting from the former, or rather, from the opposite bank of the Ohio river, in Newport or Covington, it would traverse the state of Kentucky to the Cumberland Gap, near the south-western angle of the state of Virginia, then cross the state of Tennessee, and, ascending the valley of the French Broad, in North Carolina, arrive at

Greenville, or some other point, in South Carolina, beyond the Allegheny mountains, whence it may pass down to Augusta, in Georgia, by one branch, and by another more immediately to Charleston, in the direction of Columbia. In traversing North Carolina, it might with facility, the surface of the country permitting, be connected by a lateral road, with the projected Cape Fear and Yadkin Rail-Way, which passing through Fayetteville, is to terminate at Beaty's Ford, on the Catawba river.

The distance between Cincinnati and Charleston, on a straight line, is about 500, which would probably require a road of 700 miles. South Carolina, however, has already made a rail-way, 135 miles in length, to Hamburgh, on the Savannah river, opposite Augusta, nearly in the direction of Cincinnati; and the contemplated rail-road to Paris, in Bourbon county, Kentucky, exactly in the course of Charleston, (for the construction of which there are, in the opinion of your committee, a great many weighty reasons of a local nature,) would have a length of about 90 miles, thus leaving but 475 miles to complete this new and most important communication, between the interior and the sea-board of the south.

The middle of this main trunk would be intersected by the projected rail-road from Richmond, Virginia, *via* Lynchburgh, to Knoxville, in east Tennessee, by which the Old Dominion would acquire a new channel of intercourse with her daughter Kentucky; and also with several of the states formed out of the North-Western Territory, which was once her property,— traveling from the West to southern Virginia, being thus restored to the route which it took in the infancy of our settlements.

By an extension west, to Nashville, of the Richmond, Lynchburg and Knoxville road, the whole of central and northern Tennessee would be enabled, with great facility, to communicate with the Carolinas and Georgia, by means of the southern extremity, and with the state of Ohio, by means of the northern extremity of the great highway under consideration.

From the maritime terminations, and the lateral branches of this extended trunk, let us turn our attention to the northern or continental connections which it would establish.

These would extend, both east and west, from Cincinnati, for several hundred miles, and through every intervening northern point. First, the Ohio river would connect it with western Virginia and western Pennsylvania—embracing the valleys of the Great Kenhawa, Monongahela and Allegheny rivers: Second, the Ohio and Erie canal, from Portsmouth to Cleveland, already finished; the Miami and Maumee canal, in progress from Cincinnati to Lake Erie, uniting at Fort Wayne, with the Erie and Wabash canal of Indiana; and the Madriver and Sandusky rail-road, from Dayton to the Lake, the execution of which has commenced, would connect it with the entire chain of northern lakes, from the Falls of Niagara to the Straits of Mackinac, and even Green Bay, on the western shore of Lake Michigan, including the eastern border of Wisconsin Territory, north or maritime Illinois and Indiana, the whole of Michigan Territory, a part of Upper Canada, and the centre and northern declivity of Ohio: Third, the Wabash and Erie canal just mentioned, and the rail-road from Lawrenceburg, at the mouth of the Great Miami, to Indianapolis, already begun, would carry its advantages into the depths of Indiana: Fourth, the Ohio river from Cincinnati to the Mississippi would connect it, beneficially, with south and west Illinois, Missouri, and the immense extent of unsettled territory watered by the upper Mississippi and Missouri rivers. Thus the proposed main trunk, from Cincinnati to Charleston, would resemble an immense horizontal tree extending its roots through, or into, ten states, and a vast expanse of uninhabited territory, in the northern interior of the Union, while its branches would wind through half as many populous states on the southern sea-board.

The extent of this inland communication from north to south, through the centre of the United States, would comprehend at least 15° of latitude, and could only be compared with that established by the Mississippi river. It would not indeed be limited by the continent, for, as many important islands of the West Indies are contiguous to South Carolina, *they* would, in fact, be comprehended in the new facilities of intercourse that would be established between the south and north, and should, therefore, be taken into the estimate.

Of the physical practicability of constructing the main trunk of the proposed rail-way, across the states of Kentucky, Tennessee, and North Carolina, your committee see no reason to entertain a doubt. It is true, that it must traverse many of the branches of the Cumberland and Tennessee rivers, and scale the southern extremities of the Allegheny mountains. One of the branches, however, of the latter river, the French Broad, as we have already seen, originating on the slopes of the Blue Ridge, the most southern of the mountain chains, runs to the north, traversing the western angle of North Carolina, to unite with the Tennessee, thus opening a pass through a part of the mountains, and inviting to the enterprize. Of the height of the remaining mountains, your committee cannot speak with confidence, but believe it to be less than that of the Alleghenies, where they are traversed by the rail-road and canals from Philadelphia to Pittsburg. However this may be, no decision of the question of physical practicability can be made, but by competent engineers, on an actual examination of the route.

The question of expense can of course only be settled by the same means. Assuming that the projected rail-road from the Ohio river, opposite Cincinnati, to Paris, in Bourbon county, Kentucky, will, from the considerations limited to the region of country concerned, be most certainly executed, and referring to the actual completion of the rail-road from Charleston to Augusta, the intervening section would not, as we have seen, exceed 475 miles, which, at the high price of 12,500 dollars per mile, would not amount to 6,000,000 of dollars; a sum not greater than is about to be expended by a company of capitalists, in the construction of a rail-way within the state of New York, to run nearly parallel with her grand canal, and connect the same waters with the same city.

It may be said, however, that the central part of the Cincinnati and Charleston road would run through a country but thinly inhabited, and furnishing little aid, either in the construction of the road or in swelling the amount of transportation upon it. But why *is* it so sparsely peopled? Manifestly, in part, because of all portions of our common country, it is the most inaccessible and the most destitute of facilities for the exportation of its iron, salt, coal, tar, turpentine, and other natural productions. To wait, therefore, for a denser population, as a condition for commencing a great work of internal improvement, which only can augment that density, would be to wait for the development of an effect, before resorting to the only cause that can produce it. Let the road be executed, and an instantaneous impulse will be given to improvement in that region. If, however, it were too sterile for such a result to occur, no argument against the project could arise from that fact, for the undertaking is necessary to the reciprocal exchange of the production of the states penetrated by its extremities, in which respect it would be similar to the Philadelphia and Pittsburg route, which, in a part of its course, passes over uninhabited mountains, and still facilitates an immense trade between the east and west.

Thus it is not necessary that the whole line of an artificial way should lie through a cultivated and populous country, nor need we look to the inhabitants along this or any other projected rail-road or canal, for the means of its construction. These will be furnished by the capitalists of any and every part of the country, or even by those of Europe, the moment the enterprize is authorized by the states through which it is to be carried on, and the probabilities of a profitable investment are rendered manifest. In the opinion of your committee, the states of Kentucky, Tennessee, and the Carolinas, might, in their sovereign capacity, execute this work, and make it a rich and lasting source of revenue; and, they have as little doubt, that the incorporated joint stock companies would at once be able to command the requisite capital.

Your committee are of opinion, that the strongest motives exist for the immediate execution of this great work. At least half the people of the Union, comprehended, in whole or in part, in East Florida, Georgia, South Carolina, North Carolina, Virginia, Pennsylvania, Tennessee, Ohio, Michigan, Indiana, Illinois and Missouri, are interested in its completion, as they would instantly participate in its advantages; and, as your committee believe, need only to investigate the subject, to be at once aroused to efficient action.

Would it pass, like the New York canal, or the projected rail-road from Augusta, in Georgia, to Memphis, in Tennessee, nearly from east to west, and consequently combine regions which have similar climates, and identical productions, its value would be far less. But, as we have seen, stretching boldly from north to south, and, with the present and future public works of the states between the Ohio river and the lakes, establishing a high road of communication through nearly all the climates and varieties of soil, productions, and people of the United States, it would forever stand alone and conspicuous among the public works of the Union, both in the kind and amount of commercial and social intercourse which it would promote.

The sustenance and manufactures of the corn states, from Kentucky to Michigan, would instantly pass along it to the southern consumer, of the region from Cape Florida to the Chesapeake Bay, avoiding all the delays, commissions, dangers of the river, and dangers and damages of a tropical sea voyage which belong to the Mississippi and Gulf route; and even much of the produce that might be designed for coasting or foreign exportation, would reach the sea-ports of South Carolina and Georgia, by the same channel, instead of going to New Orleans or New York. On the other hand, the tropical productions of the north-east of Cuba, and of East Florida—their spices, sugar, oranges, lemons, and figs;—and the indigo, rice and cotton of Georgia and Carolina would, by the same direct route, penetrate, in a few days, the interior of the continent, and spread among the consumers, even to the shores of Lake Superior.

Some of your committee, indeed, incline to the belief, that the same channel would, at no distant time, become an inlet for many of the productions and manufactures of foreign countries; for commerce, as far as possible, should be based upon a *direct* exchange of productions and commodities. Thus the shipping merchants of Charleston and Savannah, might barter their cotton in Europe for manufactures required by the people of the states in the Valley of the Ohio, and exchange the same for their sustenance; the whole operation, both continental and marine, being performed without the instrumentality of any other money than that employed in defraying the expenses of transportation.

Of the *amount* of the business that would, at length, be conducted on this national high-way, the committee scarcely dare to speak. To them it appears of a magnitude, which they fear the meeting and the community at the *present time* would regard as extravagant and incredible. By the existing population of the portions of country, even now connected with the work, there would be a great amount of traveling and transportation; but the extent to which it would augment the population of the zone of country through which it would pass; the impulse to agriculture it would impart; the manufacturing establishments it would set up, and the lateral turn-pikes, rail-roads and canals it would suggest, to new districts of country, from the western slopes of the Allegheny mountains to the banks of the Mississippi, from the sea to the lakes, would make it the parent of a great system of central internal improvement, and enable it to augment the amount of its articles of transportation to an indefinite degree. These immense pecuniary benefits, accruing to millions of people, should, of themselves, prompt those who are interested to an immediate attention to the work; but there are other and nobler considerations, which should not be overlooked.

No public work could contribute more powerfully to our national defence. Establishing a direct and rapid communication, between the northern and southern frontiers of the United States, separated, unlike the eastern and western, from the dominions of foreign nations by narrow sheets of water only, it would afford facilities for the transportation of troops, munitions of war, and military sustenance, from the centre to the borders, or even from one frontier to the other, with unexampled rapidity; thus favoring a concentration, requisite to national defence in time of war, which could not otherwise be effected; and which would present a new triumph of civilization over barbarism, by making civil public works, an efficient substitute for standing armies and powerful navies, which exhaust the resources and endanger the liberties of a nation.

But the most interesting and affecting consequences that would flow from the execution of this enterprize, would be the social and political.

What is now the amount of personal intercourse between the millions of American fellow-citizens, of North Carolina, South Carolina, and Georgia, on the one hand, and Kentucky, Ohio, Indiana, and Illinois, on the other? Do they not live and die in ignorance of each other; and, perhaps, with wrong opinions and prejudices, which the intercourse of a few years would annihilate forever? Should this work be executed, the personal communication between the north and south would instantly become unprecedented in the United States. Louisville and Augusta would be brought into social intercourse; Cincinnati and Charleston be neighbors; and parties of pleasure start from the banks of the Savannah for those of the Ohio river. The people of the two great valleys would, in summer, meet in the intervening mountain region of North

Carolina and Tennessee, one of the most delightful climates in the United States; exchange their opinions, compare their sentiments, and blend their feelings—the north and the south would, in fact, shake hands with each other, yield up their social and political hostility, pledge themselves to common national interests, and part as friends and brethren.

Finally, the immense summer throng of visiters which annually go up to the north, along the sea-board, would be made still greater, and turning westwardly, through the states of Virginia, Maryland, Pennsylvania, and New York, spread over the northern centre of the United States, to the shores of the lakes and upper Mississippi; concentrating on their return in the Valley of the Ohio; having seen what they now never see, and made acquaintance with what at present is unknown to them, the very heart of the Republic. On the other hand, the people of the north would, in autumn and winter, pour down upon the temperate plains of the south, in turn, studying their political, civil, and literary institutions, participating in their warm hospitality, catching a glow of southern feeling, gratifying their curiosity, and return enlarged in their patriotism and enriched in their knowledge of our common country: Thus this traveling, alone, would, at no distant day, reimburse the expenditures by which it might be created, while it would unite with the ties of business, in confining with a new girdle, states which are now but loosely connected, and thereby contribute powerfully to the perpetuity and happiness of the Union.

ADDRESS:

TO THE PEOPLE OF SOUTH CAROLINA, GEORGIA, EAST FLORIDA, NORTH CAROLINA, EAST TENNESSEE, KENTUCKY, WESTERN VIRGINIA AND PENNSYLVANIA, OHIO, MICHIGAN, INDIANA, ILLINOIS, AND MISSOURI.

FELLOW CITIZENS, The bare enumeration of the states to which you belong, shows that the extent of the public work, here proposed for your consideration, is greater than any other, as yet undertaken in the United States; but its value and importance would, we conceive, be equal to its cost. Many of you, residing in south-east Kentucky, east Tennessee, and western North Carolina, inhabit a region of country not connected with other parts of the Union, or even with the main body of the states to which you belong, by navigable rivers, canals, rail-ways or turn-pike roads; and from the natural configuration of the surface of the country, you never can enjoy the advantages of the two former to any considerable extent. You must, therefore, rely upon the two latter, of which, we believe, you ought to choose a rail-road in preference to a turn-pike.

The former is practicable through any region of country where the latter could be constructed, and for transportation especially, is far superior. An immediate rise in the value of your real estate and its various productions, with an increased value of your minerals, forests, and water-falls, must be the consequence of completing the road; while the pleasure of being able to visit, with facility, places to which you cannot now go without great difficulty and expense, would of itself reward you for a vigorous effort at its execution.

To the two great regions which spread out in different climates, at the extremities of this public work, the benefits, although not exactly of the same kind, would be manifold, and forever increasing with the population.

The cost of this work should not for a moment deter us from its prosecution. The states of Kentucky, Tennessee, North Carolina, and South Carolina, through which it must pass, might of themselves execute the different portions falling within their respective jurisdictions. They could borrow the money from eastern or European capitalists, for five or six per cent., and would, in the opinion of this committee and all whom they have consulted, find the amount of traveling and transportation such as would at an early day extinguish the loans, and leave the road a rich source of permanent income. We would earnestly but respectfully press this subject on the immediate attention of the legislatures of those states, at least so far as to induce them to appropriate the sums necessary to obtain a survey and estimate of the route.

Meanwhile we would, fellow citizens of all the states, or at least those portions immediately interested, suggest the propriety and advantage of newspaper discussion, and public meetings for inquiry and deliberation. Let committees of correspondence be appointed in all the towns lying in the general course of the proposed road, from the banks of the Ohio to the tide-waters of the south, and an interchange of opinion and feeling be promoted. Let a small amount of money be raised in each place; concentrate it at a central point in each of the states through which the road must pass; employ for each state a competent engineer, and have a preliminary survey or exploration effected at as early a period as possible—even within the present autumn. This would settle the question of practicability; and prepare the way for legislative action, the ensuing winter, either by the appointment of commissioners to make a more minute examination with estimates, or by the chartering of companies.

Should the preliminary survey, here recommended, not prove to be practicable the present fall, there might, at least, be a meeting, at some central point, of intelligent and public spirited citizens from Charleston to Cincinnati, who, in conferring together, might digest a plan of co-operative effort for the ensuing spring. Such a convention might be held on the first day of November, and would, no doubt, be well attended; but should it not be a numerous body, (from the shortness of the notice,) it would prove the commencement of that union of enterprize, which is indispensable to the success of the object. The reciprocal correspondence here proposed, between the citizens of the different towns within the region through which the road must pass, would enable them to come to an understanding as to this meeting, and should we be apprized by letter, that any tolerable number of persons are likely to assemble, representatives from Cincinnati, Newport, Covington, and Paris, would undoubtedly attend.

In conclusion, we would address ourselves particularly to South Carolina, the oldest southern member of the original Thirteen, and to Kentucky, the first-born of the Union, and ask them, whether their relative rank and seniority, do not impose on them the duty, of promptly moving in this national enterprize? If it be incumbent on youth to respect the claims of age, it is no less the duty of the latter to lead the way, and direct the energies of the former. The people of those states, from their very origin, have been distinguished for traits of character, which, in the days of external danger were most precious to their brethren; and should the same energies of feeling and action now be thrown into the arts and enterprizes of peace, results must be obtained, at once honorable to themselves and beneficial to the Union.

Introductory Lecture on the Means of Promoting the Moral and Intellectual Improvement of the Students and Physicians of the Mississippi Valley* (1844)

Drake's introductory lecture to the students of the Louisville Medical Institute in 1844 was a brief but complete summary of his thoughts on the desirability and possibility of bringing medical practice in the West up to high standards of professional excellence and scientific competence. The fruit of more than three decades of effort in this direction, his ideas and his program were based upon his personal observation of the relative advantages and disadvantages of the practice of medicine and the pursuit of science in the Mississippi Valley during the first half of the nineteenth century and were indeed the principles which he himself followed in bringing his own career to a high level of professional excellence and scientific competence. Drake himself, though his pride forbade his saying so, was the model he held before the students, to be emulated, not merely respected.

* *An Introductory Lecture, on the Means of Promoting the Moral and Intellectual Improvement of the Students and Physicians, of the Valley of the Mississippi. Delivered in the Medical Institute of Louisville, November 4th, 1844. By Daniel Drake, M.D. Professor of Pathology and the Practice of Medicine.* Second Edition. Published by the Class. Louisville, Ky. Prentice and Weissinger. 1844.

I PRESUME you are aware, that by a late regulation of the Institute, we have but one public Introductory Lecture.† Its delivery this year is confided to me, not because I am regarded as peculiarly fitted for the task, but as the next in seniority to the distinguished and venerable professor who, as the organ of the Faculty, met the class of the last session.

You will please, gentlemen, to make the distinction between a salutatory by the Faculty through one of its members, and the greeting of each professor for himself. In the latter case, it is his duty to introduce you to his special department, and raise in you an interest, without which, your attendance on his lectures will be irksome and profitless: In the former, it is the duty of him who has the honor to speak for his brethren, to press on your attention, and, if possible, impress on your hearts, those principles and precepts, which are as necessary to the successful progress of an assemblage of students, as are the laws of the Universe to the movements of a constellation of heavenly bodies. But you must not suppose, that I shall attempt to illustrate all the great principles of impulse and propriety, in which the student should be trained, for the limits of a lecture would not be sufficient; and I shall, therefore, select such as seem to me, from experience and observation, to be of most practical utility. And among these, again, I shall almost limit myself to those, which relate to INTELLECTUAL IMPROVEMENT. Some of them will apply only to your studies in the Institute, others to your whole professional lives; but they will concur in establishing the habits, without which professional excellence and fame will be found unattainable.

My first proposition is, that you should labor to conform to all the rules laid down by the managers and professors of the Institute, for the government of those who constitute its classes. A want of such regulations in some of our schools, and a practical indifference towards them in others, may fairly be ranked among the drawbacks on the progress of the whole. This indifference, sometimes amounting to contempt, has its origin in four different sources.

---

† *NOTE.* Had the Author anticipated, in a hasty composition of the following lecture, that those to whom it would be addressed, would ask for its publication, he might have written it in a less careless and declamatory style, and sought to throw into it some ideas not quite so common place, as those which make it up. His engagements do not permit him to re-write it, and he does not feel disposed to consult his reputation so far, as to withhold, from those who are desirous of carrying away, the plain spoken advice which it contains. It goes out, therefore, with all its literary imperfections, bearing the Author's hope, that, in despite of them, it may effect something, in quarters where reform is greatly needed.

First, the short duration of our sessions, raising in the minds of students the feeling rather than the matured idea, that for so brief a period, the same organization and government are not necessary, as for a longer college residence. Second: The almost total absence of discipline in the period of private pupilage; the preceptor seldom laying down any rules, and the student, for the most part, coming and going at his pleasure, not even making his intended absence known to his preceptor; reading on one topic to-day, and, leaving it unfinished, taking up another to-morrow; defacing and displacing both volumes; surrounding himself [with] acquaintances, and delivering himself over to light and excursive conversation, in the midst of the gravest inquiries; neglecting to take the requisite exercise, and at the same time establishing no regular hours of study. When students with such habits, congregate in a school, they have of course little predisposition to coalesce into a class, under the regimen of wholesome and necessary laws. Third: The almost unlimited personal liberty guaranteed by our political constitutions and laws; generating in young minds a high degree of freedom of thought and action, imparting to them ease and indifference in the presence of persons venerable from age or science, and rendering them impatient under the restraints, which in many countries have been considered essential, to family and college discipline, contributes largely to prevent the students of our medical institutions, from submitting willingly and gracefully, to rules of order; which, nevertheless, are essential to the object for which they congregate. Fourth: Another obstacle to the attainment of the same end, is the presence of so many, who have passed the age of discipline, and of a number who are quite as old as some of their teachers, and have even been private teachers themselves, and who start at the proposition, that they must submit themselves, as pupils, to the authority of an *alma mater*.

Let me, however, say to one and all of you, gentlemen, that the fruits of a ready and persevering acquiescence, in a system of rules for the preservation of order in the lecture rooms, the dissecting rooms and the hospital, will be precious to every one of you. Just as precious as they are found to be, in the common school, the academy, or the literary college. I would especially say to those who have been engaged in the practice of medicine, the seniors of the class, that it is expected of them, to set an example to their younger and more volatile class-mates, of that respectful observance of the laws of the Institute, which will secure the obedience of all; and thus maintain the state of repose and propriety, which will enable themselves, without interruption, to prosecute the studies which have attracted them within our walls. Should *they* fail in deference to the established rules of order, their juniors cannot be expected to comply; and thus there may grow up around them, a state of confusion utterly incompatible with the end for which they left their families and patients, and traveled hither from their distant homes.

A second proposition which I would submit to you is, that you should forego the pleasures and amusements of the city. They stand in direct opposition to successful progress in the study of science. The loss of the time actually devoted to them, is the least of their sinister effects. They stir up the physical sensibilities, excite the imagination, and banish that sobermindedness, which the search for facts, and the analysis of problems, absolutely demand. Thus the indulgencies of an hour will mar the acquisitions of days. The disorderly trains of thought and feeling, which they engender, may for a long time render the ear insensible to the voice of truth: Her most eloquent and impressive accents will fall unheeded upon it. The eye that is dancing with pleasure, or dull from its excess, sees but imperfectly the aspects of disease in the clinical wards, the most instructive demonstrations of the amphitheatre, or the most brilliant experiments of the laboratory. The hand that has been rendered tremulous from excess of nervous excitement, is poorly fitted to wield the dissecting knife. The love of pleasure and the love of science *may* coexist, but cannot be indulged at the same time; though in fact they are seldom found united. A student should draw his pleasures from the discovery of truth, and find his amusements in the beauties and wonders of nature. He should seek for recreation, not debauchery. The former invigorates the mind, the latter enervates it. Study and recreation, properly alternated, bring out the glorious results of rich and powerful thought, original conception, and elevated design; —dissipation wastes the whole, perverts the moral taste, and impoverishes the intellect. One makes great men—the other wild men. One creates the sun— the other the comet of the social and scientific heavens. A fixed luminary, spreading light and life on all around, is one—a wandering, flashing, and vanishing meteor, is the other. Would you shine on, with undimmed lustre, till a new generation shall bow in reverence before your mellow light, taste only of recreations: would you flare up for an hour, and then go out forever, prepare yourselves for the garish exhibition by feasts of pleasure.

A third proposition, which I would maintain, is, that the time allotted for studies in our schools, is so short, that the utmost diligence is required. How often is a student ruined by the delusion, that having secured many teachers, he may relax in his exertions! I tell you, gentlemen, that thousands of young men have been blighted in the bud, by this Sirocco of indolent minds. It is a common place remark, that the application of the students of America is small, compared with the industry of Europe: it is painful to add, that this mortifying contrast has not yet been sufficient to arouse our young men to greater effort. They are content to look up to those whom they ought to emulate. They are willing to feed on bread earned by foreign labor. Their want of industry makes them dependent. These remarks apply to the whole pupilage, often to the whole professional lives, of too many of us. We enter on the study of medicine without earnestness, and prosecute it without energy. The first feeling of fatigue closes the book: The first yawn extinguishes the lamp of study. Admirable hygienic prudence! Gentlemen, I tell you in plain, straightforward, homely phrase, that you must study hard, or you will run the risk of disappointment, in aspiring to the honors of your *alma mater*. Her ambition is to raise the standard of graduating attainment, higher and higher every year; she wishes *her* sons to walk forth into the world, on as high a level as any others. I beseech you, not to suffer the fruits of your indolence to put her to the blush. The ambition to rival your brethren of the old world, and the desire to do honor to your *alma mater,* should of themselves, stimulate you to industry; but there are still nobler motives under which you should prosecute your studies, both now and hereafter—here and elsewhere. It is your duty to leave your profession better than you found it, but if you while away your time in the Institute, how can you pay the debt you owe to those, who brought our science to the state in which you find it? Again, the lives of your friends and fellow-citizens are to be confided to your case, and how can you make yourselves worthy of the sacred trust, if you suffer the opportunities within your reach to pass unimproved? It is in the Institute, that such of you as may be constitutionally indolent, or, from bad example, have fallen into habits of idleness, can best correct them; for the circumstances of your new situation, are well fitted to stimulate hebetude into activity, and apathy into earnestness. The case of a student who comes to the school with idle habits, is not hopeless; but he who *goes from* it without their being corrected, is incurable, and should be banished from the ranks of our noble profession. No period of study however protracted, no estate, however opulent, no family connexions, however influential, no genius, however brilliant and pervasive—nay, even the whole, in one illustrious union, can set aside the simple law of human nature, that intellectual riches can only be acquired by unwearied industry.

A fourth proposition which I lay down, is, that you must proceed from the known to the unknown. But few of you have done this, under the guidance of your private preceptors, and I am sorry to be obliged to confess, that our schools are not so organized, as to correct the evil. While certain branches cannot be understood without a previous acquaintance with others, the whole are taught at the same time. Whether our universities will ever be so re-organized, as to improve our present imperfect methods, cannot be foreseen; but this is certain, that *you* must complete your courses of study on the present plan. You must take them as they are, and make yourselves sound and scientific physicians and surgeons, in despite of their defects. You must, in your first course, devote yourselves with the greatest intensity, to the branches which are introductory to others—first, the rudimental, second the practical.

If you do not understand chemistry, and to a limited extent, natural philosophy, how can you comprehend certain functions of the body? How can you know the reciprocal reactions of your medicines, and the relations of poisons and antidotes? How can you prosecute several important ætiological inquiries? If ignorant of descriptive anatomy, how can you understandingly follow the knife of the professor of surgery? If unacquainted with the tissues, and the composition and relations of the organs, how can you know their functions? how embrace the beautiful generalizations of physiology? how compare the healthy with the morbid structures? how enrich your minds with the principles and precepts of modern diagnosis? Should you proceed to the study of pathology, without a previous knowledge of physiology, how hopeless the task! It has been well said, that disease is morbid physiology. It is, in fact, a mode of action incompatible with the end, for which an organ was created. Can any thing, then, be more self-evident, than that pathology has such dependencies on physiology, that it must forever remain a sealed book to him who is ignorant of the latter? When the violations of a statute can be understood by a judge who is unacquainted with its provisions; or a departure from a given curve in mathematics appreciated by him who knows not its properties, then, and not till then, can the student who knows not physiology make himself a pathologist. How can the curative effects of medicines be studied by him who is ignorant of their modus operandi on the body in health? Medicines cannot even be classed, much less prescribed, understandingly, without such preliminary knowledge, to the acquisition of which chemistry, pharmacy, and physiology, are indispensable pre-requisites. Lastly, how is it possible to study the practice of medicine, except in downright empyricism, if deficient, in any of the branches I have enumerated, save operative surgery and surgical anatomy? It is the violation of this great truth, that has alloyed our profession with most of the quackery which disgraces it. I do not refer to the unblushing charlatanrie of the newspapers, to steaming, homœopathy, root-doctors, and therapeutick mesmerism, but to nosology without diagnosis, and prescriptions without indications. I call upon you, gentlemen, hereafter to avoid this inversion in the order of study. I recur to the proposition, that you *must* pass to the unknown through the known. If you cannot follow those, to whom practical departments are confided, in their elaborate analyses, the fault is not theirs, but yours, unless you come up to the study with the full preparation I have indicated. They could not expound, had they not passed through the course of study on which I have insisted, and of course you cannot comprehend, without the same qualification. In this matter nature is no respecter of persons. She will not let herself down to the level of ignorance or mental weakness—will not dissolve into her elements to accommodate those who cannot analyze—will not approach that she may stand within the focal distance of the near sighted. Let me then entreat those who are but little advanced in their studies, to bestow their first thoughts on what, in the order of nature, must be first studied; reserving an earnest attention to the more difficult branches for a subsequent course.

Sixth. You should make a *right* use of your text books. I have sometimes thought it would be well, if all text books were banished from our schools. The idea present in the mind of a student, that he may by reading when he goes to his room, make up for inattention to the lectures, is *productive* of inattention. Still further, when he reads he very often does it, not to bring the lecturer to the standard of the author, but to learn from the latter what he neglected to learn from the former. My opinion is, that the best mode of study would be by text books, with examinations upon them and illustrations of them, by the professor; but such is not the fashion of the world. You are to be taught by lectures, rendered intelligible by experiments and demonstration. It is your duty, then, to apply yourselves to what you hear and see in our halls; not to the books with which you may load your tables. You must not only hear and see, but remember. A defective scientific memory is almost universal in the country to which you belong. It is the natural consequence of imperfect exercise in early life. Memory, like attention, comparison, or any other operation of the mind, is strengthened by exercise. Resolve, then, that you will strengthen your memories by exerting them on the lectures. Make it the work of every evening to travel over the lectures of the day, and recollect and write down, at least, a synopsis of the whole. Thus and thus only will you make them your own—receive an equivalent for your time and money. When you have reproduced a lecture in your mind, then, and not before, proceed to compare it with what your author has said on the same subject. This is what I mean by making a right use of text books. Employed in any other way, they do harm instead of good.

Seventh. You should remain to the end of the session. It is I think universally admitted, that our sessions are too short. Every observing professor, and every inquisitive or aspiring student, feel this great truth; and every physician, who loves his profession, should labor to correct the error and extend the term. Meanwhile it would be a great step for all our schools to require pupils, not only to enroll themselves before the delivery of the first lecture, but to remain till after the delivery of the last. The absence of such a law, is a great absurdity—a practical abandonment of the principle involved in the rule, that two full courses are necessary to graduation. Why require two courses? Is not this demand an evidence, that we do not rely on mere examination into the qualifications of a candidate? Does it not say, that those who grant the degree must be assured, that the *opportunities* of the student have been adequate? If so, why not compel his attention to them? Why not require him to be present throughout the whole period declared to be necessary? Why allow him to enter a month after the course begins? Why allow him, at his own discretion, to absent himself for weeks, in the course of the session? Why permit him, after taking the tickets of the first session, to return home the next day if he choose? or at the holidays? or before the middle of February? after which, but few besides the candidates remain. I argue this point, gentlemen, to secure your influence in favor of a reform, in which the interests and character of our beloved profession are so deeply involved. You can talk it over among yourselves; you can speak to students whom you may see elsewhere; you can press it on the attention of your private preceptors, and thus contribute to the formation of a salutary public opinion. Above all, you can *stay out* the full term, on which you are now entering, and thus set a noble example, while you secure to yourselves the inestimable advantages of ample preparation.

Eighth. You should attend three courses of lectures before you attempt to graduate. In this opinion, I believe all my respected colleagues concur. For myself, I regard it as most unfortunate, that all our schools had not been organized on the principle of three courses for graduation, two only being paid for. Such as should not choose to incur the expense of mere travel, boarding, and lodging, for the sake of graduating, after a third, could go into practice at the end of the second course, as the majority now do, in a most immature and unqualified state, at the end of the first. It is a most vicious public opinion, that frightens students into candidates, in the midst of their career of elementary study. They are afraid to hold back, when they have become *technically* eligible, least it should be said they are not *intellectually* qualified. Under this contemptible cowardice, we see students who are yet young, and whose means are ample, pressing forward and subjecting themselves to the risk of being rejected! To avoid an imaginary, they plunge themselves into real danger! Who is the physician that ought to be respected —that is most likely to receive the confidence of society—is best qualified by his triumphs to maintain that confidence, and is most amply prepared to make discoveries and improvements? He who graduates in the shortest possible time, or he who makes the most patient and persevering efforts to become eminently qualified? Such questions suggest their own answers. Let me then entreat you, gentlemen, to ponder deeply on this subject, and resolve that you will rise from the path of mediocrity to that of eminence.

Ninth. I must give such of you as may not be candidates at the end of this session, a word of advice relative to your inaugural theses. By all means prepare them during vacation, for the labor of doing it in the session will divert your attention, and even detain you from many lectures. And while it produces these sinister effects, you will not be able to make your theses what they should be—specimens of experimental inquiry, or deep and logical thinking on existing facts—set forth in a pure and simple style. If you prepare them in vacation, you will have time for original observation and experiment, for reading, for natural arrangement, for conciseness and perspicuity of style. The time once was when every candidate had to write and print his thesis in the Latin language; then came writing and printing in his vernacular tongue; lastly, writing with the option to print or not. I have always regarded the last relaxation of rule, as injurious from its abolishing a motive of application. Let any one look into the collection of Theses, published in the early part of this century by a learned professor of this school, and he will be convinced, that when students were required to print their theses, they applied themselves with greater and more successful diligence. Since candidates are no longer compelled to publish they regard the composition of their theses as labor lost. But have they no desire for the approbation of their teachers? No regard to what their successors—near and remote—may say of these productions, when they look over the ponderous manuscript quartos, which are beginning to encumber our shelves? The truth is, that while all graduates ought to feel anxious and ambitious about their maiden productions, most of them present such as are any thing else than creditable—such as should make them ashamed. There are three classes of subjects from which candidates might advantageously select. *First*. The physiological and therapeutic effects—the *modus operandi* and curative powers—of the new medicines which chemistry is pouring into our magazines. *Second*. The endemic diseases of our country. *Third*. Our native medicinal plants—the pharmaceutic and clinical history of which, is so strangely neglected. The candidate for graduation should choose from among these, and make his thesis a positive contribution to the science. In doing so, he would give himself an impulse on the path of improvement, which might continue throughout his whole life; while his teachers would have it in their power, greatly to his advantage, to speak of him as a young man who had already given an earnest of future distinction.

Gentlemen, should you prosecute your studies in the manner I have pointed out, most of you will enter on the practice of the profession with adequate preparation. I cannot, however, indulge the expectation that such will be the case with the majority. True, I hope that all will be industrious and orderly, and *almost* hope that all will continue their pupilage to the end of the term; but most of you will never return to this or any other institution. I may, therefore, extend my remarks to the profession, of which you are shortly to become members and discuss some of the great principles on which you should then act.

In the FIRST place, let me apprise you, that on leaving the Institute to practise medicine, whether as graduates or not, you will enter a profession, the majority of whose members are inadequately educated, both in literature and science. It is your duty, as it should be your ambition, not to swell this catalogue, already so numerous. You must not compare yourselves with the poor in professional spirit and attainments, and being content to equal them, fall into the throng of mere sciolists. Remember that you will go forth as juniors in the midst of seniors, and that in forming your professional character, you will instinctively imitate those around you. Unfortunately this instinct does not always direct us to the proper models. Many young physicians have been ruined in their intellectual developement, by adopting low and imperfect examples. If you would improve your literature, and expand and purify your science, compare yourselves with the learned and logical—with those who are ambitious to write their recipies in plain and accurate English or in correct Latin, instead of a grotesque mixture of the two, each being incorrect; with those who attend to the rules of grammar in the directions they write for their patients, and do not begin by *addressing them,* and end by speaking *of them,* as though there was no foundation in nature for the persons and cases of pronouns; with those who can write letters without violating the first laws of orthography and syntax, and report cases without mistaking the technical terms of the science; who can spell stomach without a *u,* and bilious with a single *l;* who can stop when they are done, and be done when they stop! If you do not open your eyes to the lethargy into which so many of our brethren have sunk; if you do not cherish a livelier and purer taste for accurate learning than theirs; if you do not feel ashamed of gross illiteracy, superficial science, intellectual indolence, and contented mediocrity, you are ruined. I beseech you, gentlemen, to form the resolution—now, at this moment—that you will not add to the number of those who drag down our beloved profession, but of those who are toiling to raise it into the pure, upper regions of truth and beauty.

SECOND. When you may open offices, do not practise imitation, but guide yourselves by first principles. If it be difficult to find good models of study, it would still be more difficult to find an office so arranged and kept, that the aptest mind could study in it. I say, then, appeal to principle, and begin a new series. You may smile at the importance I am attaching, to what you will regard as a small affair, but time will rectify your estimate of things. My firm opinion is, that one of the causes which retard us, in raising our profession in the West and South, to an excellence which at present seems almost hopeless, is the style in which our offices are fitted up and kept. Who can read and think, with method or sound logic, while every thing around him is dirty and disordered? His little stock of furniture displaced, as if a riot had just passed away; his books scattered on chairs, tables, and the greasy medicine shelves; in his book-cases, volumes of different sets mixed together, some lying flat, and some, like the ideas of their reader, up side down; his skeleton exposed, and joint after joint torn off; his few injected preparations, unvarnished as my narrative, and worm eaten, as the books of an old doctor; his medicines unlabelled, and thrown into a chaos, as great as a treatise on the Materia Medica in the 14th century; bundles untied and bottles left uncorked, or stopped with plugs of paper; dead flies in the ointment within his jars, while others are wading through that which has lain so long spread over his counter, that their feet are blistered by its rancidity; his spatulas, foul and rusty; his scales tied with strings and balanced with pieces of paper; his mortar about as clean as the ancient Kentucky hommony block, which, in the same day, contained the food of the family, and the family cow and horse, as it stood convenient to all the parties, on neutral ground, near the door of the cabin; his surgical instruments oxidated and rusting away, like his mind; his study table, covered with loose papers and medical journals (even the WESTERN) with their covers torn off; his walls overspread with a tapestry of cobwebs; his windows as opake from the dust as the painted glass of an ancient cathedral; his foul candlestick standing all day on his lexicon, and his floor spotted over with the blood of his surgical patients, and his own tobacco juice!

The human intellect cannot act when thus encompassed. Ideas will not arrange themselves; nor will their foul surfaces cohere. The scene re-acts upon his mind, and a chaos within rivals that without.

Other disadvantages still result from this official derangement. 1st. When he who vegetates in its midst wants a book, a medicine, or an instrument in haste, it cannot be found. 2d. In compounding medicines he is apt to make mistakes, and all he puts up is filthy. 3d. It is attended with great waste, an effect which few young physicians are able to stand.

He who should cause every physician in the valley of the Mississippi to keep his office in order and neatness, would give a greater impulse to efficient study, than the eloquence of all its professors, and secure to himself the name and fame of a public benefactor.

In the THIRD place: You will mingle with colleagues, whose habits of *observation* are extremely imperfect. Medicine is not a science of meditation but observation. Hippocrates was an observer—Sydenham was an observer—

Hunter was an observer—Rush was an observer. But you will ask, are not all men observers? I answer yes, but *what* observers the majority are! Children, mere children, and not children either; for they lose the activity of observation which belongs to childhood, while they retain its changefulness—its superficiality—its thoughtlessness. The faculty of observation requires a training which it seldom receives—the function of observation, an amount of accurate preliminary knowledge, which very few acquire. The diagnosis of every case of disease, is to be made out by observation: the effects of medicines are matters of observation: the *post mortem* appearances in every dissection are subjects of observation. In the whole, inaccurate observations lead to erroneous conclusions. Without previous elementary knowledge, we know not what things should be looked for, and this is a great source of defective observation; but with ample preparation, our observations my be vitiated: 1st. By mistakes resulting from carelessness; 2d, by omissions. In the former case, we report and diffuse positive, in the latter, negative error. Half the reported cases which make up our clinical archives, are imperfect from one or both of these causes. To this criticism, we of the valley of the Mississippi are as obnoxious as any other physicians of the age. Who among us has not found, when a physician has been called to consult with him, or he has been summoned to consult with another, that to make out a satisfactory history of the case has been extremely difficult? Who does not know that in this extremity the imagination often supplies, from an analogy, what should have been collected by observation? Who, in reading reported cases, does not find, that facts are absent, which more activity and fulness of observation would have introduced? Many a clinical history, on the *composition* of which its author had expended great labor, has utterly failed to enrich or advance our science, because of his vague and inaccurate observation. Correct observation implies not only intelligence, but sustained attention. Look to the pilot of the steamer which brought you hither, for illustration. He knows what should be avoided, and where flow the deep waters through which he may steer in safety. This is his knowledge—the rest is observation; in the performance of which, constancy of attention is all that he exhibits. The moment he fails in this, you are liable to be wrecked. This attention, without which there can be no efficient observation, is equally indispensable to sound memory. Why do we recollect the symptoms of an *uncommon* disease for years, while those of *common* cases fade away in as many weeks, but because the new or eccentric character of the former, awakens and sustains our attention? The mind opens its bosom and receives the impression on its heart, rendered soft and susceptible by the fervent excitement which awakened attention imparts. Would you weld one piece of metal to another, heat is indispensable.

But in the FOURTH place, the best observations will in the end prove of little value, unless they be recorded; for no memory can long retain a great variety of details. As the trunk may remain after the fruit-bearing branches of a tree have perished and fallen to the ground, so the central facts of a case may be recollected, after the appurtenant have faded away; but such recollections are seldom of much value. You will enroll yourselves with physicians and surgeons, who, on the main, record as negligently as they observe. I might rather say that you go forth among those who do not record at all; for the majority of the physicians of the Mississippi valley keep no other books than those in which they write down the names and debts of their patients, now and then adding those of their diseases. At first it may be otherwise with them, but adopting no suitable method, practising no discrimination, and beginning with a record of *every* thing, they end by recording *nothing*. Their registers, moreover, are often made many days after their observations, instead of being written on the day, the hour, or at the very moment, which is the best of all. Still further, they are loosely and confusedly put down, instead of being methodised and expressed in the purity and precision of style which science requires; and hence the labor of the composition is not made to improve the literature of the writer; not does he afterwards peruse them with pleasure or profit.

In the FIFTH place: You are destined to blend yourselves with elder brethren, who will present but few examples of that reading which no physician can neglect, without retarding, and at length arresting his professional growth. If observation be the soil, reading is the manure of intellectual culture. In the Institute, as throughout your private pupilage, your reading is necessarily rudimental. When you enter on the practical duties of the profession, you can scarcely have acquired more than its elements. In reference to all its debateable departments, you will not have seen more than one aspect of its ever varying form. Much reading of the same kind in which you have been engaged, will still be necessary for the first years of professional life. Then will come up the works of the 'mighty dead;' the massive volumes of observation, experiment, and speculation, bequeathed to us by the eminent men of all ages, from which you should enrich and invigorate your minds. In addition to these, you will have the monographs, treatises, and systems of your more industrious or highly gifted cotemporaries. Lastly, there is the periodical literature of the press, which flows through the profession, in a thousand fertilizing currents, inviting every member to drink of the fresh waters of discovery and improvement. What is to be thought of the physician or surgeon who scarcely vegetates in this exuberance? Who remains a dwarf, when aliment on which he might become an intellectual giant, is placed before him?

And yet how few are to be found, who put forth their hand and pluck that which might confer upon them the power of doing what kings and conquerors cannot,—rescue the good and great from the jaws of the destroyer. Let me exhort, let me conjure you, gentlemen, to turn with deep repugnance from the examples of superficial or neglected reading, with which you will be surrounded. Labor to make industrious and systematic reading a matter of love, of ambition, of duty. Resolve that you will never allow a single day to pass, without adding something to your previous stock of professional learning. Endeavor to establish hours of study; and when they are trenched upon, by professional duties, supply the loss by the sacrifice of your pleasures. Indeed, seek to find your pleasures in your books, and not in the gossip of the world around you. Raise yourselves above it—that you may secure its admiration.

SIXTH. Do not shape your course, by the example of those with whom you will associate, in regard to publishing. Of the thousands who practise the profession, in our great valley, how few contribute by their pens to the magazines of knowledge! An immense majority leave no manuscripts behind them, but their ledgers and recipies. In vain have the most precious original phenomena passed in review before them, never perhaps to be repeated before any other observer. In vain have the arrangements of Providence, favored them with an opportunity of consigning their names to immortality. In vain has the periodical press invited their contributions. With stolid insensibility they neglect the whole, and die without leaving their profession better than they found it. The last mantling on their cheeks, should be a blush of shame for such criminal neglect. But do not conclude, that I recommend extended publications; or an indiscriminate putting forth of *all* that may fall under your observation. No, indeed, far, very far and foreign from what is in my mind, would be such advice. Publish that which *ought* to be published, and nothing else. But, you will ask, who shall decide? I answer you yourselves. Study your profession as you ought: read, to know what has been already published: think, to be assured of the bearing and value of what lies before you, and you will seldom offer that, which your brethren will not gratefully receive.

SEVENTH. Resist the western propensity for change of residence. Turn from the example of those physicians, who seek in new localities the practice and prosperity, which must flow from industry, patience and perseverance. The endless shifting of place which our western and southern profession displays, does not diminish their number, nor abate the competition; but it keeps, at all times, a multitude out of employment, and has destroyed the prospects of more than it ever built up. Select with care—then persevere with deathless constancy. Let me draw an example from the mines of Wisconsin. The mineral district, as they call it, presents two classes of miners, first, those who dig superficially, and finding no treasures, abandon spot after spot for some other, which their fancy suggests will be more productive. Second, Those who dig deeply, and perseveringly follow up the same metalic vein, till they enter its rich caverns. Make *these* your model, and you will sooner or later enjoy their reward.

EIGHTH. Avoid the example of those, who are ever ready to relinquish their profession for some other. In giving you this advice, I am apparently counteracting the interests of those whom I represent, for the practice I am reprehending, increases the number of students. Nevertheless, my colleagues look to the dignity of the profession, not less than the emoluments of teaching. He who is willing to abandon his vocation, generally looks out for the opportunity, and, as a matter of course, ceases to cultivate it. When a very young man leaves the ranks, it is a small affair; but when a physician of ripened experience goes out, society suffers. A raw recruit may desert when the battle is impending without causing defeat, but a veteran cannot be spared. The practice of which I speak, prevails in some parts of our country to such a degree, that the care of society is always in the hands of the inexperienced. There is little of the knowledge which comes from long and varied practice—the sound judgment which time only can confer—the sympathy which flows from the cherished intercourse of years, between physician and patient. An epidemic sweeps over the land—there is no veteran arm to save—no friendly eye to pity. I beseech you, gentlemen, after having through days and nights of toil, acquired your profession, not to cast it off. Regard it as you would a wife whom, through storm and sunshine, you had won by vows and prayers and pledges. Make it your bride and value her for the price you have paid. Adhere to her for the sake of her honor. Let not merchandize, nor tavern-keeping, nor office-hunting, divide your affections with her. Let not the sugar plantation boast of a conquest over her; nor the cotton field triumph at her desertion; see that she does not perish by the hemp! Confide in her and she will prove faithful to you; labour in her service, and she will amply repay you, with patients when you are in health—with sympathizing friends in your sickness and sorrow. Such a bride, never waxes either old or ugly, for every year brings out some new aspect of beauty—some brighter tint of loveliness: She will enrich you with the means of independence—'She will do you good and not evil all the days of your life,' and when you are no more, she will erect to your memory an imperishable monument.

NINTH. When you join the profession, you will become teachers, and a responsibility of a new and serious kind will devolve upon you. Do not prepare to meet it, by copying after those with whom you will associate. Do not encourage young men, who are unqualified by infirm health, or mental weakness, or want of means adequate to a proper continuance of their pupilage, to engage in the study. Suffer not your own interests to conflict with those of society; nor your thirst for the distinction of having a pupil, to render you insensible to the dignity of your profession. We want more talent and learning, more ambition, more love of knowledge, more of burning zeal in our profession; and it is only to private preceptors that we can look for their introduction. Of yourselves, I know nothing, and hope there may be none among you, whom nature fitted for some humbler calling. But this hope is not founded on experience; which testifies that in all the schools of the Union, there have been many whom no exhortation could arouse, no tuition instruct: whose dulness, fickleness, or indolence barred them from every other calling, and caused them to be drifted upon ours, in which it was assumed those defects would be no disgrace. It is time that the watchman on the walls of our professional zion, sounded an alarm. Our gates should, henceforth and forever, be closed against those who cannot be made men of science. But your duty as private preceptors is not comprehended in this exclusion of the unqualified and unworthy—you must teach as well as select. Your pupil should never be out of your grasp—never permitted to wander from the path of study which you mark out—never be allowed to loitre upon it without reproof. Whatever may be the defects of the best schools of the Union, however disqualified, their professors, justice to both requires me to declare, that the delinquences of private preceptors, are far greater; and constitute the first, the most prolific cause of imperfections, which all who love and respect the profession, unite in deploring. It will be for you, gentlemen, to arrest the downward tendency which this negligence has generated. You should do as you would be done by, not as too many of you, I fear, *have* been done by. You should spur up the indolent—cheer on the faint-hearted—clear away the doubts of the perplexed. You should travel over every book which your student reads, and by a few searching questions, on its governing propositions, ascertain that he has mastered it. But you must impart as well as extract; and illustrate every topic from your own resources, as well as interrogate. Thus, while you do your duty to him and the profession, you will not only preserve your own science, but enlarge it under the creative action of your own faculties. So true is it, of every relation in life, that a conscientious discharge of duty, benefits ourselves not less than others.

GENTLEMEN! What I have spoken on the necessity and means of intellectual improvement is of general application. In conclusion, I must say something, in which a part only will find themselves interested. A portion of you are the sons of men of wealth, and may, without inconvenience, defer the time of entering on your professional duties. To this delay there can be no objection, but the reverse; provided, you occupy yourselves on the means of increasing, diversifying, and embellishing your scientific knowledge. Apart from mere reading, there are three modes, any one, two, or all of which, you may pursue. I propose briefly to state and illustrate the whole.

First, You may visit other schools, in our own country, and attend the lectures on practical branches, or indeed on all. Thus your knowledge will be confirmed to you, without the irksomeness of listening too often, to the same professors. Some of your new teachers may, at times, go more extensively into certain subjects, than you had been inducted before. Others, in coming over debateable ground, will excite your curiosity, stir up your feelings, and give activity to your thinking. At the same time, you can visit different hospitals, and witness various modes of practice. You will also be able to examine museums of natural history and comparative anatomy, listen to lectures on those subjects, and attend the sittings of learned societies, finally, you can mingle with general society, and thus increase your knowledge of the world.

Second, Some of you may be desirous and able to visit the Universities of Europe. So far from discouraging, I would encourage this disposition. But let me warn you, gentlemen, not to substitute foreign schools for those of your own country. When you have exhausted the means of instruction afforded by the latter, then, if convenient, resort to the former. It is mortifying to see our unfledged young men, with diplomas to which they often have perhaps but an equivocal claim, depart for foreign lands, profoundly ignorant of their own, and unacquainted with most of what is taught in its schools. Why should a young man thus bring odium on his native land, and contempt upon himself? Why go abroad to acquire, what he can better learn at home, the very rudiments of the profession the first lessons of science? And that too, without knowing the tongues spoken by his new teachers! Will change of place confer knowledge, or change of teachers give understanding? To gain knowledge, a man must observe, to get understanding he must think. Let not those who have never observed nor thought on the banks of the Ohio, assume that they will do either, on the banks of the Seine or the Rhine or the Thames. The chances in fact are quite against them; in as much as the novelty and allurements of their new situation, are likely to divert them from its advantages. There is another obstacle to the improvement of our novices in Europe. They sojourn there, under the natural feeling, that opportunity may supply the place of application; and that the reputation of having studied diseases abroad, can be substituted for a knowledge of their proper treatment at home, both of which will abate their industry, and reconcile conscience to the neglect of opportunity. To reap benefit from foreign study, three things are indispensable—first, that the student should have mastered the elements of his profession in the colleges of his own land—second, that he should understand the languages to which he is to listen—third, that he should protract his stay, far beyond the time usually allotted. Under these conditions, any one or all of you, might visit Europe with advantage—in their absence you had better stay at home.

Lastly, I wish to recommend to those who have ample time and means, a third source of improvement after their graduation. I mean researches into the physical condition, manners, customs, physiology, and diseases of their native land—this great and glorious valley of ours, the Eden of the New World, from which I trust our crimes and follies may never drive us out. Who has opened his eyes and looked inquisitively and yearningly on this vast book of nature, here and there, embossed with art? Who has studied even its table of contents? Who has made an index to its countless beauties? Who has yet realized that it *might* be to all its physicians a book of wisdom? Alas, nobody.

Young Gentlemen, I invite such of you as are not under the necessity of going into practice, when you leave our *alma mater,* to turn your eyes to this mighty field. Do you not wish to know its structure, productions, and climates; and could you better study geology, natural history, and meteorology, than by connecting those sciences, with all that is dear to your childhood—with the cherished associations of school-boy days? Do you not wish to study the varieties of the human race, that your knowledge of physiology may be enlarged? Why not compare, then, the three which surround you? Do you thirst after an acquaintance with the influence of manners and customs and vocations on human health? Why not dwell on those around you? Would you know the diseases of various latitudes and localities, why not traverse and examine those which spread out from you on every side? Would you extend your acquaintance with the members of the profession to which you aspire, why not visit and sojourn among them; learn the secret of their success or failure; acquire the results of their experience, and establish with them that scientific correspondence, which they have not yet begun among themselves; but which made general would at once enlighten and exalt our young profession? I am deep in the conviction, that a couple of years spent in this way, by a thoroughly educated young physician, would do more to develop his character, and fit him for every day usefulness, than even the boasted European voyage. But he may, if he will, avail himself of both; in doing which, I would earnestly advise him not to go abroad, till he can no longer find objects of appropriate scientific interest at home. Let him carry out an ample stock of knowledge on the physical condition and diseases of the new world, and he will find easy access to the men and magazines of the old.

Young Gentlemen: Let me repeat to you, let me say to the young persons of both sexes who have honored us with their presence this evening, that travels in our own country *might* be made to yield an abundant harvest of pleasure, patriotic emotion, and general improvement, yet how few appreciate and use these precious opportunities! We are a migrating but not a traveling people. We think of our country only as abounding in residences; and pass from one to another, without inspecting anything between them. Thus, even migration bestows on us none of the benefits of travel. In our transits, all things are sacrificed to speed. We are not satisfied unless we add night to day; and when we awake in the morning, congratulate ourselves that we are a hundred miles nearer the point of attraction; although we may have passed through scenes and objects the most interesting, without having beheld the least or the greatest of them. Thus while we are wanderers, we remain ignorant of the relations and true character of all among which we roam; or know them only in connexion with hemp or cotton planting, commerce, land speculation, or the practice of law and medicine.

How few have yet realized from personal observation, the limits of this valley of a thousand streams—more comprehensive than Great Britain, France, Germany, Spain, and Italy—countries which have been the chief theatres of civilized society, since the dark ages passed away. Who has ascended its gentle elevations, rising one above another, to the region of perpetual snow; or felt its climates, from the tropical heats of southern Florida, where the orange blossom perfumes the air, and the live-oak stretches wide its gnarled limbs, to the icy fountains of the upper Mississippi? How many of the sons and daughters of this fair and favored land, have stood on its terraces and looked over its undulating prairies, where the grass waves and the flowers smile, from the close of one winter to the beginning of another? Who has admired the tangled copses of flowering shrubs in the South, the lofty forests of the Middle, the open woods of the North, or the boundless Savannahs of the unpeopled West, where in autumn a running fire lights up the darkness of the night? Few, indeed, have had their hearts thus kindled. Quite as little, do we regard the rolling waves of green water, which, age after age, have mingled their wild voices with the murmuring breeze or the boisterous tempest—amid the green islands of Michigan, or on the white sands of the Gulf. Nor do we float on our mighty rivers and their countless tributaries, with their graceful curves, their shady banks of summer green, their autumn robes of 'mourning gold;' their cane-brakes and their towering cliffs, with the admiration which ought to fill our hearts. Still less do we ponder on the rocks beneath—the tertiary formations of the South, more extensive than those of any other country; our ample limestone regions, so rich in organic remains; our drifted rocks and diluvial beds, abounding in fossil bones; our coal fields the most extensive in the world; our transition rocks; our hills of iron, and veins of lead and copper; our hot and mineral springs—chalybeate, saline and sulphurous! —But, not to dwell on the inanimate world, who bestows adequate attention on the life, the living and social movement, of which he is a part? Who for himself, compares the African with the North American—the black man with the red—in their propensities, physiology, and diseases? Who studies them in connexion with the Caucasian, whose subvarieties from the vallies of Spain to the mountains of Norway, bring hither and commingle their peculiarities of physiology, modes of living, and exercises of ingenuity? Who contemplates with a fixed and discriminating eye, the new social combinations which are starting into existence on the banks of every little river? Who gives scope to his imagination and prophecies of the future? And yet, is there not improvement in the study of nature as well as of art? Is it not as salutary to the mind, to witness growth as decay? Is it not more delightful to roam among the germs of a budding empire, than to clamber over the crumbling monuments of those which are passing away? Is not the rising more glorious than the setting sun? Does it not speak to the heart of day instead of night—of life instead of death? Let us, in idea, connect the present with the future not the past, and resolve to make that future illustrious. Let us cherish all that is American, and love our native West the dearer, *because* it is American. Let us breathe into our new-born institutions—social, literary, scientific—the breath of life, and foster them in the arms of a deathless affection: Do for them, one and all, that you have come to do for this young and aspiring *alma mater*— cheer them on with our applause, and support them with our patronage!

Young Gentlemen! You have two missions to execute—one of science and humanity, the other of freedom and national glory. In becoming physicians, you will not cease to be citizens. Most of you will, indeed, attain to both distinctions at the same time; and all should feel, and practically acknowledge the responsibilities of both characters. You have a profession to organize—a country to build up—a high destiny to fulfill. A heritage of great principles, requiring diligent cultivation, is descending to you and the generation of which you are a part. Hand it down enlarged, purified, and embellished. Do your duty, your whole duty, and nothing but your duty; and thus you will hasten on the day, when love of science, and love of constitutional liberty, and love of country, will unite in one deep and swelling emotion of soul; and the natives of every hill and valley salute each other with the anthem—'We are Americans! Our forefathers were the first of the human race to plant the seed of universal learning, christianity and freedom in the solitudes of a wilderness! We must honor their memory, by emulating their deeds!'

Introductory Lecture, Thirtieth Session

of the Medical College of Ohio*

(1849)

During the last years of his life Daniel Drake turned increasingly to retrospection. Shortly after his sixtieth birthday, in 1845, he began to set down his reminiscences of pioneer life in the Ohio Valley in the form of letters to his children.[1] These, and the *Systematic Treatise* describing current conditions in the West, would be his legacy to future generations. Early in 1849 Drake resigned his professorship at the Louisville Medical Institute (now the University of Louisville College of Medicine) and his editorship of the *Western Journal of Medicine and Surgery,* and returned to Cincinnati, expecting to make that city his home for the remainder of his life. There his family was, there the *Systematic Treatise* was to be published, there also the Medical College of Ohio had been reorganized. From the chair of special pathology, practice, and clinical medicine Drake delivered the introductory lecture in November. Quoting extensively from the introductory lecture he had delivered in 1820, Drake struck a personal and reminiscential note. Where in earlier lectures he had viewed the distance between past and present as progress, he now looked back with some longing on an earlier and younger time.

Drake's high hopes for a revitalized Medical College of Ohio yielded only disappointment, and in the spring of 1850 he accepted a position as professor of pathology and the practice of medicine in Louisville. He continued to spend his summers in Cincinnati, visiting with his family and working on volume II of the *Systematic Treatise.* Early in 1852 he delivered two discourses on the early history of medicine and the early history of medical periodicals in the West before the Cincin-

---

* *An Introductory Lecture at the Opening of the Thirtieth Session of the Medical College of Ohio. Delivered at the Request of the Faculty, November 5, 1849. By Daniel Drake, M.D., Professor of Special Pathology, Practice, and Clinical Medicine.* Published by the Class. Cincinnati: Morgan and Overend, Printers. 1849.

[1] "Reminiscential Letters of Daniel Drake, M.D., to His Children [December, 1845, to January, 1848]." MSS in Cincinnati General Hospital Library. Published as *Pioneer Life in Kentucky: A Series of Reminiscential Letters from Daniel Drake, M.D., of Cincinnati, to His Children,* Charles D. Drake, ed. (Cincinnati, 1870).

315

nati Medical Library Association, which he had helped organize the previous year.[2] Later in that spring he determined upon a return to Cincinnati, and accepted appointment to the chair he had first held at the Medical College of Ohio, in theory and practice of medicine. In July, Drake served as chairman of a committee to arrange for the reception of the body of his old friend, Henry Clay, as it passed through Cincinnati on its way to burial in Lexington. Drake's last public appearance was at a memorial meeting for Daniel Webster on October 26, where he caught a chill that sent him to his bed. Venesection, purgatives, and the services of a cupper failed to relieve the intense pain in his chest, much less restore his health. Daniel Drake died on November 5, 1852, at the age of sixty-seven.[3]

[2] *Discourse Delivered by Appointment before the Cincinnati Medical Library Association, January 9th and 10th, 1852* (Cincinnati, 1852).
[3] Cf. William Holmes McGuffey, "Particulars concerning the Last Illness and Death of Daniel Drake, M.D. [dated November 1852]," MS in Cincinnati Historical Society.

IT IS AT the request of the Faculty, as well as of a respectable portion of your own number, who attended our preliminary lectures in October, that I appear before you, this evening. I do not come to begin a course of instruction on special pathology and the practice of medicine; nor yet to greet you as students of my own class room. My mission is, to salute you as pupils of the College; and, on behalf of its Trustees and Professors, to welcome you within its walls, and give you the assurance, that whatever may be practicable will be done for your instruction and comfort.

Of the prospective and probable value of our labors to these ends, it is not my intention to speak; but I *may* remark, that whether it be much or little, a willing and earnest coöperation on your part, will be indispensable to your improvement. It will, therefore, be expected of you to be punctual in your attendance in the College and Hospital, orderly in your deportment, courteous to each other, respectful to your teachers, temperate in your habits, and unwearied in your devotion to study.

Some of you, young gentlemen, have already attended a course of lectures, in this or some other medical school; others have not, although they may have been for some time engaged in medical studies; while others have read but little, and attended no lectures. Now, no truth could be more obvious, than that students so differently advanced, should be placed in different classes. In every stage of our studies, from the infant school to the university, this is done. Medical institutions, alone, constitute an exception—an exception which establishes the necessity of the rule; for the instruction afforded by our medical schools, under the present system, is far less profitable to the student, than it would be if the pupils were properly classed. I hope that some of you may live to be the honored authors of this great improvement.

Meanwhile, let me advise those who have as yet made but little progress, to direct their attention chiefly to the first or elementary branches. These are anatomy, physiology, chemistry, pharmacy, and the natural history and classification of medicines. Until they are well understood, the more complex and practical branches cannot be mastered, any more than we can comprehend the circulation of the blood, without knowing the structure of the heart; or the movements and powers of the magnetic clock, without a knowledge of its mechanism. By neglecting this obvious truth, the student involves himself in the most discouraging perplexities; and not unfrequently finishes his first course of lectures, without having acquired an accurate or well-defined knowledge, of any of the subjects to which his attention has been called; has a superficial knowledge of many things, and a deep and exact knowledge of nothing, except the absurdity of having neglected what I am here recommending.

To protect you the more certainly from the error into which so many students fall, I must present the subject in another aspect. The phrase, "Science of Medicine," is by no means an accurate and logical expression. Strictly speaking, there is no such thing as *the* science of medicine, but there is a *profession* of medicine, to which you are aspiring; but how are you to reach it? I answer, by studying, not one, but several sciences. If medicine were a science, the advice I have just given you, would be superfluous, for you could make no progress whatever without beginning with its rudiments—its definitions and axioms—its primary and fundamental propositions; and proceeding from the known to the unknown, in the invariable order of its relations. As it is, you may study one of the sciences, which make up the profession and neglect another; you may study *first* that which should be studied *last;* or, neglecting the first, attempt to build without a foundation. A just view of the relations and dependencies of the different sciences with which the physician must be acquainted, as constituting his profession, is of the utmost importance; and should be attained by all who conduct medical education, and by every student who aims at distinction in his calling. Some of these sciences, as I have already said, must be acquired before others; I may now say, that some must be studied more *carefully* than others, because they are more intrinsic—essential—indispensable. The essential are anatomy, physiology, pathology, therapeutics, and operative surgery. You may study several of these, and yet be unqualified for the duties of the profession at large. You may be good anatomists and adroit operative surgeons, without being physiologists. You may know physiology and remain ignorant of pathology. You may be profoundly skilled in pathology and pathological anatomy, without making the slightest acquaintance with therapeutics or clinical practice. All this is abundantly obvious to every well-educated physician, and yet nothing is more common, than to see young men graduated on the basis of full attainment in some of these branches, while their knowledge of others remains deplorably superficial and imperfect.

But while the sciences I have named are essential, there are others which, to a certain extent, must be known. The most important of these is chemistry, with its applications to materia medica, physiology, and pathology. Next comes natural philosophy, especially that important and delightful branch called meteorology—then comparative anatomy, and the general physiology of organized nature—then botany and mineralogy—lastly, the philosophy of the human mind. The whole of these in varying degrees, and on various sides, connect themselves with medicine—make portions, more or less considerable, of that coalition of sciences, which constitute what is called the profession of medicine. I would not advise you to prosecute the study of the whole, nor even any single one, to its limits; but I *would* impress you deeply with the momentous truth, that if these auxiliary sciences had never been cultivated, those which are essential would have remained fatally imperfect and uncertain.

Such is "the Science of Medicine," to employ an expression which I have condemned; but which, from the want of a better, I must still employ, to indicate the branches of human knowledge, with which the accomplished and skillful physician must be more or less acquainted. You will perceive, young gentlemen, that the whole are sciences of observation and experiment. Their very existence depends on the induction of facts. With the progress of observation they began and have advanced—by the aid of future observation and experiment, they must reach a degree of perfection, of which, at present, we have but an imperfect glimpse. The earliest record of these facts is found in the writings of Hippocrates. For nearly twenty centuries those writings were the text book of all the schools of medicine in the world—the chief treasure of every medical library, public and private. All discoveries and inventions—all recorded experience—all speculative efforts of the human mind on the subject of health and disease, arranged themselves around the writings of this Father of Medicine. From him there was a lineal derivation—a legitimate descent—an unbroken succession. The fountain which he opened, although at first, of necessity, feeble and impure, but still a fountain before which there was no other, has continued to flow on, and *will* roll on, widening, and deepening, and becoming purer and purer, to the end of time. In every age the banks of this stream have been perforated, and a portion of its waters made the head spring of a rival current; but, like the waters which overflow the banks of our great Mississippi, they have returned to the channel, after having deposited the errors which ignorance and presumption had mingled with them. There is, then, but one profession of medicine—one medical literature—one legacy of medical knowledge; and, however scattered over the civilized world, but one class of laborers. All the truths which have been discovered—all that may ever be discovered—will, by an irresistible attraction, associate themselves with those already known. It is vain and idle to talk of two kinds of medical science. We might as well speak of two sciences of anatomy, chemistry, botany, or natural philosophy. All the truths bearing on the production, effects, and cure of diseases, which have been developed, belong to the same original, ancient, and honorable profession—enrich its archives, and nobly swell the list of its cultivators.

You have assembled, young gentlemen, in one of its schools, to prepare yourselves for discharging the duties which it imposes, and for handing it down to the generation that will follow you, augmented in the number of its truths, and purified from many of its errors. I hope and trust, that you will fulfill the mission, with which you will be charged, by your graduation; but, to do so, you must take expanded views, and compass, if possible, the wide range of scientific knowledge, embraced within the limits of the profession, as indicated by the sciences which have been enumerated.

The *reason* that so many branches of human knowledge unite in the medical profession, is to be found in the varied and intimate relations of man with all the objects and operations of nature, all the works of art, and all the events of society and the world. It is these extended connections, that make him, under God, its governor; that constitute him the greatest, not less than the latest, of its organized beings. In his perfect development he lives, and feels, and thinks, and acts, amid things and movements as countless in number, as the leaves of our summer woods—as diversified in aspect, as their forms and autumnal tints. The oak, on the summit of the hill, has a relation only to the soil, the winds, the rain, and the thunderbolt, which may at last rive it asunder. The flower, which blooms beneath its wide-spreading branches, is sustained by the evening and morning dew, the shower, and the sun; and, having these, it sends forth its fragrance for a season, and dies a natural death. The wild deer, of our forest, crop the natural herbage, escape from the wolf, and lie down in their grassy lair, until age puts an end to an existence, more simple than that of any human being. The passenger pigeon leaves the south, as the heat of summer arises, spends a season in the north, rears its young, and returns to the place of departure. Its relations are with food, and water, and temperature, and the air by which it performs its migrations. The dwellers in the deep, from the whale to the coral insect, have relations to the water in which they live, and to each other, on which, in part, they feed, and to these they are confined. Far different from all these limited connections and dependencies, are those which man sustains with other portions of the creation. His instincts and his wants, corporeal, intellectual, and moral, lead him into all climates. In the south he lodges under the boughs of trees; in the north in huts built of compacted snow, and on the swampy banks of lakes and rivers he drives down piles, on which to support his dwellings. Go into the sandy desert, and there you find his tents; turn your eye to the bleak mountain-top, and there you see his cabin embosomed in the cloud; look upon the ocean, and you behold it white with the canvas of his floating habitations, driven to and fro by the tempest. On the broad continents he cuts down, and burns or applies to his own use, the mightiest forests; his plowshare annihilates the natural herbage, and his sagacity and toil replace it with productions more to his taste, or better suited to his wants; in one region he lives solitary, and gives full play to all his desires and fancies; in another he struggles through the conflicting masses of a city population, where he invents new modes of industry; dooms himself to a confined and unventilated room, or a dark and damp cellar; passes much of every day in some constrained posture; breathes an atmosphere impregnated with irritating dust, or poisonous gases; watches his rivals, and seeks to wrap himself up from their scrutiny; labors to outstrip them in the career of mingled ambition and avarice; concocts schemes for their defeat, or his own advancement; delivers himself over to midnight studies, or gluts his pampered appetite with meats and drinks, drawn from the abounding storehouses of every land: He founds and conducts institutions of benevolence, literature, and science; devises enterprises which demand concert of action; adventures on gigantic projects the most visionary; and grasps riches, with the power and luxury which they bestow, or involves himself and others in hopeless ruin.

When we look into man's organization, organism, and internal functions —his capacities of body—his faculties of mind—and the emotions and desires of his heart—we find a fullness, variety, and perfection—a complexity—a quickness of irritability and a delicacy of feeling, no where else to be found in the world of organized nature; fitting him admirably for the position he occupies in that world; but, at the same time rendering him vulnerable to its thousand influences. If he act on all around, everything around reacts on him. If he shape and fashion them to his liking, they, in turn, modify his organization, impress on him peculiarities of constitution—develop one organ into excessive dimensions, and arrest the growth of another—exalt *this* sensibility, and depress *that*—break up the equilibrium of his vital functions, and derange his system with a greater variety of diseases, than are found in the whole kingdom of organized nature, of which he is at once the monarch and the victim.

Such, young gentlemen, is the being, whose diseases you will be called upon to prevent or heal. To accomplish the former, you must inquire into the action of all the influences, material and mental, to which he may be subjected, and either withdraw him from them, or fortify his constitution against their deleterious impress. But how can you do this, without that power which knowledge bestows? How can you detect the sinister agency of the water that he drinks—or the food which he eats—or the soil which he tills—or the shop in which he labors—or the climate which he breathes—or the passion by which he is consuming away—unless you have studied their nature, and know the relations which they bear to his constitution of body and mind? But, you will be called upon, still oftener, to cure, than prevent his diseases; and how can you perform that part of your mission, without a thorough and diversified knowledge, of the effects of the multitude of agents, which exert their influence on his constitution? In proportion to this knowledge, will be the richness of your resources—according to your sagacity and sound judgment, will be your success in their application.

And, now, have I not shown, that you have undertaken no ordinary task? You may feel discouraged at its magnitude, and disposed to seek some simpler and humbler pursuit; but do not falter too soon. To a chill succeeds a fever; on a momentary revulsion of feeling, from the sudden appearance of danger, there follows a reaction, which nerves the arm and achieves a victory. The morning sun may not be able to send his beams through the mists, and fogs, and clouds, which hang over us in autumn; but he rises higher, and shines with fiercer and warmer rays, until they gradually melt away. To the little child, the knowledge *you* now possess, seems vast and various. At the same age, you might have thought it beyond your reach; yet it *has* been reached, and how? By intuition? By gift? By infusion, while you slept, or lolled at ease, or reveled in luxurious pleasures? All experience answers, no. I say to you, then, array against the future, your recollections of the past; compare yourselves with what you once were, and thence draw hope and expectation, as to what you may become hereafter. Call to mind your labors, and their slow but certain results—word after word, unfolding to you its meaning—fact upon fact, infixing itself in your memory—truth upon truth, rising before you in smiling beauty—and limited generalizations, imperceptibly expanding into great propositions—universal principles—perfect sciences. You owe these conquests to toil; labor is the parent of all knowledge; constancy the secret of all success. Hold on, therefore, to the end. Every day's application will leave something behind you, conquered and brought into subjection to your will; while it lessens that which lies before you. When the engineer projects his railroad, a mountain, which cannot be turned, rises in its course; but he resolves to finish his enterprise. The pickax and the powder-blast are soon heard resounding in its bowels; the hardest rocks, by blow upon blow, at length lie in fragments at his feet; and he finally drives through the mountain in triumph. Imitate his labors, and you will be crowned with his success.

I have spoken a few imperfect words of counsel, and might say many more, but will leave them to my colleagues, who will advise you on all things which belong to their respective departments. For the remainder of my hour, I ask your attention, and that of the Trustees of the College and the Hospital, who have honored us with their presence, to a different subject; but one, I hope, in which both you and they will take an interest. If, in presenting it, I am led to speak of labors and events in which it was my destiny to bear a part, you will not, I trust, feel disposed to judge me with severity. The young man lives in the future—the old man in the past. One dwells with fond anticipation on that in which he hopes to bear a part—the other in fond recollection, of that in which he *has* borne a part; and each is true to the principles of his nature.

My object is to say something on the origin of medical schools and hospitals, in this, our own peculiar and cherished portion of the great land of liberty; and, especially to speak of your own Alma Mater.

To the state of Kentucky, the oldest of the Western sisterhood, belongs the distinction of having chartered the first University in the wilderness. Its franchises permitted the organization of every faculty, usual in the higher seminaries of learning; but a medical school was not in the contemplation of those who asked, or those who gave the charter. The town of Lexington, in early times the metropolis of the West, may justly claim the preëminence of being the first to cherish such a school. About forty years ago, Doctor Samuel Brown, and several other respectable physicians of that town, suggested the founding of a medical department in Transylvania University, but the suggestion was not then acted on; and it was reserved for Doctor Benjamin W. Dudley, in the year 1815, soon after his return from Europe, to revive the project. It was favorably received by the Trustees of the University, and four professors were appointed. In the winter of 1815-16, some of them, without concert, delivered short courses of lectures, to a small number of pupils. In the summer of the latter year, I was invited to join them; but did not do it, until the autumn of 1817; when, for the first time, the Faculty was organized. The professors of that, the pioneer medical Faculty of the interior of the continent, were Doctors Dudley, Overton, Richardson, Blythe, and myself. Our class consisted of twenty pupils, drawn chiefly from the neighboring counties. In the following spring, a public commencement was held, and the degree, to which you are aspiring, was conferred upon a single candidate—Doctor John McCullough—the first ever graduated in the West. Such was the implantation of medical institutions in the Valley of the Mississippi.

Soon after the session ended, believing Cincinnati superior to Lexington as a site for a medical school, I resigned my chair—Materia Medica—and determined on proposing the establishment of a school in this city.

As promotive of the enterprise, a course of lectures on Botany was delivered, in the following summer, to a class of ladies and gentlemen, including several students of medicine; and in the winter, ten or twelve students of medicine, were collected for instruction of a strictly professional character. On the 19th of the following January, 1819, in consequence of a personal application to our legislature, the MEDICAL COLLEGE OF OHIO, in which you have enrolled your names, was chartered; but it did not become organized until the autumn of 1820; when its first session began with a class of twenty-five pupils. Thus came into existence the second medical school of the West, and the first on the northern side of the Ohio river.

* * * * *

You may feel surprised that it should have been thought necessary to show by argument, that the period was not distant, when the West might support two respectable medical schools. But we must not disparage the limited foresight of the men of that day. They predicted the future from their knowledge of the past; and the past afforded no precedent for such a future as was then about to unfold before them. Never before, in any land or age, did a few fleeting years transform the broad face of half a continent, from the wild and savage scenery of nature, into a panorama of cultivated fields, with throngs of intelligent and busy men. At the time that appeal was made, there was in the West, as we have just seen, but one medical college. Now, there are twelve in operation, one temporarily suspended, and three in the forming stage. Thus, on an average, every three years has given birth to a new medical institution. Their distribution is as follows: Lexington, one; Louisville, one; New Orleans, one; St. Louis, two; Chicago, one; Laporte, one; Buffalo, one; Geneva, one; Cleveland, one; Columbus, one; Cincinnati, one. The one in a state of suspension, is at Memphis; and the three projected or not yet fully organized, are at Indianapolis, Evansville, and Nashville. In the twelve schools, there were, last winter, not less than fifteen hundred pupils; and yet large numbers left the West, for the institutions of Philadelphia and New York. From this evidence of the growth of your native or adopted country, you will perceive, young gentlemen, how great it must become, while you are still actors on the stage of professional life. You will realize, I trust, that you are not doomed to spend your days among rude and ignorant backwoodsmen; but in the midst of intelligent and refined communities; and that you should, by deep and protracted study, prepare yourselves for such associations;—for building up a profession worthy of such a state of society, and for performing an honorable part in a thousand works of learning, science, charity, and patriotism.

* * * * *

Since [1820], one generation has been buried and another has risen; the age of the pioneers has passed away; the ax has conquered the forest; infant villages, overshadowed by native trees, have grown into towns, with shadetrees from distant lands; towns have expanded into cities, which abound in the luxuries of every climate; new institutions of elementary learning have, from year to year, come into existence, and abounding opportunities have been brought to the door of every young man who aspires to the study of medicine; and yet, I am sorry to know, that the preparatory learning of the pupils of the West, is now scarcely superior to what it then was. All else has displayed progress—this alone shows no improvement. Young gentleman, a defect so degrading to our noble profession, should not be permitted to continue; and I most earnestly and affectionately, call upon you, now while you are in the springtime of life, to devote to the cultivation of letters, a portion of every day, from the time you leave the college. All that I propose—all that is necessary —lies entirely within your grasp. You have only to resolve, and it will be accomplished. In the year 1840, while making a tour through the central parts of this State, for the study of its diseases, I met with a young physician, who rode with me for a while, and desired to converse on the means of improving himself in the literature and science of his profession. He had entered on its study with but little preparatory learning, and after an attendance on a single course of medical lectures. The result of our conversation, was a determination, on his part, to assign a fixed portion of every day to the study of the elements of literature, and other portions to the cultivation of his profession. He left me, but was not forgotten, and the hopes he had inspired were soon realized. In three years I had the pleasure to see his name in one of our periodicals, to which he became a regular and respectable contributor. His writings brought him reputation; and, in less than seven years from our interview, without attending a second course of lectures, he received an honorary degree from one of the most distinguished schools of the West. I must return to the discourse which has suggested this narrative, but cannot do it until I have exhorted each and all, when you shall become members of the profession, to look faithfully into the literary acquirements of those who desire to become your pupils; rejecting all who are not qualified, and turning them to some other pursuit, or to the academy for further elementary instruction.

* * * * *

Thus it appears that, in its infancy, the Medical College of Ohio had a session of five months, and it must be gratifying to all who participated in the enactment of that regulation to find, that very lately, the physicians of the United States, in general convention, have repeatedly and earnestly recommended the same thing to all our schools. But few, however, have had the courage to prolong their sessions; although professors, very generally, admit the advantages that would flow from such a change. Deeply and thoroughly convinced that its general adoption would confer great benefits on the profession, and through it on society at large, I beg, young gentlemen, that you will bear the subject in mind, and use your future influence in its favor.

* * * * *

In the second month of the first session of the school, a bill was drawn up by one of the Faculty, and laid before the trustees of the township of Cincinnati. It proposed a union of the college, the township, and the state, in the establishment of a public charity. The trustees wisely gave it their sanction, and united with the Faculty in a memorial to the General Assembly, which had already been informed, by Governor Brown, of the application from the Governor of Louisiana. When I carried the memorials and bill to the legislature, they met with opposition; but, after a month of laborious explanation and personal effort, the bill became a law. The state, having in view the relief of her sick boatmen, gave a small sum of money to assist in the erection of a house, and pledged, forever, half the auction duties of the city, toward the support of the patients. The township was to supply the remainder, and the professors of the college were to be its medical and surgical attendants, with the privilege of introducing their pupils for clinical instruction; the fees of admission to constitute a fund for the purchase of chemical apparatus, anatomical preparations, and books for the college. Such was the second step taken by the State of Ohio for the promotion of medical education, the chartering of the college being the first.

An acceptance of the hospital law, by the trustees of the township, was deferred, until after the annual election in the following spring; and, during the time of delay, the subject was canvassed before the people. An argument against it was, that the plan which had been devised, however beneficial to the college, might throw the boatmen of Ohio on the city, beyond the revenues provided by the State, and thus increase the taxes for the support of the poor. To counteract the force of this argument, some of the professors moved their friends to unite with them in a bond of indemnity to the township. The bond was executed and tendered; and at the annual election, the ticket in favor of an acceptance of the charter, was carried by a vote of seven hundred to five hundred. The new trustees of the township confirmed what their predecessors had undertaken, and the hospital was erected. Five years afterward the state manifested itself a third time in favor of the school, by a grant of money with which the edifice since occupied was built.

Such, young gentlemen, was the origin of the school to which you have resorted, and the hospital in which you are to receive clinical instruction. With the increase of the city and of western population, the wards of the latter have been gradually multiplied or enlarged, and for sometime past, nearly two thousand five hundred patients have been admitted annually. Among so great a number, almost every form of disease that prevails in the country in which you are to practice the profession, will, of course, be found; and we may, without boasting, or one invidious comparison, affirm, that no where else in the West, or in the world, can better opportunities for clinical observation be enjoyed.

And here allow me to say, that all the tendencies of the age are to the study of medicine and surgery in hospitals. In them it is, that a student learns pathological anatomy, diagnosis, the art of prescribing, and operative surgery. The laboratory is not more necessary for the study of chemistry, or a garden of plants for the study of botany, than a hospital for the study of practical medicine and surgery. The time has passed by, when students will flock to men of genius (as they once flocked to Boerhaave) for the purpose of listening to expositions of theory, or to be amused with creations of imagination. The school which is not based on a hospital, may have learned and able professors; but the results of their teaching can never be satisfactory to the student, who seeks to make himself a good practical physician and surgeon. A mathematician might compose an admirable system of navigation, but you would prefer to trust yourselves, on a dangerous voyage, with one of more practical skill, though less learned. In the arrangement for the session now opening, the Faculty have made ample provision for clinical teachings, by assigning each alternate afternoon to the hospital; and I would earnestly advise all, except those who are but entering on their studies, to be punctual in their attendance on those days.

In conclusion, permit me to say something more of myself. Experience, down to the present hour, has shown, that every medical school, attempted in the West, has suffered more or less of revolution in its forming stage; the inevitable effect, perhaps, of a new and unsettled state of society. The school at Lexington had its early revolutions—that in Cincinnati was overtaken with the same infantile convulsions. After its second session, I was separated from it; and, as you are aware, have, at different periods, labored in several other institutions. They flourished; and I was happy in the consciousness of an honest fidelity to their fame and interests; but my thoughts often went back to the city, where, in the twelfth year of its settlement, I had begun the study of medicine, and in which, twenty years afterward, I had labored to found the

CINCINNATI COLLEGE.

College. My heart still fondly turned to my first love—your Alma Mater. Her image, glowing in the warm and radiant tints of earlier life, was ever in my view. Transylvania had been re-organized in 1819, and included in its Faculty, Professor Dudley, whose surgical fame had already spread throughout the West, and that paragon of labor and perseverance, Professor Caldwell, now a veteran octogenarian. In the year after my separation from this school, I was recalled to that; but neither the eloquence of colleagues, nor the greetings of the largest classes, which the University ever enjoyed, could drive that beautiful image from my mind. After four sessions, I resigned; and was subsequently called to Jefferson College, Philadelphia; but the image mingled with my shadow; and when we reached the summit of the mountain, it bade me stop, and gaze upon the silvery cloud which hung over the place where you are now assembled. Afterward, in the medical department of Cincinnati College, I lectured with men of power, to young men thirsting for knowledge, but the image still hovered around me. I was then invited to Louisville, became a member of one of the ablest Faculties ever embodied in the West, and saw the halls of the University rapidly filled. But when I looked on the faces of four hundred students, behold, the image was in their midst. While there, I prosecuted an extensive course of personal inquiry into the causes and cure of the diseases of the interior of the continent; and in journeyings by day, and journeyings by night—on the water, and on the land—while struggling through the matted rushes where the Mississippi mingles with the Gulf—or camping with Indians and Canadian boatmen, under the pines and birches of Lake Superior, the image was still my faithful companion, and whispered sweet words of encouragement and hope. I bided my time; and after twice doubling the period through which Jacob waited for his Rachel, the united voice of the Trustees and Professors, has recalled me to the chair which I held in the beginning.

The first moments of reunion are always passionate, and wisdom places little confidence in what is then promised; nevertheless, I must declare to you, that I stand ready to pledge the remnant of my active life, and all the humble talents with which the Creator has endowed me, to her future elevation; and were I to put up the prayer of Hezekiah, for length of days, it would be to devote them to her aggrandizement; and, for the pleasure of seeing her halls overflowing with inquiring pupils, attentively listening to ardent, learned, and eloquent professors. With this pledge, those who watch over her welfare, and those who govern the Hospital which she caused to be erected, are now silently mingling theirs; while you, I trust, are resolving that your own lives shall spread abroad her fame. Thus will she rise, and gracefully move onward and upward, until she stands in beauty and honorable rank, among her distinguished sisters of the Union—the pride of her sons, and a blessing to society.

Systematic Treatise on the Principal Diseases of the Interior Valley of North America*

(1850)

The publication of volume I of the *Systematic Treatise* in the summer of 1850 marked the culmination of Drake's career as citizen and scientist in the trans-Allegheny West. Directed ostensibly at physicians, it was an attempt to complete that catalogue of the endemic and epidemic diseases of the West which he had begun in *Notices concerning Cincinnati* (1810) and the *Natural and Statistical View, or Picture of Cincinnati and the Miami Country* (1815). In the process, however, as a good Baconian convinced of the necessity for gathering all relevant data prior to the construction of hypotheses, Drake included as full a description of the region as a region, as personal observation and the information provided by his correspondents would permit. It is Drake's title which is misleading. His methods were the same as those of the descriptive volumes of 1810 and 1815, and his intentions similarly were to present data systematically, not to construct a "system" which would explain the data. In the *Systematic Treatise,* in other words, Drake created at once a characteristic work of Baconian science and a kind of emigrant's guide, useful as well to citizens of the region as to outsiders. Like a cabinet of minerals or a botanical garden, the *Systematic Treatise* was one of those artificial institutions through which Nature's ways might be revealed. Like a literary convention, a railroad, or a lecture on the "history, character, and prospects of the West," the *Systematic Treatise* was one of those artificial institutions through which the natural unity of the West as a region might be recognized, and hence possibly realized.

Described in his dedication as an "attempt to lay an extended foundation" for a history of the diseases of the Mississippi Valley, Book 1 of the *Systematic Treatise* contained all that Drake knew about the geology, the climate, and the character of the population of the West, conceived as a contribution to the understanding of

---

* *A Systematic Treatise, Historical, Etiological, and Practical, on the Principal Diseases of the Interior Valley of North America, As They Appear in the Caucasion, African, Indian, and Esquimaux Varieties of Its Population.* By Daniel Drake, M.D. Cincinnati: Winthrop B. Smith & Co., Publishers. Philadelphia: Grigg, Elliot & Co. New York: Mason & Law. 1850. Pp. v–viii, 447–49, 611–13, 637–53, 703–705, 709–28, 866.

general etiology. For Drake this was the logical way to begin, but not only because of his long interest in these subjects. Book 1 was to serve as introduction to Book 2, on the febrile diseases. This would be organized under five heads: Autumnal Fever, Yellow Fever, Typhous Fevers, Eruptive Fevers, and Phlogistic Fevers, or the Phlegmasiae. "The transition from general etiology," Drake explained, "to that fever which, in its origin, has a close connection with soil and climate, is natural; and the transition from the phlegmasiae to many other forms of disease, will be found equally natural, and hence I have placed them last, although in a system of elementary pathology, or nosology, they should stand first."[1] With the exception of his hundred page essay on autumnal fever, included in the first volume of the *Systematic Treatise* in 1850, however, Book 2 was never completed. "Although he had bestowed a vast amount of care on the matter and its arrangements, evident in the numberless additions at different periods to the former, and alterations of the latter, as well as in the entire remodeling of many chapters," his editors noted in 1854, little of the three thousand manuscript pages had been left in a condition to place in the printer's hands.[2]

In one sense this was strange. Work on Book 2 was begun not in 1850 but long before, and until his final illness Drake remained in vigorous good health. Ultimately it was a failure of imagination which must be seen in his inability to examine the etiology of diseases which seemed to bear little or no connection with the environmental conditions of their occurrence. As Drake must have recognized, description alone was inadequate when induction was impossible; yet to the end he remained true to the Baconian methods which had served him so well in the past. He based his conclusions on observation, and resisted with all his will the generation of explanatory hypotheses which would have to be accepted a priori.

[1] *Systematic Treatise*, p. 702.
[2] *A Systematic Treatise . . . on the Principal Diseases of the Interior Valley of North America.* Second Series [Vol. II], S. Hanbury Smith, M.D., and Francis G. Smith, M.D., eds. (Philadelphia, 1854), p. 1.

PREFACE

THE OBJECT PROPOSED in the following work, is to give an account of the causes, symptoms, pathology, and treatment, of the principal diseases of an extensive portion of NORTH AMERICA—its INTERIOR VALLEY. In exploring it, for the purpose of collecting facts, the Author endeavored to leave behind him all opinions but the single one, that he who would observe correctly, must have no theories either to maintain or destroy.

To say that he has always been faithful to this rule of observation, would be rash; but, he *may* say, that he has sincerely and earnestly desired, to keep himself under its sway. He may affirm, still further, that it has been his constant aim, to purify from error, the facts he was collecting; and he trusts, therefore, that all the more important will be found substantially correct. Nevertheless, the country to which the work relates, is of such vast geographical extent, that he cannot doubt, but that every reader will detect some errors, in what relates to the topography, climate, or diseases of his own locality.

But while the object of this work is to embody facts, drawn, by personal intercourse, from numerous living physicians, or from publications made by them and their predecessors, and to combine the whole with his own observations, he has not been unmindful of the discoveries and improvements in etiology, pathology, and practice, of older and more enlightened countries; but sought, as far as they have become known to him, to amalgamate the foreign with the indigenous, and thus present to his brethren of the Interior Valley, a book of practice, so full on all the diseases of which it treats, as to make it a useful manual for daily reference. He is obliged to admit, however, that, while seeking after knowledge among the physicians of his own country, he could give but little attention to the writings of those who live in other countries.

Long journeys of observation, repeated through a large part of several years, with elementary teaching in winter, have much abridged the time for bibliothecal research; and, perhaps, even diminished the taste for that mode of inquiry.

Extensive as his explorations have been, large regions of country remain unvisited; and many conclusions, at which he has arrived, might possibly have been different, had the facts, which those regions could have furnished, been obtained by him. Yet, as his personal examinations were carried through eighteen degrees of latitude, and nearly as many of longitude, he trusts that facts which may, in some degree, stand as representatives of the whole, *have* been collected; and, therefore, that no general conclusion will be found radically wrong.

As announced on the title page, it is the design of this work to treat of the diseases of the Caucasian, Indian, and African Varieties of our population, in contrast and comparison with each other—the first being the standard to which the other two are brought. For this purpose, no other country presents equal advantages; since, in no other, do we find masses of three varieties of the human race, in permanent juxta-position. There is, moreover, a fourth variety, the Mongolian, represented by the tribes of Esquimaux, whose huts of snow are scattered across the northern extremity of the Valley; who subsist on a simpler diet, and live in a lower temperature, than any other known portion of the human race; and, therefore, present, in their habits and physiology, many points of interest, to which he has given such attention as the books of voyages and travels, have enabled him to bestow.

In his traveling intercourse, with his brethren and collaborators of the Great Valley, from Florida, through to Canada, inclusive, although going among them generally, without letters of introduction, he has, with very few exceptions, been received in the kindest manner, and afforded every facility in their power; for which he cheerfully makes this public acknowledgment. To designate, by name, all who manifested a high and encouraging interest in his enterprise, would be to form a catalogue too long for introduction here; but of gentlemen residing without the United States, he is not at liberty to omit the names of Professor Joseph Morrin, of Quebec, Professor Archibald Hall, of Montreal, and Captain John Henry Lefroy, of the Royal, Magnetical, and Meteorological Observatory, Toronto, as having afforded him important assistance.

While prosecuting his researches, he visited the larger part of the military and naval posts of the Interior Valley, both American and British, bearing a letter, explanatory of his object, from Major General Scott, and received, at each, such facilities as were practicable.

He desires, also, to record the names of several young gentlemen, who have rendered him various kinds of aid, in the preparation of the work for the press. They are Doctor Charles A. Hentz, Mr. Theodore S. Dana, and Mr. Charles A. Caroland, students of medicine, and Mr. David Smith; each of whom performed the part assigned to him, in the most faithful and zealous manner. Notwithstanding this, however, it is feared that, in the statistical portions, some errors may be found, though none he trusts of great magnitude.

The hydrographical map, which forms the frontispiece of the book, seemed indispensable to its plan. The reader will perceive, that it is not designed to represent civil and political divisions; but to assist in connecting what is said on medical topography, climate, and the limits imposed by latitude and altitude, on certain diseases, into one system. It was drawn by Major D. P. Whiting, U.S.A., who also drew several of the topographical maps; the remainder and larger part were from the accurate pencil of Captain C. A. Fuller, U.S. Civil Engineer. They were all executed under the author's inspection, out of the best materials he could command; for a part of which, together with many useful suggestions, he is indebted to the veteran Topographical Engineer, Colonel Stephen H. Long, U.S.A. The engravings are on stone, by a young German artist, Mr. A. Wocher, of Cincinnati, and will, the author trusts, be found not unworthy of the typographical execution, under the supervision of Mr. Charles H. Bronson; whose abilities and taste as a practical printer, have overcome many difficulties, resulting from the introduction of more than a hundred Statistical Tables, and from the absence of the Author, at the University of Louisville, during the past winter, while the work was in the press. Finally, the Author desires to express his obligations to Messrs. Winthrop B. Smith & Co., for their willingness to turn aside from their ordinary business, and become the publishers of the largest original work which, as yet, has been written and printed in the Interior Valley; thus rendering it, in all respects, an indigenous production.

The germ of this work, was a pamphlet entitled *"Notices concerning Cincinnati,"* printed for distribution, forty years ago. The greater part of the Interior Valley of North America, was at that time a primitive wilderness. Ten years afterward, the author formed the design of preparing a more extended work, on the diseases of the Ohio Valley; but being called to teach, he became interested in medical schools, which, with the ceaseless labors of medical practice, for the next twenty years, left no time for personal observation, beyond the immediate sphere of his own business. Meanwhile, settlements extended in all directions, with which the area of observation expanded; and the plan of the promised work, underwent a corresponding enlargement. He could look upon this long delay, without regret, if he were conscious, that his work had, thereby, been rendered proportionally more perfect; but he is obliged to confess, that the labors of a pioneer in many things, have not been auspicious, to a high degree of perfection, in any; and, that a new country, with its diversified scenes and objects, is not favorable to the concentration of attention, upon any one.

He expected to have introduced into the first volume, the article, Yellow Fever, but found it would swell the book to an inconvenient size. It will make the first part of the second volume; the materials for which have been chiefly collected, and considerable portions of it written, so that the author hopes it may be committed to the press in about a year.

On the manner in which the work (when finished) will be received by the profession, he does not attempt to form a prediction; but has entire confidence in the justice of those for whom it is especially designed. He has, also, no reason to doubt, that the periodical press of the country, will treat him with equal justice; and he desires nothing more. If a second edition should be demanded, the errors which may be pointed out, would be corrected, and new facts and observations introduced: If the work prove a failure as it respects public favor, the author will not be without his reward; for he has found enjoyment in the labor of producing it; and, having confidence in its general accuracy, knows that it must stand as a great collection of facts; a picture of the etiological condition and the diseases, of a newly settled country, in the middle of the nineteenth century; with which future, and more gifted, medical historians, will compare the causes, phenomena, and treatment of the maladies which may then prevail.

DAN. DRAKE, M.D.

Cincinnati, December 20, 1849.

* * * * *

[BOOK I, GENERAL ETIOLOGY.]

PART II, CLIMATIC ETIOLOGY. CHAPTER I, NATURE, DYNAMICS, AND ELEMENTS OF CLIMATE.

SECTION I. GENERAL VIEWS

I. CLIMATE OCCASIONS DISEASES. As no fact in etiology is more universally admitted, than the influence of climate in the production of disease, it follows that he who would understand the origin and modifications of the diseases of a country, must study its meteorology. The effects of climate are both predisposing and exciting. Thus, the long-continued action of a particular kind or condition of climate, may bring about such changes in our physiology as to incline us to some particular form of disease; while sudden changes often act as exciting causes to other diseases, to which we may be inclined, from agencies not connected with climate. Again, the influences of climate are both direct and indirect. The former result from the immediate action of the atmosphere on our systems—the latter from its action on the matters which are accumulated on the surface of the earth, which are thus made to send forth agents of an insalubrious character. Thus, the same state of the earth's surface which in one climate may prove highly pernicious, in another may be altogether harmless.

II. CLIMATE CURES DISEASES. But climate must not be studied with a reference to etiology only; for it can cure as well as occasion disease. It modifies the effects of blood-letting, medicines, and regimen; and although it maintains some diseases against the united powers of the most active and appropriate articles of the materia medica, it cures others in the absence of the whole. Considered as a therapeutic agent, it is, when skillfully ordered, entitled to great confidence. Its action is not often speedy, but the certainty of its salutary effects, in general, compensates for their slow development.

III. DEFINITIONS OF CLIMATE. In physical geography, the word climate expresses a zone of the earth, running parallel to the equator, of such width that the longest day at its northern limit is half an hour longer than that of its southern limit, supposing we are in the northern hemisphere; but in etiology and therapeutics, the term is used in a different sense, and simply expresses states of the atmosphere. These states involve, or consist in, varying quantities or qualities of certain elements of the air itself—its caloric, light, and electricity; its aqueous vapor, fogs, mists, and clouds; its dews, rain, hail, frost, and snow; its weight and density; its movements or winds; it factitious gases, and mechanical impurities; all of which may be very different in different times or places of the same geographical climate, and nearly the same in different zones.

IV. CLIMATE OF A GLOBE OF UNIFORM SURFACE. If the earth, with its present form and relations to the sun, had a smooth, uniform, terrestrial surface, of the same mineral composition, and were destitute of both air and water, the temperature of its crust in every latitude would bear a fixed relation to the solar influence. If, then, an atmosphere were added, winds would be created, and blow with a uniform velocity and direction, as the same seasons returned. But if mountains were anywhere upheaved, or the atmosphere should be impregnated with aqueous vapor and electricity, this uniformity would be disturbed; which prepares us for considering the proper elements of climate.

V. ELEMENTS OF CLIMATE ON THE GLOBE AS IT EXISTS. The crust of the earth is not uniform in chemical composition or surface; it abounds in mountains, plains, and valleys, distributed in a very irregular manner; portions of it are densely overshadowed, while others are destitute of forest; the larger part is covered with oceans, lakes, rivers, and swamps; an elastic atmosphere rests upon the whole; and every part—solid, fluid, and aeriform,—is permeated by electricity. Were the earth, with this surface, removed from the influence of the sun, the phenomena of climate would be annihilated; in that luminary, then, reside the dynamics on which they depend; and the rays of light and heat, are the efficient agents by which its quickening influence is exerted on the earth. When they reach its surface, their effects are, substantially, according to the angle of incidence; but falling on material elements so diversified, a vast variety of movements are generated, and results or phenomena the most complicated, are incessantly developed. Thus, unequal degrees of heat are accumulated in portions of a continent having the same latitude, but different elevations; or, as they are covered with forests or destitute of shade; the heating and cooling of the land and water do not proceed according to the same laws; aqueous vapor is raised into the air from the oceans and transported over the continents, by winds, generated by the unequal heating of the atmosphere, to be condensed and precipitated, on regions remote from those in which the evaporation took place; in the condensation of the vapor, caloric is liberated,—by the evaporation of the fallen water it is absorbed; the clouds intercept the rays of the sun and limit their effects upon the surface, but, at the same time, arrest and throw back much of the caloric which radiates from the surface; dead calms and hurricanes rapidly succeed each other; electrical phenomena are generated; the luminous solar rays are decomposed by the clouds, which they tinge with various colors; finally, different gaseous exhalations, from decomposable matters lodged on the surface of the earth, ascend into the atmosphere.

VI. THE ELEMENTS OF CLIMATE NOT THE SAME IN DIFFERENT PARTS OF THE EARTH. It results from what has been said, that the elements of climate are not precisely the same in any two regions of the globe; and, therefore, that the climate of every region, even in the same latitude, must possess some peculiarities; the causes of which are to be sought in the physical geography and hydrography of the region itself, and of those by which it is immediately surrounded. For this aid to the study of the Climate of the Interior Valley of North America, the necessary facts have been given in Part I; but as they are scattered through several hundred pages, it will be useful to collect and condense them into one section.

* * * *

CHAPTER V, ELECTRICAL PHENOMENA: DISTRIBUTION OF PLANTS AND ANIMALS.

SECTION I. ATMOSPHERIC ELECTRICITY, THUNDER STORMS, HURRICANES

I. ATMOSPHERIC ELECTRICITY. It would be absurd to question the value, to a physician, of a thorough knowledge of the temperature, winds, weather, and moisture of the country in which he is to ascertain the causes, and prosecute the cure of diseases. If the study of its electrical condition and phenomena, cannot, in the present stage of our knowledge, be shown to possess an equal importance, it is by no means to be neglected. The mysterious, but apparently, intimate relations between light, caloric, magnetism, galvanism, and electricity, suggest that they are, perhaps, but one agent in different states or modes of action, the whole of which should be studied in connection. The manifest part which electricity plays in the systems of certain aquatic animals, as the *gymnotus electricus,* and the influence it may be made to exert on the nervous system of man, still further point to it as an agent, upon which the physician is bound to direct a portion of his attention. After the experimental demonstration, by our great countryman, Franklin, of the identity of electricity and lightning, by which the immense amount of that fluid in the atmosphere was made known, many physicians vaguely indulged the opinion, that it performs an important part in the animal economy. The subsequent discoveries of Galvani and Volta, gave a new impulse to these speculations; and suggested many experiments, on the effects of galvanic electricity upon the living body, both in health and disease. No satisfactory results have, however, been reached. The subject, nevertheless, is not likely to be relinquished by the imaginative; and, even the most sober-minded must admit, that an agent so powerful, so universally present, and so operative in many of the secret processes of inorganic nature, can scarcely fail to perform some important part in the living body. Perhaps the proper mode of studying it, in reference to that body, has not yet been discovered. That atmospheric air performed a vital part, in the kingdom of organized nature, was known from the beginning; but it was reserved for the last century to discover its mode of action. We are, perhaps, at this time, in a similar condition, in reference to electricity; and some future generation may devise instrumentalities, by which modes of action and effects, of which, at present, we can form no conception, will be rendered plain.

Electricity exists at the surface of the earth, and in the atmosphere at all times, and is forever circulating between them. Experiment has shown, that during combustion, the æriform products which escape are positively electrified, leaving the residuum negative; but the process by which the greatest amount of electricity makes its way into the atmosphere, is solar evaporation; the vapor, which arises from every wet or watery surface, in which any saline matter is held in solution, being positively electrified. Wherever, then, there is a high temperature with a high dew point, the atmospheric, electrical phenomena, are of a striking character; and where the temperature is low, and the evaporation feeble, they are correspondingly reduced. The condensation of vapor into fog, rain, or dew, appears to increase atmospheric electricity, setting it free, in conjunction, if we may so speak, with the latent caloric of the vapor. The tension or quantity of atmospheric electricity, as measured by the electrometer, is greater, in a tranquil state of the atmosphere, in winter than summer; it would seem that, as the vapor of the atmosphere is more and more deposited by increasing coldness, its electricity is disengaged and left behind. There is, also, a diurnal variation in the amount of electricity at the same place, when the weather is serene.

* * * * *

I am not in the possession of any series of experiments on the electrical condition of the atmosphere, at any place in the Interior Valley. Two questions may be here proposed: *First*. What are the effects, if any, on the human constitution, of a highly positive or highly negative state of the atmosphere, and of the sudden transition from one to the other? *Second.* In what manner, if at all, does electricity contribute to the production or spread of epidemic diseases? I shall not undertake a reply to these questions; but dismissing the consideration of electricity as it exists in *equilibrio,* or a neutral state, say something of its phenomena and effects when in a state of perturbation [*i.e.* thunder storms].

* * * * *

PART III, PHYSIOLOGICAL AND SOCIAL ETIOLOGY.

CHAPTER I, POPULATION.

SECTION I. DIVISION INTO VARIETIES

THE INTERIOR VALLEY OF NORTH AMERICA, embraces four of the five varieties, into which naturalists have commonly divided the human race. In reference to their numbers, civilization, and interest to the physiologist and physician, they stand in the following order: *First,* the CAUCASIAN; *Second,* the AFRICAN; *Third,* the NORTH-AMERICAN INDIAN; *Fourth,* the MONGOLIAN. The three former, existing in contiguous or intermingled masses, present opportunities for studying the comparative physiology and diseases of different races, which we should not neglect. The last, known under the name of *Esquimaux,* are but a handful, compared with either of the other varieties, and live contiguous to but one of them—the Indian.

The CAUCASIAN races are found in large numbers, within the tropics, to the latitude of forty-seven degrees north. Beyond that latitude, they are met with but in trading establishments, missionary stations, and other small settlements, on the rivers and shores of Hudson Bay, and Davis' Straits, up to the fifty-eighth or sixtieth degree. The most populous zone is between the thirty-fourth and forty-fourth degrees; the line of greatest length and density of population being near the latitude of thirty-nine degrees. A great majority of the whole reside east of the Mississippi.

The AFRICAN, or NEGRO variety—nearly all of whom are natives of this continent, though many were born out of the VALLEY—have a more southern residence. Extending upward from the tropics, they gradually become more numerous in proportion to the whites, to the latitude of thirty-two or thirty-three degrees; when they begin to decrease, and above the thirty-ninth degree are found chiefly in large cities; though single families, or small settlements, are to be met with, beyond Lake Erie and Lake Ontario, as far north as the latitude of forty-four degrees. Like the whites, they are most numerous on the eastern side of the Mississippi.

The NORTH-AMERICAN, or INDIAN variety, on the other hand, nearly all reside west of the Gulf of Mexico and of the Mississippi, up to the forty-fourth degree of latitude; beyond which they are found over the interior of the continent generally, but are much more numerous to the west than east. After passing the fiftieth parallel, the number diminishes rapidly, and very few are found within the Polar Circle.

The MONGOLIAN variety, of which the Esquimaux are the representatives, succeed in the north to the Indian, and are found on the entire polar margin of the Valley.

Intending to make the diseases of the African, Indian, and Esquimaux varieties, respectively, the subjects of special dissertations, I shall dismiss them, until the history of the diseases of the Caucasian races is finished.

SECTION II. CAUCASIAN VARIETY—HISTORICAL, CHRONOLOGICAL, AND GEOGRAPHICAL ANALYSIS.

I. CURVES OF MIGRATION FROM EUROPE. Western Europe, either directly or indirectly, has given our Valley its Caucasian population. The emigrating zone of that continent, extends from the south of Spain, in latitude thirty-six degrees, to the middle of Sweden and Norway, in the sixtieth parallel; and the emigrants from it have settled chiefly between the eighteenth and forty-eighth degrees. The principal emigration, however, has been from the region lying between the forty-fifth and fifty-fifth parallels; to that part of this continent comprehended between the thirty-fifth and forty-fifth degrees. Thus, the curve of emigration bends southwardly, and in traversing the Atlantic Ocean, has sunk, on an average, ten degrees to the south. The different lines of emigration, have, moreover, not often crossed each other; and hence, those who resided furthest north in Europe, now reside furthest north in America. The greatest exception to this remark, was the emigration of the French to Canada, while the English, from a higher latitude, were emigrating to Virginia and the Carolinas. On the other hand, the Spaniards, from the southern shores of Europe, settled around the Gulf of Mexico; and the existing emigration from beyond the Baltic, is to the regions west of Lake Michigan. Let us now look at these migrations in the order of their occurrence.

II. THE FRENCH. The beginning of the settlement of our Valley was in the North, and the first immigrants were French. As early as 1506, only fourteen years after the discovery of America, they had made a map of the Gulf of St. Lawrence, and in 1534, Cartier entered the St. Lawrence river. In 1535, he ascended it to the island of Montreal. After a long period of suspended operations, Quebec was founded by Champlain, in 1608. Emigration from France was then recommenced, and continued through the St. Lawrence, for more than one hundred and fifty years; that is, till Canada was ceded to Great Britain, in 1763. The settlers planted themselves on either bank of the St. Lawrence and its tributaries; and extended along the Lakes, chiefly upon their northern shores and intervening straits, to Mackinac and Lake Superior; always arranging themselves in open villages. In 1674, Father Marquette, and M. Joliet, a trader, entered the basin of the Mississippi, by Wisconsin River; and soon afterward their countrymen began settlements, which, at length, spread as far south as the latitude of thirty-eight degrees. Peoria, Cahokia, and Kaskaskia, in the western part of the State of Illinois; Carondelet, St. Charles, and St. Louis, on the opposite side of the Mississippi, in the State of Missouri; Vincennes, in the State of Indiana, and Fort Du Quesne—now Pittsburgh—are among the fruits of this early enterprise, among the upper and eastern tributaries of the great river.

A few facts connected with this, the earliest colony of our Valley, belong to its medical history. *First*. The pursuits and modes of life of the immigrants and their descendants, have always been remarkably simple. *Second*. They have not dispersed and intermarried, to any great extent, among the immigrants from other parts of Europe. *Third*. Their long canoe voyages up the great lakes and their tributary streams, gradually produced a peculiar class of men, generally called *Voyageurs,* of whom more hereafter.

At the present time, the chief portion of our northern French population is found, as indeed it has always been, in Lower Canada.

In 1683, De La Salle undertook to descend to the mouth of the great river, discovered by Father Marquette. This he accomplished, and returning by the same route, departed for France; where he promoted a southern emigration. Appointed to the command of the first expedition, he missed the mouths of the Mississippi, the object of his voyage, and landed, in 1685, on the shores of Matagorda Bay, in Texas, where he built Fort St. Louis. In 1687, he was assassinated by one of his own men, and no permanent settlement followed. In the same year, Tonti, from Canada, descended the Mississippi to Arkansas River, on which he established a post. After the unsuccessful expedition of La Salle, nothing more was done by sea, until 1699, when a settlement, under M. D'Iberville, was effected, on Biloxi Bay, whence excursions were made into the interior. In 1717, New Orleans was founded by the same leader. Colonists continued to arrive, and the Mississippi was ascended, until the settlements on the Illinois River were reached; and thus a curved zone of French population, extending from the estuary of the St. Lawrence to that of the Mississippi, traversed the great Valley, through seventeen degrees of latitude; before any other European inhabitants had entered it, except a few Spaniards in the south. In addition to their settlements on the Mississippi, the French made others, to a more limited extent, on the Arkansas, Red, and Mobile Rivers; in all cases confining themselves to the banks of the streams. A small and often interrupted current of immigration, continued until 1769; when Louisiana passed into the possession of Spain. Several years before that event, a considerable colony of French migrated from Arcadia—Nova Scotia—and settled on that part of the Mississippi, which is called the Arcadian Coast.

The French of the lower Mississippi, and their descendants, called Creoles, like their brethren on the St. Lawrence, lead simple, and, in the main, temperate lives; their pursuits are chiefly agricultural and commercial; they intermarry with the surrounding population rather more than those of Canada.

It would be interesting to trace out the influences of climates so distant, and soils so different, as those of Quebec and New Orleans, on people of the same national blood; but my intercourse with them has been too limited to justify the attempt.

III. THE SPANIARDS. In 1528, the Spaniards, under Narvaez, effected a landing in Appalachee Bay, Middle Florida, and made an incursion into the interior, coming out to the Gulf of Mexico, at Pensacola Bay, West Florida, of which they were the discoverers; but they left no permanent settlement. In 1538, De Soto effected a landing at Spiritu Santo, now Tampa Bay; and wandered into the interior as far as the Mississippi, of which he was the discoverer. Passing far west of that river, he returned to be buried beneath its waters; and left no permanent settlement in the country. In 1540, Pensacola Bay was again visited by the Spaniards, from Cuba; after which, for a long period, West Florida was neglected. Meanwhile, however, Spanish settlements, in the character of Roman Catholic, Missionary stations, were extended from Mexico, through Texas, to Red River; of which Nacogdoches and Natchitoches were the most important.

In 1689, attention was again turned to Pensacola Bay, and a Fort, called the Barrancas, was built near its mouth. This was followed, in 1693, by a settlement, where the town of Pensacola now stands. Thus Spanish immigration into West Florida was begun, and continued until 1763, when the whole of Florida was ceded to Great Britain; an event which was followed by the emigration, to Cuba and Louisiana, of a large portion of the inhabitants. After the lapse of twenty years, it was restored to Spain; when these people returned in considerable numbers; but, on the sale of Florida to the United States, in 1822, the greater part of them left it.

The Spanish immigration to the banks of the Mississippi, and its tributaries, began with the transfer of Louisiana to Spain, in 1769, as already mentioned. The new-comers extended their settlements up the great river to the Missouri; but on the retrocession of the country to France, in 1802, and its sale to the United States, in 1803, most of the Spanish population again emigrated to Florida, Cuba, and Mexico. Thus the Spanish Creoles, at last, make but a feeble element of our population. In modes of living and physiology, they resemble the French, much more than they do the Anglo-Americans; and have intermarried with the former more extensively than with the latter. If we pass from West Florida and Louisiana, to the Valley of the Rio del Norte, we find a larger Spanish population extending up to Santa Fe and the Valley of Taos, and westwardly to the Sierra Madre.

IV. THE BRITISH. The next immigration into our Valley was British, and took place, both on its northern and southern borders. Immediately after the cession of Canada, in 1763, the English, Scotch, and Irish began to enter it by the St. Lawrence, and these streams of immigration have continued ever since; that from Ireland being, in the latter years, much greater than both the others. In Canada East, the immigrants from Great Britain and Ireland, with their descendants, make a large porportion of the whole population; and, in Canada West, there are few others, except immigrants from the United States. They have spread out, in detached trading establishments, to the north-west, as far as Lake Winnipeg and Hudson Bay; in fact, to the west of both, up to the Rocky Mountains. More enterprising and diversified in their pursuits than the Canadians—as the French are called—they are, at the same time, more addicted to a full diet and intemperate drinking.

The British emigration to West Florida, during the twenty years which England held that Province, was not very great. Yet Pensacola and Mobile were once English towns; and the first notice we have of the medical topography and fevers of those localities, is to be found in the well-known Essay on the Diseases of Hot Climates, by Dr. Lind, an English naval surgeon. After the restoration of Florida to Spain, in 1782, most of the British population withdrew.

Having thus traced out the only *direct* emigrations from Europe to our Valley, which preceded the *indirect,* or that of Europeans and their descendants from the Atlantic states, we are now brought to the latter which, as we shall see, make up the mass of its population.

V. IMMIGRANTS FROM THE ATLANTIC STATES. Before proceeding to speak of the peopling of our Valley, by emigrants from the Atlantic States, it is proper to give the dates of the settlements of those states, which, in the order of time, are as follows: Virginia, 1607; New York, 1615; New England, 1629; Delaware, 1630; Maryland, 1632; Pennsylvania, 1643; New Jersey, 1650; North Carolina, 1660; South Carolina, 1670; Georgia, 1733.

The first advances of population to this side of the Appalachian Mountains, were from the colonies of North Carolina, Virginia, and Pennsylvania, into a region extending from the Tennessee River to Lake Erie. Permanent settlements were begun in East Tennessee, as early as 1761; in Western Virginia and Pennsylvania soon afterward. The settlement of Kentucky, then a part of Virginia, began in 1774; that of Ohio, by immigrants from Massachusetts, in 1788; that of Indiana and Illinois, about 1795; that of Mississippi and Alabama, at a still later period; that of Florida, in 1822; of the states beyond the Mississippi, in 1804; of Western New York, including the coasts of Lakes Erie and Ontario, about 1788; and that of Texas, by Americans, in 1822. Thus the oldest Europo-American settlements in the Valley, are those of Eastern and Middle Tennessee, North-west Virginia, South-west Pennsylvania, and Kentucky.

The majority of the people of Western New York are either from the eastern portions of that state, or from New England; those of Western Pennsylvania, from the middle portions of that state, embracing a large proportion of Irish, Scotch, and Germans; those of Western Virginia and Kentucky, from the eastern and middle regions of the former, and from Maryland; those of Tennessee, from North Carolina; those of Alabama and Florida, from South Carolina and Georgia; those of Mississippi, Louisiana, and Arkansas, from North Carolina, Virginia, Kentucky, and Tennessee; those of Texas, from all the states just named; those of Missouri, from North Carolina, Kentucky, Tennessee, and Virginia; those of Ohio, from New Jersey, Massachusetts, Connecticut, and other New England States, New York, Pennsylvania, Maryland, Virginia, and Kentucky; those of Indiana, from Ohio, Kentucky, and Pennsylvania; those of Illinois, from Ohio, Kentucky, and New York; those of Iowa, from Kentucky, Ohio, Indiana, and New England; those of Michigan and Wisconsin, from New York and New England.

These statements must be received in the most general sense; for, though various regions have a great predominance of people from some of the old states, still the intermingling in every part of the Valley has been very great. South of the Ohio River, both east and west of the Mississippi, the chief elements of this intermixture, are from the slaveholding, Atlantic States, south of Pennsylvania; while north of that river, they are derived from the non-slaveholding states, including the one just named. As we advance westwardly from the Alleghany Mountains, into the newer states of the Valley, the elements of variety display a regular increase; and thus the later the settlement of any portion of the Valley, the more is it compounded; which is especially true of the non-slaveholding states. It is scarcely necessary to add, that these immigrants, together with their descendants, constitute the greater part of our population, from the Lakes to the Gulf of Mexico. There are, however, many immigrants from Europe, who have reached us, and are still arriving, through the Atlantic states, and deserve a passing notice.

VI. LATE AND PRESENT IMMIGRANTS FROM EUROPE. For the last quarter of a century there has been a direct and increasing immigration of northern Europeans into the Valley; the majority of whom have settled north of the thirty-ninth degree of latitude. Nearly all have come from kingdoms northeast of France, that is, above the fiftieth degree of latitude. In the order of their numbers we may begin with,

1. *The Germans*. They are from various parts of Germany, and outnumber the immigrants from any other part of Europe. But few of them settle in the slave States, with the the exception of Missouri. They are most numerous between the Lakes and the Ohio River. While many remain in the larger towns and cities, the majority disperse into the country, where they amalgamate readily with the existing population.

2. *The Irish*. In number they stand next to the German immigrants. More disposed to inhabit cities than the country, and less offended by slavery, they are found in large numbers in New Orleans and Mobile, not less than in St. Louis, Louisville, Cincinnati, Pittsburgh, Buffalo, Montreal, and Quebec. In the two latter cities, and, generally, north of the Lakes, they are much more numerous than the Germans.

3. *The English*. They, no doubt, rank next in number, but scattering, and bearing a close resemblance to their brethren of the Valley, are soon confounded with them. In Canada and the lead-mine regions of Illinois and Wisconsin, they are numerous.

4. *The Scotch*. Less numerous, perhaps, than the English, and found both in town and country. They are chiefly from the Lowlands. More numerous in Canada than the United States.

5. *The Welsh*. More clannish than the last two, they have settled chiefly in Cincinnati, and in the south-eastern portions of Ohio; where many of them are employed in its iron mines and coaleries.

6. *The Norwegians*. These make a new stream of immigration. Its termination is in northern Illinois, Wisconsin, and Iowa, about the forty-third degree of latitude.

7. *The Poles*. The revolutions of Poland, have dispersed a considerable number of the people of that country over the Valley. They are chiefly men, and abide in towns and cities.

8. *The Jews*. Mostly English, German, and Polish; they prefer the cities,

and are found from Quebec to New Orleans. They are, perhaps, more numerous in Cincinnati than in any other city.

VII. NATIONAL GENEALOGIES. Such are the principal elements of our population. We have seen that a vast majority have been derived from the Atlantic States; and it remains to inquire, whence they or their ancestors originally emigrated to those States. The answer is, almost entirely from Great Britain and Ireland—above all, from England. Emigrants from the last, were almost the exclusive settlers of New England; contributed liberally to the settlement of New York and New Jersey; still more largely to that of Pennsylvania and Maryland, and composed nearly the whole population of Virginia, the Carolinas, and Georgia. But there were, among the early immigrants into the Atlantic colonies, several communities, which deserve to be mentioned. The most numerous were the Low Dutch, in New York and New Jersey; Swedes and Germans, in Pennsylvania; and French Huguenots, in South Carolina.

When we trace up these streams of European emigration to their sources —distant in point of time and space—we have the following results. *First.* The Irish, Welsh, Highland Scotch, Normans, Low Dutch, and French, land us among the Kimmerian or Celtic nations—the earliest known inhabitants of western Europe; and distinguished by different appellations in different countries, as the Kelts, Kimbri, Belgæ, Erse, Cimbri, Britons, Scots, Caledonians, and Gauls. *Second.* The English, Lowland Scotch, French, High Dutch, Swiss, Danes, Swedes, and Norwegians, carry us to the Scythian or Gothic nations; of which the Saxons, Germans, Angles, Jutes, and Franks, were the principal tribes.

Laborious researches have convinced the ethnogrophers, that both these classes of nomadic barbarians, entered Europe south of the Baltic, from beyond the Euxine or Black Sea; and were, in fact, wanderers from the Caucasian Mountains, and the plains of Asia lying north of them; but that the migration of the Celtic, was many centuries before that of the Gothic hordes.

Let us change from analysis to synthesis, and thus obtain a fuller view of the composition of our society. The amalgamation of tribes, by which the main stock of our population was formed, began in England. During the time that island was held by the Romans, the Celtic population must have received an infusion of Pelasgic, or southern European blood, not less than of civilization. Then came the conquests by the Saxons and Angles, and the establishment of the Anglo-Saxon nation—with which, however, the Celtic population must have become more or less blended—although many were destroyed, and many driven into the mountains of Wales. The conquest, and long occupation of the country by the Danes, contributed another, though kindred element; as they had descended from the Jutes, a Gothic race. Lastly, the Norman conquest introduced another element, which, from its magnitude, must have greatly changed the blood and national character of the conquered. Thus, we see that the compound term, Anglo-Saxon, is not an accurate expression for the present English race; but an arbitrary epithet for a compound of Celts, Romans, Angles, Saxons, Jutes or Danes, and Normans; in which the predominating elements are those which have imposed their names upon the mass. Emigrants from this mass, peopled the Atlantic States; where they absorbed a portion of Swedish, Low Dutch, French, German, and Irish blood; then ascended the Alleghany Mountains, spread—and are still spreading—over the great Valley, and constitute the basis and bulk of our population.

SECTION III. PHYSIOLOGICAL CHARACTERISTICS

The modifications of physiology, consequent on the immigration and intermingling of western Europeans in America, may be considered as prospective rather than present. Races do not change their type in a single generation, and some preserve it for long and indefinite periods; though placed under conditions which, *a priori,* might be expected to work out rapid changes. In comparing the circumstances which surround the people of the Interior Valley of North America, with those which surround their European brethren at home, we may refer them to several distinct heads; and, although not able to appreciate the exact influence of either, they must be regarded as causes, of which the effects will, in due time, be so far developed, as to merit the attention of the practical physician. The causes to which I refer, may be included under the following heads: *First.* Intermarriage: *Second.* Change of climate: *Third.* Change of Food: *Fourth.* Change of political, moral, and social condition. Let us consider them separately.

I. INTERMARRIAGE. As western Europe was peopled by nomadic and barbarous tribes, of a warlike spirit, civilization found them divided into many kingdoms; between which, until within the last thirty years, that is, since the downfall of Napoleon, there was but little intercourse, and, consequently, but little intermarriage.

The density of population, has even prevented much intestine change of place, by the people of the same kingdom; who, in its different counties or departments, have continued, through a long period of time, to intermarry with each other; and thus perpetuate their characteristics, corporeal and mental; which perpetuation has been negatively promoted, by the subsistence of each nation, for many generations, on the same kind of diet; in climates which continued without alteration from time; under the influence of forms of government that underwent no important modification, and with usages and manners which varied quite as little. In short, from the remote period, when the Celtic nations were conquered by the later Asiatic hordes, chiefly known under the names of Goths, Saxons, Germans, and Franks, down to the present century, all the circumstances favored the full development of well-defined varieties of constitution, in the different kingdoms of that continent.

The history of the settlement of our Valley, as sketched in the preceding section, shows how much its people differ from their brethren of the old world, and even of the old states of the Union, on the subject of intermarriage.

1. Our frontiers, from Quebec, round by the Lakes and Hudson Bay, to the Gulf of Mexico, beyond the Rio del Norte, present a mixed race of whites and Indians; which is gradually lost, in the population residing immediately *within* that boundary. Thus Indian blood is, as it were, absorbed by the surface of the new nation. The readiest amalgamation with the people of that race, is by the northern French, and the southern Spanish Creoles; but the Anglo-American immigrants from the Atlantic States, and their descendants have, at all times, when war did not prevent it, shown a propensity of the same kind.

2. Wherever there is a negro population, bond or free, the same coalescence is displayed; so that in all our towns, from Mobile to Montreal, and from Pittsburgh to St. Louis, the streets are more or less thronged with mulattoes, quadroons, and other mixed breeds; all pressing upward, that is, ambitious of intermarriage with those whiter than themselves; and thus our Caucasian blood is constantly, though slowly, acquiring an African element. In the willingness for this commingling, the Spanish Creoles of Florida, Louisiana, and Mexico stand first; next come the French Creoles of the Lower Mississippi; then some of the classes of the modern emigrants from Great Britain, Ireland, and Germany; lastly, the native Anglo-Americans.

3. In Canada, intermarriage between the French and British population, although limited by the prejudices of race, and the aversion of the conquered toward their conquerors, is by no means uncommon, especially in the towns and cities, and hence the process of assimilation is going on; in the west, as along the middle portions of the Mississippi, marriages of the French with Anglo-Americans are so common, that the former element is fast disappearing. On the lower Mississippi and the northern coasts of the Gulf of Mexico, the same union has been occurring ever since the cession of Louisiana; and, as the ratio is on the increase, a copious infusion of Franco-American blood into the Anglo-American, will mark the final absorption of the French Creole race.

4. In the same region, we have long been receiving, directly, or indirectly through the French, a tincture of Spanish blood.

5. Intermarriages by the English and Irish immigrants with the indigenous population of the Valley, are familiar events; and if the language of the

German immigrants oppose, for a while, the same kind of union, the work of amalgamation is only deferred. In other words, they are not destined to form distinct and permanent communities any where in our Valley, as they once did in Pennsylvania. It will, no doubt, turn out in the same way with the Norwegians, now pouring into the regions west of the great Lakes.

6. Finally, immigrants and their descendants, from all the Atlantic States, here intermarry, unrestrained by any kind of prejudice; for, on reaching the Valley, their sectional feelings are soon moderated, and many of their antipathies become extinct. Even in the extreme south, among the immigrants from the Carolinas and Georgia, the introduction of New England, New York, Pennsylvania, and Ohio blood, is constantly going on. A large proportion of the emigrants, of both sexes, from those states, are unmarried persons—young men who go out as overseers or superintendents of plantations, clerks, mechanics, watermen, merchants, teachers, physicians, lawyers, and divines; and young women who teach, or act as governesses. The marriages of these classes are not, in general, among themselves, but with the children of the resident population; and thus the north mingles with the south in the lower parts of the Valley; while in the upper, the immigration of families from Virginia, Tennessee, and Kentucky, brings out the same result.

From all this it follows, that the world has not before witnessed such a commingling of races. Those of England and the Atlantic States, the most complete of modern times, bear no comparison with ours; and if we ascend to the earliest historic period, no case of equal complexity is met with. The Roman Empire, it is true, was greatly compounded; it was, however, an assemblage of distinct nations, between which there was but little, in many cases no, social, nor even commercial intercourse. It was an aggregate; ours is a living compound, as yet in the forming stage. Three out of five varieties of the human species, with all the important races which belong to one—the Caucasian or overruling element—cannot fail in the end to give a new physiological and psychological development.

In their western migrations from the sources of the Tigris and Euphrates, to the banks of the Mississippi, and the shores of the Gulf and Lakes, tribes and nations have been governed by a law of increasing social amalgamation. The head of the Mediterranean Sea presented greater diversity than the plains of Chaldea—Greece still greater—Rome went beyond her—the population of western Europe is still more compounded—that of our Atlantic States, diversified in a degree yet higher—that of our Valley beyond all. Thus the union and living coalescence of nations, have been in the direct ratio of time and distance, from the birth-place of the species. The course has been westward, bending to the north in Europe, but again, as we have seen, inclining to the south, to reach America. Dr. Robert M. Patterson, of the United States Mint, Philadelphia, has investigated its direction in the United States for fifty years. In 1790 the center of population was near Baltimore, Maryland; and in 1840 it was in Morgan county, Virginia, both in the same latitude. Thus it appears, that the curve of migration for the United States, still runs nearly from east to west.

But the influence of the latter element is not at an end. The GREAT CENTRAL VALLEY OF NORTH AMERICA is the *last* crucible into which living materials, in great and diversified streams, can be poured for amalgamation. The double range of mountains which separate it from the Pacific Ocean, leave too little space for an empire on the shores of that sea; and the detached communities which may there grow up, will be but derivatives from the homogeneous millions, with which time will people the great region between the Appalachian and Rocky Mountains, which is thus destined to present the last and greatest development of society.

II. CHANGE OF CLIMATE. While we recognize intermarriage as the greatest agent in transforming the races of mankind, we should not overlook external influences, of which climate must be regarded as one. Nor must we reject it because some naturalists, in their attempts to explain too much by it, have assigned it an influence too limited. The immense predominance of a dark or black skin, eyes, and hair, in the inter-tropical regions of Africa and Asia, with the equally uniform prevalence of the two latter, in connection with a swarthy complexion in the south of Europe and the corresponding latitudes of Asia, while the middle and northern parts of the temperate zone present an equal predominance of fair complexions, and light hair and eyes, would seem to leave no doubt, as to a general influence of climate on those parts of the body; but to suppose its agency limited to these, would be a most unphilosophical restriction, for many other physiological modifications may escape our observation.

As we have already seen, the people of western Europe, in emigrating to North America, have generally made ten degrees of southern latitude. If the mean annual heat of this country, and that, on the same parallels, were equal, this would be, as an emigration from Philadelphia or Cincinnati to New Orleans; but, in fact, they live in an average yearly temperature, so much the same as that to which they had been used, that but little influence can be ascribed to this element. There is, however, a striking difference between the summer and winter temperatures of western Europe, and those of the Atlantic States and the great Valley of the Interior; the extremes of the American, being far greater than those of the European continent. Sudden and extreme variations of temperature are, moreover, far more common in this portion of the new, than in that part of the old, world. Lastly, we have a dryer climate, and a more electrical atmosphere.

To these climatic conditions we are bound to admit a modifying influence;

which, if I mistake not, is perceptible in the loss of a ruddy complexion, in a diminution of the capillary and cellular tissues of the face, and a consequent reduction of the convexity of the cheeks, with an increasing tendency to darker hair; in short, the production or further development of a bilious temperament. Without insisting on the accuracy of these special observations, I am convinced that whatever tendency exists, is not in the direction of the sanguine temperament.

III. CHANGE OF FOOD. As a general fact, the inhabitants of America, and especially those of the Interior Valley, live on a fuller diet, than the masses of Europe. Their food differs in two respects: *first,* it is here, far more complex; *second,* the animal portion is much greater. The wheat of England, the oats of Scotland, the potatoes of Ireland, and the rye of Germany, are, in this country, represented by wheat, maize, rye, and buckwheat; and our hot summers permit the cultivation of a great number of culinary vegetables, and some fruits, which those countries cannot produce. The abundance and comparative cheapness of animal food—beef, mutton, and pork—in the great Valley, originates, however, the greatest dietetic distinction between the two countries. Of the natural desire for animal food, no observing man can entertain a doubt. It is among the earliest preferences of infancy; and the immigrants from Europe, who might have seldom tasted it, begin to indemnify themselves for their past privation, as soon as they arrive among us. Whether slow or fast to adopt other customs, they never fail to come into this; and like the indigenous inhabitants of the Valley, very generally eat it three times a day. This inordinate indulgence is often injurious, on their first arrival. Thus, Professor Brainerd, of Chicago, has informed me, that the Norwegian immigrants, on landing at that city, often sicken under the combined influence of meat and whisky. As the time is indefinitely remote, when the density of our population will limit the supply of animal food, it will long continue to enter inordinately into our diet; and, mingled with a great variety of vegetables, unskillfully cooked, indiscriminately mixed, imperfectly masticated, and rapidly swallowed, will constitute our national feeding. That such fullness and crudeness of diet, through successive generations, must work out peculiarities of constitution, and tendencies to some forms of disease, while it gives protection from others, can scarcely be doubted; but these things have not yet been made subjects of accurate observation.

IV. CHANGE OF POLITICAL, MORAL, AND SOCIAL CONDITION. In barbarous states of society, the influence of the mind over the body, is very small. In civilized communities, it becomes great, and bears a proportion to the degree of refinement. As our immigrant ancestors, not less than the people arriving from Europe, were civilized; and as the arts, if not the science and moral sentiments, which should animate and dignify civilization, are increasing among us; we must not overlook the modifying influence of moral causes on our national physiology. If these were the same, in the old and the new world, no change of constitution would result from a change of continent; but they are not.

1. Transplanted from the depths of a compact population, to one of great comparative sparseness, the immigrant has experienced a change, not unlike that of the individual who escapes from a crowd, to associate with a small and open company. His feelings, of both mind and body, undergo a modification by the change.

2. He passes from the midst of ancient works of art, to a new country, where natural objects, scenery, and events replace the artificial.

3. Instead of being compressed on every side, and limited to a small spot, beyond which he seldom passed, he finds ample space for locomotion, and, under the influence of slight motives, makes long journeys or frequent removals; thus, seeing many new objects, and forming new associations.

4. Leaving a state of society which doomed him to one and the same pursuit through life, he finds himself where freedom and facility of change are promoted, and extensively practiced; where new plans of business and exciting enterprises call him out, and inspire him to adventurous and novel efforts; in which he engages with a fearlessness proportionate to the facility with which, in this country, failures, which in an old state of society would be ruinous, may be repaired.

5. While in his native country, his thoughts, in reference to property, might never have risen above his daily bread; but he here sees many ways to wealth laid open before him, and has the love of property, with the comforts, luxuries, and influence, of which wealth is the instrument, awakened or quickened in his heart.

6. When in his native land, he saw, perhaps, but a single aspect of Christianity—one form of public worship, and one variety of worshipers; but he here finds himself surrounded by many. Freedom from legislative restrictions, permits an unrestrained manifestation of his opinions and feelings; he speedily sympathizes in some form of religious worship, and finds himself under the influence of more lively religious feelings; or drawn into the controversies which inevitably arise, in proportion as the superincumbent weight of an ecclesiastical *establishment* is thrown off.

7. But above all, in relation to the young, and to men, the immigrant from Europe is born into a political existence, in becoming naturalized in America. There he was governed, here he assists in governing; there, in feeling, he stood in opposition to the government, here, in practice, he seeks either to modify or preserve its administration. In ordinary circumstances, he was passive and obedient, he is now active and aggressive; he connects himself with a party,

harkens to its tocsin, rallies to its standard, listens to expositions of its doctrines and objects, yields up his heart to its exhortations, and bends his will to its dictation; the servant becomes a civil officer—the peasant, a party politician. The variety and amount of emotion, the excitement of passion, and the activity of thought, developed by this new condition, are great in proportion to his previous torpor—as vegetation advances more luxuriantly after a cold than after an open winter. This is the true reason why our immigrant population are so eager to plunge into the party strifes which are forever heaving the bosom of our society.

Such are some of the new, social circumstances, under which the transplanted population of western Europe, live in our great Valley; and the physiologist cannot doubt, that the mental states, intellectual and emotional, generated and permanently sustained by them, will, in successive generations, work out changes of innervation; and coöperate with the physical causes which have been discussed, in creating a type of constitution different from that of Europe, or any that has gone before it in any country.

SECTION IV. STATISTICAL PHYSIOLOGY

For this portion of the physiological history of the Valley, our stock of materials is very small. They cannot be collected to any great extent, in the ordinary practice of medicine, and no societies, or even individuals, have as yet instituted the requisite courses of observation and experiment. I will give a few facts, under several different heads, in the hope of prompting those who have time and opportunity, to more extended inquiry.

I. STATURE AND WEIGHT. When upon the Northern Lakes, in 1842, I was enabled, through the accommodating politeness of the late Major Martin Scott, of Fort Mackinac, Captains Lynde and Thompson, of Fort Gratiot, and Captain Drain, of Detroit Barracks, to ascertain the stature of three hundred and sixteen soldiers of the United States Army, which I have condensed into the following table.

The numbers in this table are too small to justify any general conclusions; and it should be recollected, moreover, in reference to weight, that most of these persons were not yet of middle age.*

---

* If any one make such inquiries, he should be aware, that he should not combine his averages with those of the table, and take the mean term of the two sums, unless the number of individuals were the same as in the table; for the result would be erroneous. He must multiply the number of individuals, composing the class in the table which he wishes to enlarge, by the mean hight or weight, and to the product add the aggregate hight or weight of those he has examined; when, on dividing the whole amount by the whole number of men, he will obtain the desired result.

NO.	NATION.	MEAN HIGHT.			MEAN WT.		TALLEST.			HEAVIEST.
		Ft.	In.	Lns.	Lbs.	Oz.	Ft.	In.	Lns.	Lbs.
155	American,	5	7	7	148	9	6	1	6	189
82	Irish,	5	8	4	144	11	6		6	192
17	English,	5	8	4	147	2	6	2		183
10	Scotch,	5	8	2	146	8	5	10	9	167
45	Germans,	5	6		146	1	5	10	6	176
7	Danes and Poles,	5	6	5	143	7	5	10		165
316	The whole,	5	7	8	146	13	6	2		192
155	Americans,	5	7	7	148	9	6	1	6	189
109	Islanders,	5	8	4	145	3	6	2		192
52	Continentals,	5	6	4	145	12	5	10	6	176

As these soldiers had reached their full stature, their hight represents that of the nations to which they belong, so far as their number goes; but as most of them were between twenty and thirty years of age, they had not yet reached their full weight; and can only serve as representatives of their respective nations in early manhood. Nevertheless, as the different groups were composed of persons in the same periods of life, they may be compared with each other. In bringing each class to the standards or mean terms of the whole, we find, in reference to stature, that the Americans are at, or rather, within one line below it; the English, Scotch, and Irish rise above it; and the Germans and other continentals fall below it. In reference to the standard of weight, the Americans are above, the rest below it.

The following statement of the relation between stature and weight, shows how much of the latter is due to an inch of the former:

The whole,	one inch gives	34.60 oz. avd.
Americans,	" " "	35.17 "
English,	" " "	34.45 "
Irish,	" " "	33.88 "
Germans,	" " "	35.41 "
Scotch,	" " "	34.38 "

From this table it appears, that the English and Scotch approach within a few decimals of the standard of the three hundred and sixteen individuals of the preceding table, in the weight which belongs to an inch of their stature; the Irish fall seventy-two hundredths of an ounce, or two *per cent* below it; while, on the other hand, the Americans rise fifty-seven hundredths, or one-sixth *per cent,* and the Germans eighty-one hundredths, or two-thirds *per cent* above it. Thus the Germans are the heaviest of the whole, in proportion to their hight, and next to them stand the Americans. This difference argues, either greater development of flesh, or a longer trunk of body compared with the lower extremities. The ratio between the stature and weight of different races, in the corresponding periods of life, being ascertained, the weight of any considerable number of individuals of that race, might be inferred from a knowledge of their stature, *et vice versa.*

In the absence of more conclusive statistics, I may say, that English, Scotch, and Irish immigrants, are so near the stature of the native inhabitants of the Valley, as, in that particular, to be identified with them; that the Germans, the Jews, and the French, both of Louisiana and Canada, are regarded as smaller. The Norwegians, whom I saw in Illinois and Wisconsin, appeared to be taller than the Germans. It is a current opinion that, of the natives of the Valley descended from British and Irish ancestors, the largest men are those of western Virginia, and the eastern and middle portions of Kentucky and Tennessee; where they breathe a salubrious air, abound in sustenance, and take exercise enough to preserve health; but do not perform sufficient labor to carry out of the system, by copious perspiration and increased pulmonary exhalation, a great amount of solid matter.

In our towns and cities, many young men, who grow up without much exercise or labor, and spend their time chiefly within doors, fail to reach the employments, have a larger development of the system, and, consequently, standard size; but those who follow laborious, mechanical, or miscellaneous greater bulk. On the whole, however, the people of the country attain a greater size than those of our cities.

Family and individual eccentricities of size are, of course, not uncommon among a people so diversified in origin, and advanced in civilization—a state which develops greater anatomical and physiological varieties, than the state of barbarism. In different parts of the Valley, families of remarkable stature are met with; and occasionally, a single member of a family rises above the rest, and overreaches the tallest around him.

II. STRENGTH. No experiments on the strength of the native or immigrant races of the Valley, have yet been published. It is a current opinion, that as we advance south, from the middle latitudes, it diminishes; in other words, is inversely to the mean temperature. Comparing all the inhabitants of one of our towns or cities, with an equal number in the surrounding country, the aggregate strength of the latter, would, I have no doubt, be found much greater, except in malarial districts. In a country long and thickly settled, where the labors of the people consist largely in stirring the loose soil with the plow and hoe, in pruning hedges and orchards, training vines, gathering in crops, and in the care of domestic animals, this might not be the case; but in a new country, where overshadowing forests are to be subdued, shrubs and bushes grubbed up, fields inclosed with heavy rails mauled out of the trunks of trees, log houses erected, stone quarried, roads opened, bridges built, and canals excavated, the labors are, in kind and degree, well-fitted to develop large, compact, and powerful muscular systems. Such from the beginning of immigration, have been the labors of a majority of our people. In the older settled regions, they are less than formerly, but in all the new States, they still continue.

But in latter years, a great number of men have been called to new labors, requiring, and therefore, developing, great muscular power. I may briefly enumerate some of these: The erection of cities, such as Toronto, Buffalo, Chicago, Pittsburgh, Cincinnati, Louisville, St. Louis, and New Orleans; the preparation and embarkation of the agricultural staples—flour, corn, pork, beef, hemp, tobacco, and cotton—the lumber trade of Canada, of the mountains of New York, and of the cypress swamps of Louisiana; the preparation of fuel for our five or six hundred steamboats between New Orleans and Quebec, and the labors upon those vessels of firemen and deckhands; lastly, the incessant use of the oar or paddle, and the carrying of heavy burdens by trappers and *voyageurs* on the rivers of the north-west.

Many other classes might be named; but these will serve to show that our country abounds in employments, which cannot be prosecuted without developing great muscular strength.

* * * * *

[BOOK II, FEBRILE DISEASES.]

PART I, AUTUMNAL FEVER. CHAPTER I, NOMENCLATURE, VARIETIES, AND GEOGRAPHICAL LIMITS OF AUTUMNAL FEVER, TOGETHER WITH THE TOPOGRAPHICAL AND CLIMATIC CONDITIONS UNDER WHICH IT PREVAILS.

SECTION I. NOMENCLATURE, VARIETY, IDENTITY

I. NOMENCLATURE. In different parts of the Interior Valley, the fevers, which we are about to study, are known under the names—autumnal, bilious, intermittent, remittent, congestive, miasmatic, malarial, marsh, malignant, chill-fever, ague, fever and ague, dumb ague, and, lastly *the* Fever. So great a variety of names suggests two facts; first, diversity of type; second, wide geographical range of prevalence. I shall use the epithet autumnal, as involving no etiological or pathological hypothesis; and, at the same time, including every modification; but, in speaking of diversities, other terms will find their appropriate places.

II. VARIETY AND IDENTITY. The varieties of autumnal fever are numerous, and often seem widely separated. Thus, the difference in phenomena between a simple tertian and an inflammatory or a malignant remittent, is greater than the difference between measles and scarlatina; in some years nearly all the cases that occur are intermittent, in others remittent; finally, although the former seem to be but mild grades of the latter, they often prove suddenly fatal; and, that too, without assuming a remittent type. Nevertheless, all the varieties must be regarded as making but a single species; as appears from the following facts: *First*. They prevail at the same times and in the same places. *Second*. Under much variety of aspect, they possess many deep-seated analogies and identities. *Third*. They frequently change from one type to the other. Thus an intermittent turns into a remittent, and the latter, assuming the type of the former, is often seen to become, first, a quotidian, then a tertian, and finally, a quartan. A simple intermittent may, in the third or fourth paroxysm, take on the character of a fatal congestive; and that which begins with an aspect of malignity, sometimes emerges into simplicity and mildness. *Fourth*. Vernal agues attack those who in autumn had suffered under remittent fever, not less than those who had experienced the intermittent form. *Fifth*. The *sequelæ* of all the varieties are almost identical. *Sixth*. The same treatment, with certain modifications, is applicable to the whole. Thus they are manifestly the offspring of the same specific, remote cause; and when no particular variety is in view, may be designated by one epithet.

SECTION II. GEOGRAPHICAL LIMITS

Being an endemic of all hot climates, we need not look to the shores of the Gulf of Mexico for a southern limit to our autumnal fever. Its base is, in fact, within the tropics; and prevailing, of course, in Havana and Vera Cruz, it is found wherever there are inhabitants, on the northern coasts of the Gulf, between those two cities. In ascending all the rivers, which discharge their waters into the northern arc of that closed sea, from Cape Florida round to the Panuco river, it is still met with; and, sometimes, as we shall hereafter see, from the influence of local causes, displays greater prevalence and malignity, than it shows further south and on a lower level.

In every direction than the south, this endemic has its geographical limits. To the east, its barrier is the Appalachian Mountains, into the very gorges of which, however, it ascends by the valleys which penetrate their flanks. But as that chain is not found south of the thirty-third degree of latitude, it has, below that parallel, no eastern limit but the Atlantic Ocean. To the south-west the Cordilleras of Mexico, and the southern Rocky Mountains, constitute its boundaries; while, in higher latitudes, it ceases on the great plains of our western desert, long before we reach those Mountains. From what can be collected out of the travels and expeditions of Lewis and Clark, Pike, Long, Catlin, Fremont, and Gregg, not less than from fur traders and Santa Fe merchants, it is almost unknown at the distance of three hundred miles from the western boundary of the states of Missouri and Iowa, and above the latitude of 37° N. To the north it does not prevail as an epidemic beyond the forty-fourth parallel, and ceases to occur even sporadically at about the forty-seventh.

The observations from which these limits are deduced have been made on the resident inhabitants included within them; on travelers into portions of country as yet unsettled; and on the soldiery of the American and British posts. From these army returns I have, with all possible care, constructed two tables, which may properly be introduced at this place. The American returns* purport to be for ten years; but this is true of a few only; and many of the others vary from each other, in the number of years through which they run, whereby the conclusions deducible from them, are entitled to less confidence than if an equal number of observations, in the same years, had been made at each post. As the number of troops was never the same at two different posts, nor during two years at the same, one thousand has, in the returns, been assumed as the mean strength of the whole; and the number of attacks of Fever, and the actual mean strength, have both been brought to that standard. The results offered in the table, then, are not what any post did afford, but what any or all would have given, had the actual strength been at all times one thousand men. At several of them, it will be perceived, the number of attacks exceed the number of men, implying that some individuals experienced several, in the course of the year. The returns are quarterly, but the quarters are those of the calendar year, and therefore, do not exactly correspond with the seasons.

The observing reader will perceive, that this table affords a variety of information; such as the decrease of the Fever in the north—its relative

* Forry's Statistical Report of the Sickness and Mortality of the Army of the United States; prepared under the direction of Thomas Lawson, M.D., Surgeon General, Washington, 1840.

prevalence at different posts in the same latitude—the proportionate number of intermittent and remittent cases, and the comparative prevalence of both, in different seasons of the year.

* * * *

SECTION III. CONDITIONS WHICH IMPOSE GEOGRAPHICAL LIMITS, AND GIVE UNEQUAL PREVALENCE TO AUTUMNAL FEVER

I. SOIL. Under this term I include all that composes the surface of the earth, apart from its waters. The loose upper stratum of our Valley consists, as far as its mineral elements are concerned, of the *debris* of the rocks beneath, or of deposits of the *debris* of other rocks, spread over the surface by ancient inundations. There are tracts of country, however, in which the rocks themselves appear at the surface. None of these conditions favor the production of autumnal fever; but, on the contrary, it prevails least where they are most perfectly developed; and hence there is no reason for referring the disease to emanations from a purely mineral surface.

The soil, however, may have another element than the mineral—dead organic matter, both animal and vegetable; and this is its general character throughout the Valley. The amount of this element is very different in different places, for its production depends, *first,* on the fertility of the surface; *second,* on temperature; and, *third,* on moisture. Where these conditions are all present, the growth of organic matter is redundant; where any one or more of them is wanting, it will be correspondingly limited. Thus it is small in quantity in the pine woods of the south (if we except the trees themselves), from the sandiness of the surface; in the desert, beyond the Mississippi, from the same cause, and also from the want of moisture; in the far north from the want of heat, yet it is abundant even beyond the limits to which the Fever extends; on the Appalachian Mountains from that deficiency in part, and from their rocky surface. Dead organic matter is, also, unequally distributed; for the rains wash it down from the hills, and deposit it in the valleys; where, adding to their fertility, it rapidly augments itself, by promoting more luxuriant crops of vegetation.

Now, it is a safe generalization to affirm that, all other circumstances being equal, autumnal fever prevails most where the amount of organic matter is greatest, and least where it is least. A diligent study of the topographical descriptions of *Book I, Part I,* will sustain this conclusion, and demonstrate that decaying organic matter is *one* of the conditions necessary to the production of autumnal fever. As to the mode in which it coöperates, two opinions may be entertained: *First.* It may supply the material out of which a poisonous gas is formed; and, *Second,* It may be a nidus or hot-bed of animalcules or vegetable germs. In either case, we may presume, that all kinds of decomposing organic matter, are not equally favorable to the production of the cause of this fever; but, although I have sought for facts bearing on this question, a sufficient number has not been found to justify their presentation here. I hope the subject may attract the attention of others.

The first breaking up of the soil appears, from a variety of observation, scattered through our topographical descriptions, to be frequently followed by autumnal fever; and, on the other hand, long-continued cultivation is accompanied by diminution of that disease; the element which contributes to its production becoming exhausted.

II. LIVING VEGETATION. Forests have been thought to modify the conditions which generate autumnal fever. Our medical topography supplies several facts, which go to show, that those who first penetrate our woods, and establish themselves in cabins, closely surrounded by trees, remain comparatively exempt from autumnal fever, till the clearing is extended. On the other hand, it is a disease of the country, and especially of newly-settled parts; where the amount of forest is so great, as to maintain a high degree of humidity. Our cities and larger towns, it is well-known, seldom suffer, and they are to be considered, as in some degree, presenting the very opposite condition from our woodlands. Again, trees have been thought to arrest the spread, of that gaseous agent, whatever it may be, which is said to be the true cause of the Fever; but in what manner they do it, no one can tell. It has been conjectured, that their leaves absorb the noxious exhalation; and also that they mechanically arrest the dissemination of the aerial poison. In harmony with the former hypothesis, is that of Dr. Cartwright in reference to the *Jussieua grandiflora,* and some other aquatic plants, in the delta of the Mississippi; which, he supposes, absorb the agent that produces autumnal fever. I have already expressed the opinion, that the facts do not establish that hypothesis; and must here, in conclusion, remark, that living vegetation is so mixed up with other conditions, necessary to the production of the Fever, that, in the existing stage, of observation, its effects cannot be correctly estimated.

III. SURFACE WATER. In the maritime parts of Florida, Alabama, Mississippi, Louisiana, and Texas, surface water is abundant, for one side of each rests on the gulf, which has many inlets and little bays, the banks of which, are inhabited. The rivers, moreover, are numerous, and as they approach the gulf, expand into broad estuaries or deltas. The delta of the Mississippi, abounds in lakes, lagoons and bayous. As we ascend this, and the smaller rivers, wide cypress and liquid-amber swamps, annually replenished, skirt both sides. The intervening plains, are cut up by smaller streams, which have wide alluvions, often subjected to inundations; and the country between them abounds in swamps; from which even the sandy, pine plateaus are not entirely free. This continues to be their condition, till we reach the flanks of the Cumberland Mountains, on the east, and those of the Ozark hills, to the west. As we ascend the Mississippi, to the mouth of the Missouri, we find its annual floods leaving small lakes, ponds, swamps, and lagoons; which in the aggregate, are of great extent, and but partially drained or dried up, before the next inundation. Now, as we have seen, the whole of this region is infested with autumnal fever, beyond any other portion of the valley.

In North Alabama, Tennessee, and Kentucky, swamps are almost unknown, except along the few rivers, which have wide bottom-lands, most of which, moreover, are exempt from inundation. The rivers, however, are sinuous, and in summer, sluggish and pondy; and it is in their vicinity, chiefly, that autumnal fever prevails. In the states of Illinois, Indiana and Ohio, the rivers generally flow through wide valleys, many of which, are liable to be overflowed. Small lakes, ponds and swamps, are also frequent, in certain portions of those states; and it is precisely these localities, which are most infested. To the east of all the states mentioned, as we climb the mountains, the surface water is no longer found in basins; and the streams, generally, have a rapid current, down narrow and rocky channels; and here, autumnal fever nearly disappears; or, when present, is confined to the valley of some stagnating stream. Everywhere, west of the states of Arkansas, Missouri, and Iowa, surface water is scarce; the declivity of the plain which stretches from the Rocky Mountains, favoring its escape; while the subjacent sand almost absorbs, even considerable rivers. Thus, as we advance into that desert, we come at the same time to the limits of surface water, and of autumnal fever. In the north there is no deficiency, for the whole country is essentially lacustrine; and up to a certain latitude, the Fever prevails. Thus the shores of Lake Ontario and Lake Erie, with those of the southern extremity of Huron and Michigan, are infested, and suffer far more than the dryer lands which surround them. But beyond these limits, on the shores of the two latter lakes, and on those of Lake Superior, the Fever, as we have seen, is never epidemic, although water is abundant; and still further north, where small lakes, and their connecting streams, exist in countless numbers, the disease is unknown; showing that, while water is essential to the production of this Fever, other causes must coöperate to give it power.

Let us inquire into the *modus operandi* of this agent in the production of the disease under consideration.

1. Under the influence of solar heat it impregnates the air with vapor,

giving a high dew point; and, other circumstances being equal, the evaporation is greatest where the heat is highest. This, of course, is in the southern part of the Valley, and there, as we have seen, the Fever prevails most.

2. Surface water not only contributes largely to the production of a luxuriant vegetation, destined annually to perish, but is indispensable to the decomposition of what it has aided in producing. Hence, without its agency, none of the deleterious gasses, which are supposed to be thus generated, could have an existence. But its presence in any or all quantities, will not answer equally well. If there be too little, the molecular movements of fermentation are arrested for want of a solvent—if too much, the atmosphere, indispensable to the process, is excluded; or the evolved gases are absorbed and retained.

3. Its presence is essential to those chemical actions, in certain soils, which are believed, by some writers, to generate exhalations that occasion the Fever.

4. It is equally indispensable to the production of both animalcules and microscopic plants.

5. Both evaporation and condensation are known to be accompanied by electrical perturbations.

Thus water is a necessary element, in all the hypotheses which have been framed to account for autumnal fever.

But a contrary and salubrious influence has been ascribed to water; for it is held by many that this fluid absorbs the noxious gas or gases, which they believe to produce the Fever, and thus limits its prevalence. According to this opinion, the deep waters in the center of a basin, may imbibe and retain the noxious gases which the shallow waters of its margins have contributed to generate; and, in support of the hypothesis, it has been affirmed that the vicinity of cataracts and rapids is more unhealthy than the banks of the rivers in which they occur. The absorbed gases are supposed to be there liberated by the agitation of the water. The medical topography of *Book I,* presents several facts bearing on this hypothesis. Thus Wetumpka, at the foot of the long rapids of the Coosa river; Louisville, at the falls of the Ohio River; and Maumee City, at the termination of the rapids of the Maumee River, are all infested with autumnal fever; but other towns, on the same rivers, are likewise scourged with that disease; and Oswego River, which drains the Montezuma swamps of western New York, has at its mouth a great number of mills, yet the inhabitants suffer but little from that disease. It prevails still less at the Falls of Niagara; and finally, at Zanesville, where a natural waterfall has been augmented by artificial means, and on the Kentucky River, where there are series of pools and dams, there is no special prevalence of the Fever. Thus the facts furnished, by our Valley, do not prove that waterfalls eliminate a gas which is the cause of the disease under consideration.

IV. TEMPERATURE. The fact that autumnal fever prevails perpetually and virulently, within the tropics, but ceases long before we reach the polar circle, demonstrates that a high temperature is one of the conditions necessary to its production. Should it be ascribed to heat alone? The answer must be in the negative; for places having the same temperature, but varying in other conditions, are very differently affected with autumnal fever. Thus the people on Mobile Bay suffer greatly, while those who live on the adjoining oak and pine terrace escape; and the summer heat of the southern portions of the great desert is intense, but those who traverse it, and keep at a distance from its water courses, pass the season unaffected. It cannot be affirmed, that the direct action of a hot atmosphere on the body, does *not* contribute to the production of the Fever; for, on the contrary, where it prevails as an epidemic, exposure to the noon-day summer sun is often followed by an attack; but such exposure, in a different locality, will not produce it; and, therefore, we may conclude that in its direct action, heat is merely an exciting cause, on which it is not necessary to expatiate in this place; and I will therefore proceed to trace out its indirect effects.

Our army statistics furnish some instructive facts on this point. The posts which lie along the Mississippi, are placed nearly under the same conditions, in everything but temperature, which varies according to their latitude. They are, therefore, well fitted to indicate the influence of this climatic condition in the production of the Fever. Its relative prevalence at these posts, which extend through more than thirteen degrees of latitude, is presented in connection with the annual and quarterly mean heat, in the first part of the following table, while the second offers a comparison of two posts in the region west of the Mississippi, and the third of two on the Lakes.

To show, by a comparison of localities, the exact relation between temperature and autumnal fever, the conditions of the different places should, in all other respects, be alike, which is not often the case; nevertheless, the medical topography and hydrography of the posts, compared together in the foregoing table, will be found substantially the same, and they show, that with the decrease of yearly and summer heat, other conditions continuing unchanged, there is an abatement of the Fever. It is, however, with the heat of summer, and not that of the year, that autumnal fever is connected; and the question here arises, what summer temperature is necessary to the production of the Fever? This question cannot be rigorously answered; for the number of observations hitherto made, in the proper region, is too small to justify a positive conclusion; we may, however, assume, that a summer temperature of sixty degrees, is necessary to the production of the Fever; and that it will not prevail as an epidemic, where the temperature of that season falls below

	POSTS	North. Lat.	Annual No. of cases, in 1000 mean strength	Annual mean temperature	Mean heat of Winter	Mean heat of Spring	Mean heat of Summer	Mean heat of Autumn
		° ′		°	°	°	°	°
Along the Mississippi River	Baton Rouge	30 36	824	67.56	52.68	68.72	81.48	67.38
	Jefferson Barracks	38 28	475	56.93	33.98	56.55	76.19	54.38
	Fort Armstrong	41 32	307	50.65	25.15	50.82	74.57	52.07
	Fort Crawford	43 03	301	47.35	20.69	48.25	72.38	48.09
	Fort Snelling	44 53	62	45.15	17.29	46.56	71.16	45.59
W. of Miss. River	Fort Gibson	35 48	1435	61.07	42.50	61.26	79.17	61.53
	Fort Leavenworth	39 23	629	52.34	27.60	53.38	74.00	54.39
On the Lakes	Fort Dearborn	41 50	251	46.14	24.31	45.39	67.80	47.09
	Fort Brady	46 30	44	40.62	18.06	38.17	62.14	44.13

sixty-five; finally, that if the other conditions favoring its production are deficient, it will cease before those reductions of temperature have been reached.

According to these conclusions, the Fever will occur in winter, at all places where that season has a mean temperature of sixty degrees or upward; as at Vera Cruz, Tampico, Havana, Key West, Tampa Bay, and Fort King, as may be seen in the table; and it is well known that cases do occur at those places, in that season; but at the two latter posts, where the winter heat barely rises over sixty, they are few in number. At New Orleans, and generally under the thirtieth parallel, where the mean winter heat is as low as fifty, the Fever is suspended. But the seasons are made up of months, and we are here brought to consider its connection with their respective temperatures.

Up to Tampa Bay, every winter month rises above sixty degrees; but at New Orleans, or the thirtieth parallel, only the nine months from March to November, have that temperature; and as we advance to the north, the number of months having it constantly decreases. Thus, at St. Louis, it is attained by five months only—from May to September, inclusive; at Fort Snelling, by four; at Fort Brady, by three; at Montreal, by four; at Quebec, by three. In advancing further north, June and September fall below it; and, finally, in the distant north, July and August, or the entire year. Long before this reduction is reached by those two months, however, the Fever ceases; and therefore it results, that a continuance for more than two months of a heat equal to sixty degrees, is necessary to the development of the Fever. Hence we can understand, why it prevails more in October than April, although their mean temperatures are nearly the same; in November than June, notwithstanding, the latter is much the warmer month, and in September and August, than July—the hottest month of the year. The greatest prevalence in every latitude, is indeed, generally some weeks, after the hottest month; showing that the effects of temperature are cumulative. It appears from all that has been said, that within the tropics, autumnal fever may occur throughout the year; and that as we move northerly, the duration of its prevalence shortens, by its beginning later in spring, and terminating earlier in autumn. March and November first escape; then April and May on the one hand, and October on the other—lastly June and September.

In contemplating the climatic relations which exist between autumnal fever, and certain aspects of vegetation, we find that in the tropical regions they are the same throughout the year, and that when we attain the thirty-third parallel, which constitutes the northern limit of several southern trees and plants, the prevalence of the Fever is for a much shorter period; that its disappearance is nearly at the same curve, at which the miscellaneous vegetation of the middle latitudes, gives place to the terebinthinate trees and birches, of the north; finally, that maize or Indian corn, which grows all the year round, in the tropical regions, finds the summers too short for the ripening of its grain, in nearly the same curve of summer temperature, at which autumnal fever is arrested.

If change of latitude, by diminishing the heat of the atmosphere and that of the earth's surface can, as we have just seen, arrest the production of autumnal fever, an increase of elevation above the level of the sea, may likewise do it. Thus the Fever which scourges the *tierra caliente* of Mexico, near the level of the sea, is almost unknown in and around the city of Mexico, at an elevation of seven thousand four hundred and fifty feet, although the latitude remains the same. The inhabitants among the sources of the Kenawha and Tennessee Rivers, on the Appalachian Mountains, at a medium elevation of nearly three thousand feet, are almost exempt, while those who occupy the valleys, under the same parallels, are affected; and, further north, at half that elevation, where the Alleghany and Genessee Rivers have their sources, the disease is almost unknown, while on the shores of Lake Ontario, directly north, it prevails. In traversing that mountain terrace, which has a mean summer temperature of sixty-three degrees, I witnessed a frost, on the night of the second of August which destroyed the Indian corn; but, on descending into the valley of the Genessee, which, although a degree further north, is infested with the Fever, the fields of maize were uninjured. Finally, the constantly increasing elevation of the desert to the west of the Mississippi is, no doubt, one cause of the disappearance of the Fever under the same parallels, in which it prevails on the banks of that river.

Having established the paramount influence of high temperature in the production of autumnal fever, it remains to inquire into the modes in which it may operate. I have already referred to its effect as an exciting cause, but this view is too limited, and others must now come under consideration.

1. The long-continued impress of summer heat upon the surface of the body, occasioning copious perspiration, and through the nerves of the skin sympathetically affecting the internal organs, more especially the abdominal, may predispose to this form of fever; and the cool nights of early autumn, acting on the same surface, may still further derange the economy. That such nights, and occasional sudden changes of temperature, are often followed by an immediate development of the Fever, is well known.

2. Heat promotes great evaporation from all moist and watery surfaces, thus giving to the atmosphere a high dew point.

3. It favors the fermentative decomposition of organic matter, and the production of new compounds.

4. It facilitates the multiplication of minute but visible animals, and cryptogamic plants, and may be presumed, therefore, to multiply the microscopic—both animal and vegetable.

5. It evaporates the superfluous water of ponds, swamps, marshes, and lagging streams; thus bringing them into a condition favorable to the more rapid decay of the organic matters which they contain or cover over, and thereby promoting the extrication of gases.

6. It dries the surface of the ground after the rains of spring and summer; and may (as has been asserted) cause it, in the act of desiccation, to send forth deleterious exhalations, different from those generated in deposits of decomposing organic matter.

7. It disturbs the equilibrium of the electricity of the atmosphere; hence summer thunder storms are of almost daily occurrence, on the coasts of the Gulf of Mexico; but on the shores of Lake Superior they are rare.

Thus solar heat plays an indispensable part, in every hypothesis which has been proposed to explain the origin of autumnal fever; answering equally well for the advocates of combined heat and moisture—miasmatic exhalations—microscopic beings, and atmospheric electricity.

We have now reviewed all the obvious conditions which seem to concur in the production of our autumnal fever, and endeavored to assign the *modus operandi* and influence of each. We have seen the necessity of their concurrence, from the fact that the absence of any one puts an end to the prevalence of the Fever. These conditions are dead organic matter, resting on or blended with the mineral elements of the soil; water, not in any, but a certain quantity; and temperature, above the sixtieth degree, continuing for at least two months. And here we might stop, but for the instinctive propensity of the human mind to arrive at the knowledge of a single efficient cause; to which, therefore, a chapter must be devoted.

CHAPTER II, SPECULATIONS ON THE EFFICIENT CAUSE OF AUTUMNAL FEVER.

SECTION I. METEORIC HYPOTHESIS

It has been suggested, and, indeed, is believed by some physicians, that while the three conditions recognized in the last chapter, are present wherever autumnal fever prevails, but two of them—heat and moisture—exert an influence in its production. Under the joint influence of these elements, vegetation will of course flourish and decay; but not contribute to the production of the Fever. The advocates of this opinion, of course, deny the existence of a special poison; and ascribe the disease to the direct, combined action of a hot, humid, and electrical atmosphere. The discussion of this hypothesis, necessarily involves, to some extent, the discussion of the question of a special agent; for but the two opinions can be held. The Fever prevails extensively, is often epidemic, and is not contagious; it must have a cause, and if that cause be not some conjunction of the ordinary elements and sensible qualities of the atmosphere, it *must* be a poison, dissolved or suspended in it. If it should appear, then, that the Fever does not depend on the former, we may affirm that it *does* depend upon the latter.

I have already shown, that neither heat nor moisture, by itself, can produce the Fever, and will now proceed to state certain objections to the hypothesis that it results from their combined influence.

1. It is well known, that autumnal fever seldom appears on board of vessels which cruise in the Gulf of Mexico, although the air, at the temperature of eighty, is nearly saturated with vapour.

2. The inhabitants of Key West, who breathe a similar atmosphere, are much less afflicted with the Fever, than those on the Peninsula of Florida, several degrees further north. Now, although that little island supports considerable vegetation, its swamps are filled with the waters of the Gulf in every high tide, and when strong winds prevail.

3. The sandy banks of Pensacola Bay; from its entrance, up to the town of Pensacola, suffer but little; while, at the head of the bay, where extensive

alluvial deposits have been made, the Fever has been so constant and fatal as to prevent permanent settlements. Yet the temperature and moisture of both localities are the same, for they are but ten miles apart.

4. The pine woods around the Gulf of Mexico, at the distance of only two or three miles from the estuaries of the rivers, are places of retreat from the Fever, although there is a sea and land breeze, which tends to equalize the humid atmosphere.

5. The inhabitants of the Balize, suffer less from the Fever than those along the rivers of the interior of Louisiana, two or three degrees further north; notwithstanding they are immersed in an atmosphere of great heat and vapour. Vegetation is as luxuriant at the Balize as above; but when it dies, it falls upon a soil impregnated with sea salt, and is often wetted by the waves of the gulf.

6. In many parts of Kentucky and Tennessee, where the surface is dry and ridgy, and the streams narrow and tortuous, the Fever occurs upon the former, although the atmospheric humidity is small.

7. It is well known that a family may settle down in the forest, and cultivating but a small spot, remain free from fever; but when several families arrive, and an extensive breaking up of the soil takes place, it immediately begins to prevail, although the heat and moisture are not thereby increased.

8. Dr. Winter gave me the following fact. On Cedar Creek, a tributary of Cumberland River, a mill dam had been erected about sixteen feet high. After twenty-two years, the basin above having become filled up with silt and drift, the dam was torn down, and the perpendicular face of the deposit, exposed to the action of the sun and air, in the month of August. The consequence of this was, that nearly all the men who performed this labor, were seized with severe autumnal fever, and one of them died. There was no pond above, nor any marsh in the neighborhood; and the people generally were healthy at the time. Here there was no combined agency of heat and moisture; and hence the facts afford strong evidence of a developed aerial poison.

9. On Paint Creek, Ohio, a millpond was generally drained the first of June, and the rains of that month, washed away the silt and dead plants, and animals; so that the people of the adjoining village of Washington, suffered but little from the Fever; the draining was postponed till July, and no rains followed to wash out the basin. Then there immediately followed an epidemic autumnal fever, which prevailed most on the side of the village next the pond. More than a fourth of the population suffered an attack, and nearly three per cent. of the whole number of inhabitants perished.

10. It has frequently happened, that individuals who have lodged for a single night in certain localities, have after several days, or even weeks, been

taken down by the Fever. More than this, persons, living in places where it never originates, have been seized in the spring with intermittents, after having in the preceding autumn, traveled where the Fever prevailed. Now it is in no degree characteristic of heat and moisture, to produce *remote* effects. A catarrh, a pleurisy, or a rheumatism, comes on soon after exposure, or not at all. The development of the disease, at a distant time from that at which the remote cause was applied, clearly suggests, that the cause was something else, than a particular condition of the sensible properties of the atmosphere.

11. At our different salt works, the operatives spend their lives in a hot atmosphere saturated with vapour; and, yet, on the whole, are more exempt from fever, than the surrounding population.

12. Lastly, in some of our manufacturing establishments, the in-door artisans and operatives, labor in a heated atmosphere supersaturated with vapor, but remain free from autumnal fever.

These facts seem to me conclusive in their bearing against the meteoric hypothesis; except so far as certain atmospheric conditions may act as exciting causes; and we are therefore, thrown upon the alternative—a deleterious agent, diffused in the atmosphere; the positive existence of which, seems to me to be established, by the facts which have been cited.

Now this agent may be either one, of two kinds—inorganic or organic—and both have a *prima facie* advantage over the hypothesis we have examined, in demanding the concurrence of all the conditions—heat, water, and dead vegetable and animal forms—which have been shown to be always present, wherever autumnal fever prevails; while the last is left out of account by the meteoric hypothesis. We must first inquire into the origin and nature of the inorganic poison.

SECTION II. MALARIAL HYPOTHESIS

I. It is unnecessary to inquire into the nature of the gases, which may be exhaled from an *earthy* surface, consisting of *nothing* but the fragments and powder of the subjacent rocks, and the different salts, or oxides, formed by their decomposition, under the influence of heat, water, and atmospheric air; for no such surface exists in our Valley. Whenever the rocky strata are thus exposed, they begin to crumble; and the pulverulent layer then immediately becomes the *nidus* of some kind of plant; thus, lichens overspread the hardest rocks, and, by their death and decay, add to the thin layer of mineral matter, an organic element, at once vegetable and animal in its composition. In this way, the spot becomes prepared for a vegetation of a higher order, which, in turn, augments the amount of organic matter; while the rock beneath, by continued disintegration, continues to contribute new mineral substances. Thus it is, that the loose upper crust of the earth is accumulated; and the nearer we come to the actual surface, the greater, in proportion, are organic elements, or those fixed compounds which are formed by its decomposition. The soil thus formed may vary exceedingly in its depth; for where the rock has undergone rapid disintegration, or the *debris* have favored a luxuriant vegetation, the soil will be much deeper, than in opposite circumstances; but there is still another source of inequality. The soils thus formed are not fixed, and consequently are liable to be drifted about by currents of water. In ancient times, great portions of the Valley, on the north side of the Ohio River, were deeply covered with this kind of drift or diluvium; and down to the present time, every considerable rain or dissolving snow, but especially the former, washes a portion of the soil, with its superincumbent dead plants and animals, into the valleys, where they are speedily deposited.

But the soil of every inhabitable part of the Valley has, at all times, resting on its surface, a layer of dead and decomposing organic matter; which is abundant in proportion to its fertility, and its favorable exposure to rains and the heat of the sun—that is, to those conditions which cherish the growth of animals and vegetables.

Now, in the study of medical topography, with reference to autumnal fever, our attention has been generally directed to this layer only; and as there may be some physicians who even doubt the existence of those organized and decaying forms, in the soil beneath, supposing that they suffer decomposition when they disappear from the surface, it may be well to say something more on this subject.

The soil all contain organic matters, which, in one, more than equal all the inorganic substances. One of the specimens examined was silt, taken from a point ten feet below the surface, in New Orleans; and Professor Riddel found, that nearly one fourth consisted of "organized matters, such as the sporules or germs of algæ, animalcules, and their ova;" and at the depth of sixteen feet, in sinking the gas tanks of that city, wood was found, which had the texture of cheese, when the spade passed through it. The length of time required for the Mississippi to deposit the sixteen feet of superincumbent silt, must have been indefinitely long. Again: In parts of Ohio, where there is a deep diluvial or post-diluvial deposit, when wells are dug, plants unknown in the neighborhood often appear upon the earth which has been thrown out, and doubtless spring from seeds, which had lain buried for an immense length of time. Still further: Where the upper crust is composed of sand, but produces the kinds of vegetation that can grow on such a surface, the decaying organic matter is washed into the ground by rains. Thus it is that the manure or mold, that is spread on the white sands of the gardens of the navy yard at Pensacola Bay, rapidly disappears. In this way, a spot which seems destitute of dead organic matter, may have an admixture of that element below the surface. From these facts, we are warranted in reaffirming, that the soil and subsoil, of all parts of the Interior Valley, contain organic matter, in every stage of decomposition.

II. We come now to consider the dead and decaying organic matter deposited on the surface. This does not consist of vegetable forms merely, as we too often suppose, but likewise of animal. An inspection with the naked eye, and still more with the miscroscope, reveals to us that innumerable insects, and other minute animals, live and perish among vegetables. Many tribes, moreover, find their sustenance and abode in the decaying remains of plants. Still further, the surface and superficial parts of the ground teem with small quadrupeds, reptiles, and worms; while the trunk of every fallen tree, in a certain stage of its decay, abounds in various kinds of grubs or larvæ. From the moldering remains of trees and other vegetables, moreover, spring mushrooms, algæ, lichens, and other cryptogamic plants, which abound in nitrogen beyond the higher order of vegetables, and have, in fact, nearly the same chemical elements with animals. Finally, wherever there are pools, or swamps, or running streams, there are fishes, moluscæ, and crustaceæ, which multiply and perish, and whose bodies then float and dissolve, or sink to the bottom, or are thrown upon the shores, and mingled with the remains of land animals and plants. Thus, a vegeto-animal layer overspreads the surface of the country; and under the combined influence of water, heat, and air, when the two former are in the right proportions, is constantly undergoing decomposition, and originating new chemical compounds.

III. But the organic covering of the surface is, by no means, of the same nature in every locality. We cannot tell what kind of plants and animals, in past ages, left their remains on what now makes our subsoil; but the existing forms are subjects of observation, and, in the investigation which occupies us, should not be entirely overlooked.

1. The trees, in what are called the pine woods of the south, are chiefly resinous, and abound in hydrogen. Vegetable matters having such a composition are little disposed to pass into fermentation, but are decomposed by the slow combustion of several of their principles, by the oxygen of the atmosphere; and if the efficient cause of autumnal fever be a gas, formed during the fermentative decomposition of organic matter, we have here one explanation of the comparative absence of that fever in those woods.

2. The *gramineæ, equisetacea* and, indeed, all kinds of grasses, contain in their culms and blades a great quantity of silicate of potash, and in their seed much phosphate of magnesia and lime. They undergo decomposition very

slowly, and the results cannot be the same as those of plants widely differing from them in composition. In describing the medical topography of the Balize, the extensive and luxuriant growth of the *Phragmites communis, Typha latifolia,* and *Scirpus lacustris,* was mentioned; and I have already conjectured that their falling, when dead, into brackish water, may modify their mode of decomposition; but we may also believe that their composition exerts an influence; and that, on the hypothesis that the Fever is the offspring of the decomposition of organic matter, one cause of its milder prevalence, at the final termination of the Mississippi, than along the same river above, may be the peculiar composition of its reigning vegetation. Again: the vegetation on the grand prairies, beyond the Mississippi, is chiefly gramineous, and to this, on the same hypothesis, we might, perhaps, consistently attribute some portion of their exemption from the Fever.

3. The oak tree abounds in tanno-gallic acid, and is often the governing tree in considerable tracts of forest; which, I think, are less infested with the Fever than localities having a diversified, arborescent vegetation. At all events the *exuviæ* of such a forest might be expected to afford the elements for gaseous exhalations of a different sort from those of pine, or of trees not abounding in that acid.

4. The *leguminosæ,* including all kinds of pulse, as peas, beans, and lentils, contain very little potash, silica, or the earthy phosphates, while they abound in nitrogen, and must, therefore, while under decomposition, yield gases of a very different kind from the *gramineæ.*

5. The extensive natural family of plants called the *cruciferæ,* embracing the radish, mustard, turnip, and cabbage, contain sulphur and nitrogenized ingredients, fitting them to give out, in decomposition, gases varying from the last.

6. Not to pursue the subject any further, the *fungi, boleti,* and other cryptogamic plants, which abound in dark and shaded woods, have a composition almost animal, and cannot, in their spontaneous decay, afford results of the same kind with plants of a widely different composition.*

IV. The facts which have been cited teach us that there is, mingled with the soil or resting upon it, a great amount and endless variety of organic matter, both animal and vegetable, to the decomposition of which, and to the resulting new compounds, the malarialists look for the efficient cause of autumnal fever. In doing this, a special stress may, with great propriety, be laid on a few unquestionable facts.

1. That, all other circumstances being equal, the Fever prevails most where the organic matter is most abundant, in or resting on the soil.

---

* Liebig: Chem. applied to Agricul. and Phys.

2. That where the surface is not moist enough to favor the decomposition of organic matter, the Fever has but little prevalence.

3. That a temperature of sixty degrees of Fahrenheit, or above, is necessary to fermentation and putrefaction, and that the Fever ceases, in going north, when we reach a summer temperature below that degree.

4. That particular localities have experienced the Fever, in an epidemic form, when a surface abounding in organic matter has been newly exposed to the action of the summer sun.

5. That under long cultivation, which exhausts the organic matter of the soil, and prevents its accumulation on the surface, the Fever almost ceases to appear.

V. These facts undeniably establish a connection between a certain condition of the surface and autumnal fever; but they do not prove the existence of malaria, or a *gas,* which is the efficient cause of the Fever, and to this point we must now give attention.

1. The observed aeriform products of this decomposition are carbonic acid, carbonic oxide, carbureted hydrogen, sulphureted hydrogen, and carbonate of ammonia. Now, there is not a single fact going to show that either of these gases can produce autumnal fever. On the contrary, as the result of experience, it may be safely affirmed, that they do not; for the effects which follow on exposure to them are of a different kind. But it can be said that, in the endless variety of new compounds, which nature may form out of the ultimate elements of plants and animals, there may be many which have not yet been detected, and that some one of these is the efficient cause of the Fever, and this cannot be denied. But we must not forget that it is an assertion without proof —a mere suggested hypothesis—a proposition to be proved.

2. It is well known to us all, that there are sickly and healthy seasons at the same place, and sometimes over large portions of our Valley, while the amount of organic matter remains unchanged; and, as yet, it has not been shown that this can be explained by a reference to varying degrees of heat and moisture, though the subject has not received sufficient attention to show that it cannot.

3. The Fever occasionally appears in limited localities, from which it is in general entirely absent; the surface meanwhile remaining, to all observation, precisely the same.

4. All the known gases are either simple bodies, as hydrogen and chlorine, or binary compounds of two simple elements, as carbonic acid, ammonia, and carbureted hydrogen, and their principles are united in definite proportions, giving to each a uniform and peculiar character. If we may depend on analogy, the assumed undiscovered gas, called malaria, must be of the same

character; and, therefore, at all times and places be productive of the same effects. Now, although autumnal fever is a disease of intrinsic uniformity, it shows modifications which have not been explained by the assignment of modifying causes; and without such causes, its diversities constitute an objection to the existence of a single agent of an unchangeable character.

On the whole, therefore, I must repeat, that while the conditions under which our autumnal fever appears, are sufficiently clear to observation, the existence of a special gaseous agent, resulting from them, remains to be proved.

SECTION III. VEGETO-ANIMALCULAR HYPOTHESIS

I have united two words to express an hypothesis which ascribes autumnal fever to living organic forms, too small to be seen with the naked eye; and which may belong either to the vegetable or animal kingdom, or partake of the characters of both.

In the year 1832, I published in the Western Medical and Physical Journal, of which I was the editor, a series of papers on Epidemic Cholera, which were afterward collected and enlarged into a small volume;* in which an attempt was made to show, that the mode in which that disease spreads, was more fully explained by the *animalcular* hypothesis than any other which had been proposed. The brief investigation then given to the subject, reinspired my respect for the opinion long before expressed, that autumnal fever, and many other forms of disease, might be of animalcular origin; and the discoveries since made by the Ehrenberg school, have seemed to render that doctrine still more probable. But I have neither had time nor means for experimental or bibliographical inquiry; and do not propose to dwell very long upon the subject in this place.

As applied to Epidemic Cholera, I regard the hypothesis of animalcules as more plausible than that of vegetable germs; but in reference to autumnal fever, either may be assumed; and in support of the assumption, I proceed to make the following observations:

1. The microscope has revealed the existence of a countless variety of organic forms, which surround and penetrate the bodies of larger animals and plants, whether living, or dead and decaying, inhabit all waters, salt and fresh, and swarm in the atmosphere; buoyed up and moving by their own organs, or sustained by their levity, and wafted about by currents of air. The difficulty of detecting them in the atmosphere is greater than in water, or when attached to

* A Practical Treatise on the History, Prevalence, and Treatment of Epidemic Cholera. By Daniel Drake, M.D. Cincinnati: 1832. Pp. 180.

solid substances; but to my own mind, it seems probable that they exist in the aerial ocean in greater multitudes than elsewhere. For, *first,* minute particles of matter, organic and inorganic, are at all times floating in that ocean, and may serve as their food or resting places; and, *second,* as the surface of a body becomes greater, in comparison with its weight, the more it is reduced in size, it follows that living, organic forms, both animal and vegetable, may be of such size, as to float permanently in the air. The power of reproduction, possessed by these microscopic creatures, is still more wonderful than their minuteness. It exceeds, indefinitely, all examples presented by the visible organic kingdom; where, however, we see the government of the same law, for, in both plants and animals, the small multiply more rapidly than the large. In contemplating the invisible living world, in which the visible is, as it were, immersed, the mind becomes bewildered, as in meditating on the infinite, and requires to fall back upon obvious facts. Now one of these facts is, that whole rock formations, of great thickness and extent, have been found, under the microscope, to be composed entirely of the silicious shells or coverings of animalcules. In such beings, the increase seems to be merely by secretion from, or division of the parent body.

2. Among visible plants and animals, there are species that form no poison, and others which secrete that, which applied to, or inserted in our bodies, produces a deleterious effect, which is generally of a definite kind. Thus, the venom of the rattlesnake produces a disease of definite form; cantharides another; certain fish are poisonous when eaten; wasps and bees instill a venom; and the smallest visible gnat, as that which inhabits the forests of the middle latitudes, and that which is known under the name of sand fly on the shores of the Gulf of Mexico, inflames the skin; while the juice of stramonium, the exhalations of the rhus toxicodendron, and the fungus which grows beneath its shade, excite peculiar diseases. It seems justifiable to ascribe, by analogy, to microscopic animals and plants, the same diversity of properties which we find in larger beings, differing from them, as we may presume, in nothing but size and complexity of organization. We may suppose, then, that while many species of this minute creation are harmless, there are others, which can exert upon our systems a pernicious influence. This, moreover, is in accordance with what we know of gases, some of which, as nitrogen, are inert, while others are deleterious. Under this head, moreover, we must not forget the fact, that nearly all the animals and plants which secrete a poisonous fluid, grow in the southern regions, and we may, analogically, suppose that the microscopic beings in those regions are more pernicious than those of higher latitudes. Now it is in the warmer portions of our Valley, that autumnal fever has its greatest prevalence.

3. We know that water is essential to the support of those animal and vegetable forms which are matters of observation by the unassisted eye; and may conclude, therefore, that it is equally necessary for the tribes which are invisible. Indeed, it is known of many, as the *rotiferæ*, that if deprived of moisture, they seem to die, but may be revived many years afterward by the application of water. Now we have seen that, in the western part of the Valley, where great aridity prevails, the Fever is almost unknown; while it prevails with greatest frequency and violence, other conditions being the same, where there is adequate humidity.

4. A high temperature is favorable to the development of animal and vegetable life. In the southern parts of the Valley, animal forms, especially of the lower order, are greatly multiplied, and vegetation is luxuriant. If this be true of the visible, why may we not conclude that it is equally true of the invisible. Now, it is precisely in those regions, that the Fever, other circumstances being equal, displays its greatest prevalence and malignity. When we look to the north, we find that, after reaching the parallel which has an isotheral curve of sixty degrees Fahrenheit, the amount of visible organic life is much diminished, and continues rapidly to decrease; we may therefore presume, that the same is true of microscopic plants and animals. But we have already seen, that where the summer temperature falls below sixty degrees Fahrenheit, autumnal fever is unknown.

5. In the visible organic world, we find animals subsisting on plants, or on other animals that have fed on vegetables. Again: the decomposing remains of one generation of plants, favors the growth of another; and thus the soil gradually acquires the ability to bring forth a more luxuriant crop. Organic matter is, then, the proper, though not sole nutriment of organized beings. Such being the law, we may presume that, *cæteris paribus*, where dead organic matter is most abundant, microscopic tribes will be most multiplied. It is a familiar fact, that such matter abounds, through almost every stage of its decomposition, in visible beings, which subsist upon it. Thus flesh has the larvæ of the green and many other flies; rotten wood its grubs; vinegar, as the result of decomposition, its eels—sometimes visible to the naked eye; cheese its visible and invisible inhabitants; and bread its mold, a cryptogamic plant. Finally, all vegetable infusions, when exposed to the air, have their *infusoria*. It is impossible, then, to doubt, that myriads of microscopic beings swarm around, and enter the interstices of all dead organic matter; and thus we have reason for believing, that they prevail most, where such matter is most abundant; and it is in the same localities, other circumstances being equal, that we find the greatest prevalence of the Fever.

6. By the vegeto-animalcular hypothesis, we can explain the concentrated

prevalence of the Fever in certain places, as rationally as by the malarial hypothesis. Thus, its virulent reign at the head of Pensacola Bay, where there are extensive deposits of river alluvion, may be referred to the multiplication of animalcules or germs, where they find abundance of nutriment; and in the case of the exposure of the face of a deep stratum of silt by the removal of a mill-dam on Cedar Creek, we have only to suppose, that they immediately began to multiply upon the denuded surface.

7. It has, often, been observed, that the Fever has suddenly increased after rain; and this might have arisen from the resuscitation of organic forms rendered torpid by previous drought.

8. It may be, that cold produces a state of suspended animation in these as in many larger animals, and in numerous plants; and that the first warm weather of spring revives and sets them to multiplying; when they generate, what are called vernal intermittents (or at least, a part of such cases); the origin of which cannot be rationally ascribed to malaria developed at that time.

9. Microscopic observation and analogy render it probable, that in the invisible, as well as the visible province of the organic kingdom, there are distinct species, which constitute, by their union, natural families or orders. We know that in each natural assemblage of the larger plants and animals, the species resemble each other in many internal qualities, as well as in their forms. Thus, an astringent principle pervades the various kinds of oak; a resinous principle the linear evergreens; an aromatic oil, the peppermint, and other didynamous herbs; a poisonous principle, the different species of rhus; and that a narcotic principle pervades a large assemblage of plants. We know, also, that these various active principles in each group, are in general analogous, but not identical; whether we examine them by their sensible properties, with chemical reagents, or observe their effects upon the living body. Now, may it not be, that two distinct species of the same natural order of microscopic beings, may produce autumnal fever? May not one be the cause of intermittents—the other of remittents? may not both act on the system at the same time? and may we not thus explain diversities, which are inexplicable on the malarial hypothesis? Every practical physician knows, that while the juice of a variety of plants will produce the pathological condition called narcotism, the symptoms of that state, when induced by different agents, differ as widely from each other, as the symptoms of the different forms of autumnal fever.

10. In discussing the meteoric hypothesis, it was said, that the pathological effects of a certain condition of the principles of the atmosphere, are always immediate; and it might have been remarked, when treating of the malarial hypothesis, that as far as we know, the effects of gases are likewise immediate;

but we are certain that autumnal fever often begins many days, and even weeks or months, after an exposure to its remote cause. Now we know, as a general fact, that many animal poisons do not develop their effects, till after the lapse of a greater or less length of time. Thus, two weeks may elapse before small pox will appear, after exposure; and two years have passed away, before hydrophobia has followed on the bite of a mad dog. On this point, then, the vegeto-animalcular hypothesis, has an advantage over both the others.

11. It has been already stated, that autumnal fever prevails very unequally in different years; and that, in the same locality, it may, in one autumn, be malignant and epidemic, and in another, mild and sporadic. This can, perhaps, be better explained on the hypothesis we are now discussing, than on either of the others; for we know, that throughout the visible organic domain, reproduction is by no means uniform. A year of great abundance, may be followed by one unproductive, in the vegetable kingdom; and in the animal, one summer and autumn will be infested by insects far beyond another. It has often happened, that musquitoes have been absent, from the banks of the middle portion of the Ohio river, for a year, and in the next appeared in immense numbers. We have but to suppose insect forms of a parallel size, to live under corresponding laws, and the hypothesis now before us, offers an explanation of sickly and healthy seasons.

12. It is well known that the long-continued cultivation of the soil, and the building of towns and cities, diminishes the prevalence of the Fever. Now this cultivation implies the drying up of a great deal of surface water; the burning up of the natural vegetation, and the gradual decomposition of that which has been mingled with the soil. Summer crops, as those of wheat and hay, are also removed, and not suffered like the natural herbage to accumulate on the surface; and those of autumn are either removed, or in the course of the winter consumed, to prepare the fields for new planting. Thus the food of microscopic beings is destroyed, and their reproduction arrested.

13. We are familiar with the fact that many persons never sicken with autumnal fever, while others around them will have repeated attacks. This is ascribed to difference of susceptibility, and of exposure to exciting causes. Such ascription is no doubt correct; but the vegeto-animalcular hypothesis offers, from analogy, an additional explanation. It is well known that certain visible insects prey on some individuals much more than others—seem to be attracted by one and repelled by another—and we have but to grant to the invisible the same tastes and instincts, to understand that some persons may always draw swarms around them, while others escape their depredations.

14. People who inhabit houses built on the hills adjoining valleys, are said to suffer more than those who reside below. Now every breeze may waft and

lodge in such habitations the microscopic beings which multiply in the rich and humid valley-soil. It has also been observed, that a grove of forest trees between an inhabited house, and what is called a sickly spot, gives comparative immunity from the Fever; and may not the leaves of such trees as successfully arrest animalcules, or vegetable germs, as they can absorb a gas not designed for their nourishment?

From what has been said, it appears obviously, I think, that the etiological history of autumnal fever, can be more successfully explained by the vegeto-animalcular hypothesis, than the malarial. But both, in the present state of our knowledge, must stand *as mere hypotheses*. Neither can claim the rank of a theory; nor will it be entitled to the confidence of the profession until many additional facts are brought to its support.*

IV. VALUE OF THE DISCOVERY OF THE EFFICIENT CAUSE OF AUTUMNAL FEVER. I cannot, *a priori,* attach much practical importance to a discovery of the *efficient* cause of autumnal fever; and have devoted several pages to its discussion, from deference to my brethren, much more than from my own conviction, of the value of the discovery to which so many minds are directed. Did we know the particular meteoric condition, the gas, or the organized microscopic species which produces the Fever, we should not probably be able to defend ourselves against it, by any precautions, but those which experience has already established; nor should we be able to destroy the efficient cause, without annihilating the conditions under which it is generated. Those conditions are already well known. The individual exposed to them is liable to an attack—he who keeps away remains exempt. The people of the country escape the vesicular eruption produced by the *rhus toxicodendron* or the *rhus vernix,* by keeping beyond the sphere of exhalation. They know nothing of the nature of the poisonous emanation, and yet their means of protection are as perfect, as those of the chemist would be, who might analyze the poison and give it an appropriate name. Nor is it probable that the discovery of the efficient cause would throw any light upon the treatment. It was not a knowledge of its cause that taught us the cold treatment of small pox; we know the cause of hydrophobia and yet cannot cure it; we do not know the cause of goitre, but have discovered that iodine is an efficient remedy.

Ignorant, however, as we are of any definite, efficient cause for autumnal fever, I am a full believer in its existence, and shall speak of it as a specific agent, known only by its effects on the living body. These effects constitute the

* When this article was about to be sent to the press, a friend handed me Professor Mitchell's Lectures on the *"Cryptogamous origin of Malarious and Epidemic Fevers,"* which I had not before seen. The array of facts made by the learned author, seems almost irresistible; and, from his distinguished reputation, it will, no doubt, lead many others into new courses of observation and experiment.

disease we have been studying in its etiology; and are now to contemplate in its symptomatology, pathology, and therapeutics. In proceeding to do this, the first inquiry naturally is, into the manner in which the assumed agent makes its impress on the system. In doing this, I wish it understood, that if I should, at any time, use the word malaria, it is merely to designate the remote cause, *whatever* it may be.

* * * * *

The article AUTUMNAL FEVER is now brought to a close. It has extended through many pages; but a smaller number would not have sufficed, to present, even an outline, of its etiological, and therapeutic history; through so wide a geographical range, as that of the southern half of our Interior Valley; in almost every part of which, it is an annual endemio-epidemic. Of all our diseases, it is the one, which has the most intimate relations with soil and climate—that, in which, peculiarities, resulting from topographical and atmospheric influences, are most likely to appear. Hence it was chosen, to stand next to the Book of General Etiology; as illustrating, better than any other disease, the importance, of the facts which make up that Book. It is, moreover, the *great* cause of mortality, or infirmity of constitution, especially in the southern portions of the Valley; and, therefore, entitled to severe and patient attention. What I have collected and presented, has required more labor, than many of our brethren might suppose; and, yet, they will not, perhaps, realize so fully as I do myself, how much must be added—how many errors corrected—before the pages through which they have traveled, can be entitled to universal acceptance. Meanwhile, if what has been written, should stir up a single young physician, to a more diligent observation of the Fever, or save the life of one individual, who might otherwise have become its victim, my labor will not have been in vain.

A Bibliography of the Writings of Daniel Drake

No comprehensive bibliography of the writings of Daniel Drake has previously been published, although lists of his major writings have appeared in Thomson, *Bibliography of Ohio* (1880), Juettner, *Daniel Drake and His Followers* (1909), the Library of Congress *Catalogue of Printed Cards,* and more indirectly in the references of Horine, *Daniel Drake.* In preparing this list, we have had occasion to visit not only the excellent libraries of Cincinnati but also collections in Lexington, Louisville, Philadelphia, and Washington, D.C., and we wish to acknowledge the assistance of the librarians in those cities who gave so willingly of their time and knowledge. Although in a few instances it has been impossible to do so, we determined at the start to include no item which we had not personally seen. In our search for Drake material, as a consequence, we have literally read complete runs of the following periodicals: *Academic Pioneer and Guardian of Education* (Cincinnati), *American Journal of Science* (New Haven), *Boston Medical and Surgical Journal, Cincinnati Chronicle and Literary Gazette* and its predecessor, *Saturday Evening Chronicle, Eclectic Repertory and Analytical Review* (Philadelphia), *Educational Disseminator* (Cincinnati), *The Hesperian, or Western Monthly Magazine* (Columbus and Cincinnati), *Louisville Journal of Medicine and Surgery, New Orleans Medical and Surgical Journal, North American Medical and Surgical Journal* (Philadelphia), *Philadelphia Journal of the Medical and Physical Sicences, Philadelphia Medical and Physical Journal, Port Folio* (Philadelphia), *Proceedings of the Medical Convention of Ohio, Transactions of the Western Literary Institute and College of Professional Teachers* (Cincinnati), *Transylvania Journal of Medicine and the Associated Sciences* (Lexington), *Western and Southern Medical Recorder* (Lexington), *Western Journal of the Medical and Physical Sciences* (Cincinnati), *Western Journal of Medicine and Surgery* (Louisville), *Western Lancet* (Cincinnati), *Western Medical Gazette* (Cincinnati), *Western Medical and Physical Journal* (Cincinnati), *Western Monthly Magazine* (Cincinnati), *Western Quarterly Journal of Practical Medicine* (Cincinnati), and *Western Quarterly Reporter of Medical, Surgical, and Natural Sciences* (Cincinnati).

Items in this bibliography are arranged chronologically, in order of composition, or more often, when this is not known precisely, in order of publication. Contemporary reprints appear following the original; reprints issued in the twentieth century have been excluded from this list. For the sake of consistency, we have dated all articles printed in the *Western Journal of the Medical and Physical Sciences* or the *Western Journal of Medicine and Surgery,* which Drake edited, as if publication occurred on the first day of the indicated month. We have ascribed to Drake all unsigned articles in these journals, as well as several initialled, or pseudonymous pieces appearing in other journals which seemed to us to be his handiwork. The latter are indicated by bracketed question marks.

"Subjects Discussed by the Cincinnati Debating Society [July–October, 1804]." MS in Tucker Collection, University of Cincinnati College of Medicine Library. 36 p.

"Medical Diary or Common place Book [1805–06]." MS in National Library of Medicine. 169 p.

"Duty of Periodical Essayists." The Literary Magazine, and American Register (Philadelphia) 6: 265–66 (July 1806).

"Some Account of the Epidemic Diseases which Prevail at Mays-Lick, in Kentucky. In a Letter to the Editor, from Dr. Daniel Drake. [July 22, 1807.]" Philadelphia Medical and Physical Journal 3, part 1: 84–90 ([March] 1808).

"[Address] delivered before the Cincinnati Lyceum, 1807 [on Debating Societies and the Obligations of their Members]." Photostat in Cincinnati Historical Society. 7 p.

"Library." Liberty Hall and Cincinnati Mercury, January 5, 1809.

Notices concerning Cincinnati. Cincinnati: Printed for the Author, at the Press of John W. Browne & Co., 1810. [Pt. I published May 1810; pt. II published May 1811.] 60 p.

"Strictures on Volney's 'View of the Soil and Climate of the United States.'" Port Folio (Philadelphia) [3d series] 4: 587–91 (December 1810; dated August 26, 1810; [3d series] 5: 320–24 (April 1811; dated February 20, 1811); [3d series] 6: 203–209 (September 1811; dated May 15, 1811).

"To Isaac G. Burnet, Esq." Liberty Hall and Cincinnati Mercury, July 18, August 15, September 17, 1810.

"Climate." Western Spy (Cincinnati), June 15, 1811.

"Medical Society, District No. I." Liberty Hall (Cincinnati), June 9, 1812.

"Statistical View or Picture of Cincinnati and Its Environs." Liberty Hall (Cincinnati), September 28, 1813.

"Cincinnati Lancaster Seminary." Western Spy (Cincinnati), April 23, 1814.

Anniversary Address, Delivered to the School of Literature and the Arts, at Cincinnati, November 23, 1814. Published by Order. Cincinnati: Looker and Wallace, 1814. 12 p.

"Anniversary Address, Delivered to the School of Literature and the Arts at Cincinnati, November 23, 1814." Liberty Hall (Cincinnati), December 27, 1814.

"Anniversary Address, Delivered to the School of Literature and the Arts, at Cincinnati, November 23, 1814." National Intelligencer (Washington, D.C.), April 16, 1815.

Natural and Statistical View, or Picture of Cincinnati and the Miami Country. Illustrated by Maps. With an Appendix Containing Observations on the Late Earthquakes, the Aurora Borealis, and Southwest Wind. Cincinnati: Looker and Wallace, 1815. 251 p.

"Medical Topography of Cincinnati [Extract from Natural and Statistical View, or Picture of Cincinnati and the Miami Country]." Eclectic Repertory and Analytical Review (Philadelphia) 6: 137–49 (April 1816).

A Systematic Catalogue of Books Belonging to the Circulating Library Society of Cincinnati. To which are Prefixed an Historical Preface, the Act of Incorporation, and By-Laws of the Society. Published By Order of the Board of Directors. Cincinnati: Looker, Palmer & Reynolds, [October] 1816. 36 p.

"Biographical Sketch of Dr. William Goforth." Western Spy (Cincinnati), June 13, 1817.

"Biographical Sketch of Dr. William Goforth, who Died in Cincinnati, May 12, 1817." Ohio Medical Repository (Cincinnati) 1: 78–79 (January 3, 1827).

"[Letter to Jose Correa de Serra, on the Geology of the Ohio Valley.]" October 1, 1817. MS in American Philosophical Society Library. 26 p.

"Geological Account of the Valley of the Ohio: In a Letter from Daniel Drake, M.D., to Joseph Correa de Serra. Read November 7, 1818." In Transactions of the American Philosophical Society n.s. 2: 124–39 (1825).

"Extract from Professor Drake's Valedication, delivered to his Class at the Close of the Session." Kentucky Reporter (Lexington), March 11, 1818.

"An Address to the People of the Western Country [Prospectus of the Western Museum]." Liberty Hall and Cincinnati Gazette, September 15, 1818.

An Address to the People of the Western Country [Prospectus of the Western Museum]. Handbill. Cincinnati, September 15, 1818. 2 p.

"An Address to the People of the Western Country [Prospectus of the Western Museum. September 15, 1818]." American Journal of Science (New York & New Haven) 1: 203–206 ([October?] 1818).

"Valedictory to the Class at Lexington in the Transylvania University. March 1, 1818." Last page missing. MS in Cincinnati General Hospital Library. 16 p.

An Appeal to the Justice of Intelligent and Respectable People of Lexington. Cincinnati: Looker, Reynolds & Co., [July 10,] 1818. 23 p.

A Second Appeal to the Justice of the Intelligent and Respectable People of Lexington. Cincinnati: Looker, Reynolds & Co., [November 6,] 1818. 34 p.

"Extracts from Dr. Drake's Introductory Lecture to a Course of Materia Medica and Practice of Physick." Inquisitor Cincinnati Advertiser, November 17, 24, 1818.

"Observations on Temulent Diseases. In a Letter from Daniel Drake, M.D., of Cincinnati, Ohio, to Samuel Brown, M.D., of Philadelphia." American Medical Recorder (Philadelphia) 1: 59–65 (January 1, 1819).

"Observations on the Means and Importance of Preserving Fruit and Forest Trees." Read at the First Meeting of the Cincinnati Society for the Promotion of Agriculture, Manufactures and Domestic Economy, September 28, 1819. Published by Order of the Standing Committee. Western Spy (Cincinnati), October 23, 1819.

"Medical College of Ohio. [Dr. Samuel Brown rejects offered Professorship]." Liberty Hall and Cincinnati Gazette, October 19, 1819.

"An Explanation [Concerning Choice of Faculty for the Medical College of Ohio]." Liberty Hall and Cincinnati Gazette, October 26, 1819.

"An Introductory Lecture on the Utility and Pleasures of the Study of Mineralogy and Geology. Delivered in the Western Museum, December 18, 1819." Liberty Hall and Cincinnati Gazette, December 28, 1819.

"On the Utility and Pleasures of the Study of Mineralogy and Geology." Port Folio (Philadelphia) 4th series, 9: 86-93 ([January, February, March] 1820).

"To the Editors of the Spy." Western Spy and Cincinnati General Advertiser, January 22, 1820.

"To the Editors of the Western Spy." Western Spy and Cincinnati General Advertiser, January 29, 1820.

"To the Editors of the Western Spy." Western Spy and Cincinnati General Advertiser, February 19, 1820.

An Anniversary Discourse, on the State and Prospects of the Western Museum Society: Delivered by Appointment, in the Chapel of the Cincinnati College, June 10th, 1820, at the Opening of the Museum. Cincinnati: Printed for the Society, by Looker, Palmer and Reynolds, 1820. 36 p.

Circular. Medical College of Ohio. Handbill. [Cincinnati:] August 20, 1820. 3 p.

"Medical College of Ohio.—'Circular' [August 20, 1820]." Western Spy and Literary Cadet (Cincinnati), August 23, 1820. Repeated, August 31, 1820.

An Inaugural Discourse on Medical Education; Delivered at the Opening of the Medical College of Ohio, in Cincinnati, November 11th, 1820. Cincinnati: Looker, Palmer and Reynolds, 1820. 31 p.

"Valedictory to the Medical College of Ohio, April 4, 1821 ['reproduced from the original manuscript']." In Otto Juettner, Daniel Drake and His Followers. Cincinnati: Harvey Publishing Co., 1909. Pages 52-55.

"Introductory Lecture for the Second Session of the Medical College of Ohio. November, 1821." Originally 30 p. Includes revisions and additions made in 1832 ("Gravitation, Affinity, Vitality. Address to the Mechanics Institute. December 8, 1832") and 1840 ("Address to the Louisville Medical Society. November 27, 1840"). MS in Cincinnati General Hospital Library. 50 p.

A Narrative of the Rise and Fall of the Medical College of Ohio. Cincinnati: Looker & Reynolds, 1822. 42 p.

"To the Physicians of the Western States [Concerning Plans for Systematic Study of Western Diseases]." Western Quarterly Reporter of Medical, Surgical, and Natural Science (Cincinnati) 1: 307-11 (September 1, 1822).

"To Debtors and Creditors." Liberty Hall and Cincinnati Gazette, May 16, 1823.

An Introductory Lecture, on the Necessity and Value of Professional Industry; Delivered in the Chapel of Transylvania University, November 7th, 1823. Published by Request of the Class. Lexington: William Tanner, 1823. 31 p.

"Address Read to the Lexington Medical Society [on the Purposes of Scientific Societies]. November 14, 1823." MS in Cincinnati General Hospital Library. 11 p.

"Valedictory Address to the Kappa Lambda Society of Hippocrates [at Transylvania University]. February 26, 1824." MS in Cincinnati General Hospital Library. 14 p.

"[Letters on] The Presidency [supporting Henry Clay's candidacy]." Liberty Hall (Cincinnati), April 20, 27, May 4, 7, 14, June 4, 25, July 2, 23, August 10, 13, 17, 27, 1824. Signed "Seventy-Six."

"Okumanitas." Cincinnati Literary Gazette, July 3, 1824. Signed "D." [?]

"A Fragment [on Intemperance]." Cincinnati Literary Gazette, July 10, 1824.

"Republican Gratitude." Cincinnati Literary Gazette, September 25, 1824.

"[An Introductory Lecture on the Principles and Practical Rules of the Profession]." For the Medical Department of Transylvania University, November, 1824. MS in Cincinnati General Hospital Library. 8 p.

"On Philosophical Truth [Extracts from an Introductory Lecture at Transylvania University]." Cincinnati Literary Gazette, November 27, December 4, 11, 1824.

"Drake's Lectures on Materia Medica. [Transylvania University, November–December, 1824.]" MS student notes, in Transylvania College Library. 96 p.

"Table IX. Synoptic View of Mineral Poisons; With their Appropriate Tests and Antidotes; Translated with Additions, from the Tables of Dr. De Salle, published in Paris, 1824." In Robert Best, Tables of Chemical Equivalents, Incompatible Substances, and Poisons and Antidotes. Lexington: W. W. Worsley, [January] 1825. Pages 42–59.

"Note on the Vegetable Poisons." In Robert Best, Tables of Chemical Equivalents, Incompatible Substances, and Poisons and Antidotes. Lexington: W. W. Worsley, [January] 1825. Pages 60–74.

"To the Editors of the National Intelligencer [March 21, 1825]." National Intelligencer (Washington, D.C.), April 4, 1825.

"The Late Presidential Election. Letter to the Editors of the National Intelligencer. [March 21, 1825.] Niles Weekly Register (Baltimore), April 9, 1825.

"To the Citizens of Cincinnati [Concerning the Cincinnati Library]." Saturday Evening Chronicle (Cincinnati), December 30, 1826.

"A Doctorate Address Delivered at the Medical Commencement of Transylvania University, March 16th, 1827." MS in Cincinnati General Hospital Library. 31 p.

"History of a Case of Neuralgia Facialis, with Reflections." Western Medical and Physical Journal (Cincinnati) 1: 13–29 (April 1827).

"[Letter Soliciting Facts and Observations for a History of the Diseases between the 'Gulph and the Lakes.']" Saturday Evening Chronicle (Cincinnati), April 21, 1827.

"[Notice:] Eye Infirmary." Western Medical and Physical Journal (Cincinnati) 1: 126–27 (May 1827).

"To the Editor of the Saturday Evening Chronicle [Concerning the Western Quarterly Review]." Saturday Evening Chronicle (Cincinnati), June 30, 1827. Signed "D."

"History of a Case of Congenital Chorea Sancti Viti." Western Medical and Physical Journal (Cincinnati) 1: 193–97 (July 1827).

"Observations on the Modus Operandi and Effects of Medicines." Western Medical and Physical Journal (Cincinnati) 1: 249–63 (August 1827).

"Editorial Remarks on Tissot on Trepanning in Epilepsy." Western Medical and Physical Journal (Cincinnati) 1: 276–77 (August 1827).

"Practical Remarks on Cramp of the Stomach from Tartar Emetic." Western Medical and Physical Journal (Cincinnati) 1: 297–304 (September 1827).

"Thoughts on Modern Travelling; designed for Valetudinarians." Western Medical and Physical Journal (Cincinnati) 1: 305–10 (September 1827).

"To the Editor of the Evening Chronicle [Concerning the Western Quarterly Review]." Saturday Evening Chronicle (Cincinnati), September 1, 1827.

"Practical Observations on the Typhoid Stage of Autumnal Fever." Western Medical and Physical Journal (Cincinnati) 1: 381–91 (October 1827).

"[Review:] Principles of Dental Surgery . . . By Leonard Koecker." Western Medical and Physical Journal (Cincinnati) 1: 328–49 (September 1827); 406–22 (October 1827).

"[Review:] A Review of the Diseases of Dutchess County, New York, from 1809 to 1825 . . . By Hunting Sherrill." Western Medical and Physical Journal (Cincinnati) 1: 458–75 (November 1827).

"Notes on a Case of Cellular Inflammation." Western Medical and Physical Journal (Cincinnati) 1: 545–50 (January 1828).

"An Account of the Death of a Man, from a Lacerated Wound of the Jejunum, the Abdominal Parieties Remaining Entire." Western Medical and Physical Journal (Cincinnati) 1: 550–55 (January 1828).

"Hepatica Triloba." Western Medical and Physical Journal (Cincinnati) 1: 595–97 (January 1828).

"Epidemic Quackery." Western Medical and Physical Journal (Cincinnati) 1: 600 (January 1828).

"To the Editor of the U.S. Telegraph." Daily Cincinnati Gazette, January 16, 1828.

A Discourse on Intemperance; Delivered at Cincinnati, March 1, 1828, before the Agricultural Society of Hamilton County, and Subsequently Pronounced, by Request, to a Popular Audience. Cincinnati: Looker & Reynolds, 1828. 96 p.

"A Discourse on Intemperance; delivered by Appointment, at a Public Meeting of the Agricultural Society of Hamilton County, Ohio, March 1st, 1828, and Subsequently Pronounced, by Request, before a Popular Audience." Western Journal of the Medical and Physical Sciences (Cincinnati) 2: 11–34 (April 1828); 65–91 (May 1828).

"[Review:] Formulary for the Preparation and Employment of Many New Medicines . . . By F. Majendie, . . . Translated . . . revised, and augmented, by John Baxter, M.D." Western Journal of the Medical and Physical Sciences (Cincinnati) 2: 35–50 (April 1828); 92–108 (May 1828).

"[Review:] Observations on the Efficacy of White Mustard-seed . . . By Charles Turner Cooke." Western Journal of the Medical and Physical Sciences (Cincinnati) 2: 50–54 (April 1828).

"Quarterly Report of the Weather and Diseases of Cincinnati." Western Journal of the Medical and Physical Sciences (Cincinnati) 2: 60–64 (April 1828).

"Winter Quarterly Report on the Diseases of Cincinnati, Concluded." Western Journal of the Medical and Physical Sciences (Cincinnati) 2: 116–20 (May 1828).

"An account of a Rupture of the Pylorus, with the Appearances after Death." Western Journal of the Medical and Physical Sciences (Cincinnati) 2: 126–30 (June 1828).

"Notices of the Principal Mineral Springs of Kentucky and Ohio." Western Journal of the Medical and Physical Sciences (Cincinnati) 2: 142–67 (June 1828).

Doctor Daniel Drake's Report of the Harrodsburg Springs, Published Originally, in the Western Journal of Medical and Physical Sciences, edited by Daniel Drake [With the Addition of Testimonial Letters]. N.p., n.d. [Cincinnati, 1828.] p. 8.

"Report on the Weather and Diseases of Cincinnati, in the Spring of 1828." Western Journal of the Medical and Physical Sciences (Cincinnati) 2: 174–76 (June 1828).

"Report on the Diseases of Cincinnati in the Spring of 1828, Concluded. Ague and Fever." Western Journal of the Medical and Physical Sciences (Cincinnati) 2: 216-18 (July 1828).

"Fractures of the Thigh." Western Journal of the Medical and Physical Sciences (Cincinnati) 2: 218-21 (July 1828).

"Our Board of Health." Western Journal of the Medical and Physical Sciences (Cincinnati) 2: 221 (July 1828).

"Splendid Engravings of American Birds." Western Journal of the Medical and Physical Sciences (Cincinnati) 2: 221-23 (July 1828).

"Dr. Drake on Intemperance." Saturday Evening Chronicle (Cincinnati), July 19, 1828.

"Domestic Restrictions: From Dr. Drake's Discourse on Intemperance." Saturday Evening Chronicle (Cincinnati), July 26, 1828.

"To the Physicians of the Western States." Western Journal of the Medical and Physical Sciences (Cincinnati) 2: [225-30] (August 1828).

"Observations on Mr. Guthrie's Double Cataract Knife." Western Journal of the Medical and Physical Sciences (Cincinnati) 2: 231-40 (August 1828).

"Practical Observations on the Uterine Hoemorrhage, connected with Abortion and Parturition." Western Journal of the Medical and Physical Sciences (Cincinnati) 2: 241-49 (August 1828).

"Observations on the Reciprocal Morbid Influences of the Abdominal and Thoracic Viscera." Western Journal of the Medical and Physical Sciences (Cincinnati) 2: 281-90 (September 1828).

"Wounds of the Heart. Translated from the Dictionaire de Sciences Medicales, Vol. 43." Western Journal of the Medical and Physical Sciences (Cincinnati) 2: 333-37 (October 1828).

"Some Account of a Fatal Case of Softening of the Heart with Apparent Emphysema of the Lungs." Western Journal of the Medical and Physical Sciences (Cincinnati) 2: 337-41 (October 1828).

"Editorial Note Containing an Account of the Analysis of . . . [Urinary] Calculi [in an Ox]." Western Journal of the Medical and Physical Sciences (Cincinnati) 2: 451-53 (December 1828).

"Remarks by the Editor [on Epilepsy]." Western Journal of the Medical and Physical Sciences (Cincinnati) 2: 457-58 (December 1828).

"[Review:] De la Lithontritie . . . Par le Docteur Civiale." Western Journal of the Medical and Physical Sciences (Cincinnati) 2: 570-95 (February 1829).

"An Account of a case of Empyema, Consequent upon Measles, Cured by Paracentesis." Western Journal of the Medical and Physical Sciences (Cincinnati) 2: 617-20 (March 1829).

"Notes of a case of Empyema from Chronic Pleurisy, in which the Operation Gave Relief." Western Journal of the Medical and Physical Sciences (Cincinnati) 2: 620-23 (March 1829).

"A Sketch of the Life and Character of Dr. Thomas Hinde." Western Journal of the Medical and Physical Sciences (Cincinnati) 2: 625-34 (March 1829).

"The New Canal Boat." Cincinnati Chronicle and Literary Gazette, May 23, 1829.

"Practical Essays on Medical Education and the Medical Profession, in the United

States. I. Selection and Preparatory Education of Pupils." Western Journal of the Medical and Physical Sciences (Cincinnati) 3: 13–27 (April, May, June, 1829).

"Diagnostic Notices of a Case of Fungus Haemotodes; Designed as a Contribution to the History of that Disease." Western Journal of the Medical and Physical Sciences (Cincinnati) 3: 34–40 (April, May, June, 1829).

"Medical Jurisprudence—Report of a Trial for Murder, in which the Culprit was Defended on the Ground of his Laboring under Mania a Potu, or Delerium from Intemperance." Western Journal of the Medical and Physical Sciences (Cincinnati) 3: 44–65 (April, May, June, 1829).

"From Dr. Drake's Western Journal of the Medical and Physical Sciences. Medical Jurisprudence." Cincinnati Chronicle and Literary Gazette, July 4, 1829.

"The London University." Western Journal of the Medical and Physical Sciences (Cincinnati) 3: 149–51 (April, May, June, 1829).

"Small Pox." Western Journal of the Medical and Physical Sciences (Cincinnati) 3: 155–56 (April, May, June, 1829).

"[Communication Concerning the Annual Examination of The Messrs. Kinmont's Academy.]" Cincinnati Chronicle and Literary Gazette, August 22, 1829. Signed "D."

"Notes on a Case of Spina-Bifida." Western Journal of the Medical and Physical Sciences (Cincinnati) 3: 210–12 (July, August, September, 1829).

"Medical Jurisprudence. Sequel to the Case of John Birdsell, with Remarks on Feigned Insanity." Western Journal of the Medical and Physical Sciences (Cincinnati) 3: 215–21 (July, August, September, 1829).

"[Review:] A Treatise on the Diseases of the Chest . . . and . . . Acoustick Instruments . . . Translated from the French of R. T. H. Laennec . . . By John Forbes; A Short Treatise on the Different Methods of Investigating the Diseases of the Chest. Translated from the French of M. Collin. By W. N. Ryland, M.D." Western Journal of the Medical and Physical Sciences (Cincinnati) 3: 67–99 (April, May, June, 1829); 3: 222–60 (July, August, September, 1829).

"Wounds of the Heart." Western Journal of the Medical and Physical Sciences (Cincinnati) 3: 295–99 (July, August, September, 1829).

"Silliman's Journal." Western Journal of the Medical and Physical Sciences (Cincinnati) 3: 313–14 (July, August, September, 1829).

"Practical Essays on Medical Education and the Medical Profession, in the United States. II. Private Pupilage." Western Journal of the Medical and Physical Sciences (Cincinnati) 3:317–40 (October, November, December, 1829).

"The People's Doctors." Western Journal of the Medical and Physical Sciences (Cincinnati) 3: 393–420, 455–62 (October, November, December, 1829).

The People's Doctors; A Review, by "The People's Friend." Cincinnati: Printed and Published for the Use of the People, 1830. 57 p.

"The Gentleman. Begun January 31, 1830. Finished February 13. Delivered September 15." MS in Cincinnati General Hospital Library. 20 p.

"[Review:] A Treatise on the Scrofulous Disease, by C. G. Hufeland, Physician to the King of Prussia . . . Translated from the French of M. Brusquet, by Charles D. Meigs, M.D." Western Journal of the Medical and Physical Sciences (Cincinnati) 3: 545 (January, February, March, 1830).

"Medical Jurisprudence. Mania a Potu. Birdsell's Case." Western Journal of the Medical and Physical Sciences (Cincinnati) 3: 598–602 (January, February, March, 1830).

"Remarks by the Editor [on a Case Resembling Dry Gangrene]." Western Journal of the Medical and Physical Sciences (Cincinnati) 4: 31 (April, May, June, 1830).

"Note by the Editor [on a Disease Called Charbon]." Western Journal of the Medical and Physical Sciences (Cincinnati) 4: 37–38 (April, May, June, 1830).

"History of Two Cases of Burn, Producing Serious Constitutional Irritation." Western Journal of the Medical and Physical Sciences (Cincinnati) 4: 48–60 (April, May, June, 1830).

"[Review:] An Essay on the Remittant and Intermittant Diseases, Including, Genrically Marsh Fever and Neuralgia . . . By John Macculloch, M.D., F.R.S." Western Journal of the Medical and Physical Sciences (Cincinnati) 4: 66–104 (April, May, June, 1830); 213–56 (July, August, September, 1830).

"Obituary [notice, Dr. John D. Godman]." Western Journal of the Medical and Physical Sciences (Cincinnati) 4: 158 (April, May, June, 1830).

"History of a Case of Exophthalmos or Protruded Eye Ball." Western Journal of the Medical and Physical Sciences (Cincinnati) 4: 197–212 (July, August, September, 1830).

"Female Fortitude [extract from 'History of a Case of Exophthalmos, or Protruded Eyeball']." Cincinnati Chronicle and Literary Gazette, November 6, 1830.

"Lectures on the Theory and Practice of Medicine, delivered by Daniel Drake [at the Jefferson Medical College, Philadelphia]. 1830-1." MS student notes [Lawrence D. Henderson?] in National Library of Medicine. 2 v., 772 p.

"[Notes:] Discourse II: Gluttony. November 21, 1830." Chapel Lectures, Jefferson Medical College, Philadelphia. MS in Cincinnati General Hospital Library. 4 p.

"Discourse 3d: Of Intemperance. Sunday, November 28, 1830." Chapel Lectures, Jefferson Medical College, Philadelphia. MS in Cincinnati General Hospital Library. 25 p.

"Discourse VI: On the Value of Time. Sunday, December 19, 1830." Chapel Lectures, Jefferson Medical College, Philadelphia. Incomplete. MS in Cincinnati General Hospital Library. 8 p.

"[Notes:] On the Duties of Physicians to Their Patients. Sunday, January 16, 1831." Chapel Lectures, Jefferson Medical College, Philadelphia. MS in Cincinnati General Hospital Library. 2 p.

"On the Duties of the Physician to his Patients, delivered January 16, 1831." In "Lectures on the Theory and Practice of Medicine, delivered by Daniel Drake [at the Jefferson Medical College, Philadelphia]. 1830-1." MS student notes [Lawrence D. Henderson?] in National Library of Medicine. Pages 751–72.

An Oration on the Intemperance of Cities: Including Remarks on Gambling, Idleness, Fashion, and Sabbath-Breaking. Delivered in Philadelphia, January 24, 1831. Philadelphia: Griggs and Dickinson, 1831. 30 p.

"[On Industry and Idleness. February? 1831.]" First pages missing. MS in Cincinnati General Hospital Library. 12 p.

An Oration on the Causes, Evils and Preventives of Intemperance. Delivered and Published by Request, in the Town of Columbus, Ohio. February 12th, 1831. Columbus: Olmstead & Bailhache, 1831. 21 p.

"[Extracts from An Oration on the Causes, Evils, and Preventives of Intemperance,

Delivered on the 12th of February . . . in Columbus.]" Cincinnati Chronicle and Literary Gazette, March 5, 1831.

"Biographia Medica [Dr. John D. Godman and Dr. Robert Best]." Western Journal of the Medical and Physical Sciences (Cincinnati) 4: 596-616 (January, February, March, 1831).

"Practical Essays on Medical Education and the Medical Profession, in the United States. III. Medical Colleges." Western Journal of the Medical and Physical Sciences (Cincinnati) 5: 9-23 (April, May, June, 1831).

"[Review:] 1. Journal of Health. Periodical. Philadelphia, 1829. 2. Rules for the Preservation of Health and Vigor of the Constitution from Infancy to Old Age. Philadelphia, 1831." Western Journal of the Medical and Physical Sciences (Cincinnati) 5: 100-18 (April, May, June, 1831).

"[Review:] The Pharmacopoeia of the United States of America. Published by Authority of the National Medical Convention, held at Washington, A.D., 1830." Western Journal of the Medical and Physical Sciences (Cincinnati) 5: 119-21 (April, May, June, 1831).

"[Review:] A Dissertation on the Remote and Proximate Causes of Phthisis Pulmonalis . . . By Andrew Hammersly, M.D." Western Journal of the Medical and Physical Sciences (Cincinnati) 5: 121-28 (April, May, June, 1831).

"[Practical] Essays on Medical Education and the Medical Profession, in the United States. IV. Studies, Duties, and Interests of Young Physicians." Western Journal of the Medical and Physical Sciences (Cincinnati) 5: 169-77 (July, August, September, 1831).

"History of a Case of Hydrophobia, Not Arising from an Ascertained Cause." Western Journal of the Medical and Physical Sciences (Cincinnati) 5: 199-205 (July, August, September, 1831).

"Notes on a Fatal Case of Quiescent Gall Stones." Western Journal of the Medical and Physical Sciences (Cincinnati) 5: 205-10 (July, August, September, 1831).

"[Review:] Elements of Physics, or Natural Philosophy, General and Medical, Explained Independently of Technical Mathematics . . . By Neil Arnott, M.D. . . . With Additions by Isaac Hays, M.D." Western Journal of the Medical and Physical Sciences (Cincinnati) 5: 211-33 (July, August, September, 1831); 397-423 (October, November, December, 1831).

"[Review:] Select Medico-Chirurgical Transactions . . . Edited by Isaac Hays, M.D." Western Journal of the Medical and Physical Sciences (Cincinnati) 5: 233-57 (July, August, September, 1831).

"Report of the Visitors of T. Hammond's Juvenile Seminary." Cincinnati Chronicle and Literary Gazette, September 10, 1831.

"[Letter on the Healthiness of Covington, Kentucky.]" Cincinnati Chronicle and Literary Gazette, October 8, 1831.

"[Review:] Elements of Chemistry . . . By Benjamin Silliman." Western Journal of the Medical and Physical Sciences (Cincinnati) 5: 361-72 (October, November, December, 1831).

"[Review:] Researches, Principally Relative to the Morbid and Curative Powers of the Loss of Blood." Western Journal of the Medical and Physical Sciences (Cincinnati) 5: 372-96 (October, November, December, 1831).

Practical Essays on Medical Education, and the Medical Profession, in the United States. Cincinnati: Roff & Young, [January 19,] 1832. 104 p.

"[Practical] Essays on Medical Education and the Medical Profession, in the United States. V. Causes of Error in the Medical and Physical Sciences.—Legislative Enactments." Western Journal of the Medical and Physical Sciences (Cincinnati) 5: 503-29 (January, February, March, 1832).

"Observations on the Use and Abuse of Emetics." Western Journal of the Medical and Physical Sciences (Cincinnati) 5: 543-59 (January, February, March, 1832).

"Epidemic Cholera:—Its Pathology and Treatment. An Eclectic Review." Western Journal of the Medical and Physical Sciences (Cincinnati) 5: 593-616, 652-64 (January, February, March, 1832).

"Resignation." Western Journal of the Medical and Physical Sciences (Cincinnati) 5: 664 (January, February, March, 1832).

"A Sketch of the Climate of the Valley of the Mississippi." Western Journal of the Medical and Physical Sciences (Cincinnati) 6: 9-22 (April, May, June, 1832).

"Notices of the Influenza and Measles, as They Appeared at Cincinnati, in 1831-2." Western Journal of the Medical and Physical Sciences (Cincinnati) 6: 45-52 (April, May, June, 1832).

"Epidemic Cholera—Its History and Aetiology. An Eclectic Review." Western Journal of the Medical and Physical Sciences (Cincinnati) 6: 78-120 (April, May, June, 1832).

"Climate and Diseases." In Robert Baird, View of the Valley of the Mississippi, or The Emigrant's and Traveller's Guide to the West . . . Philadelphia: H. S. Tanner, 1832. Pages 55-75. 2 ed., 1834. Pages 67-87.

"The Cholera." Cincinnati Daily Gazette, June 23, 25, 27, 1832.

"The Cholera. Reprinted from the Cincinnati Daily Gazette, June 23 & 25, 1832." Cincinnati Chronicle and Literary Gazette, June 30, 1832.

"The Cholera [Reprinted from the Cincinnati Daily Gazette, June 27, 1832]." Cincinnati Chronicle and Literary Gazette, July 7, 1832.

A Practical Treatise on the History, Prevention, and Treatment of Epidemic Cholera, Designed for both the Profession and the People. Cincinnati: Corey & Fairbank, 1832. 180 p. Dated July 23.

"Cholera." Cincinnati Chronicle and Literary Gazette, July 28, 1832.

"Editorial Note [to An Inquiry into the Anatomical Condition of the Alimentary Canal, in its Healthy State. Extracted from the Pathological Anatomy of Professor Andral]." Western Journal of the Medical and Physical Sciences (Cincinnati) 6: 169-70 (July, August, September, 1832).

"Editorial Note [on the Use of Digitalis in Amenorrhoea, Paramenia Obstructionis]." Western Journal of the Medical and Physical Sciences (Cincinnati) 6: 188 (July, August, September, 1832).

"Notice of the Epidemic Constitution, of the Summer of 1832, at Cincinnati." Western Journal of the Medical and Physical Sciences (Cincinnati) 6: 198-210 (July, August, September, 1832).

"[Review:] A Treatise on the Venerial Diseases of the Eye. By Wm. Lawrence, F.R.S." Western Journal of the Medical and Physical Sciences (Cincinnati) 6: 229-44 (July, August, September, 1832).

"[Note:] Elements of Chemical Philosophy, on the Basis of Reid, Comprizing the

Rudiments of That Science. By Thomas D. Mitchell, M.D." Western Journal of the Medical and Physical Sciences (Cincinnati) 6: 320 (July, August, September, 1832).

"[A Short History of the Medical College of Ohio.] September 29, 1832." In Proceedings and Correspondence of the Third District Medical Society of the State of Ohio, in Reference to the Medical College of Ohio. Published by Order of the Committee Charged with That Subject. N.p., December, 1832. Pages 4-37.

"Cholera in Cincinnati." Cincinnati Daily Gazette, October 8, 9, 1832.

"Prevention of Cholera." Cincinnati Chronicle and Literary Gazette, October 13, 1832.

Cure of Cholera. Cincinnati Chronicle—Extra. Handbill. [Cincinnati,] October 13, 1832. 1 p.

"Cholera in Cincinnati." Cincinnati Chronicle and Literary Gazette, October 13, 27, 1832.

"Prevention of Cholera." Cincinnati Daily Gazette, October 15, 1832.

"Cholera and the Steam Doctors." Cincinnati Chronicle and Literary Gazette, October 27, 1832.

"Cholera in Cincinnati. Relapses." Cincinnati Chronicle and Literary Gazette, October 27, 1832.

"Epidemic Cholera in Cincinnati." Western Journal of the Medical and Physical Sciences (Cincinnati) 6: 321-64 (October, November, December, 1832).

Epidemic Cholera, as it Appeared in Cincinnati. Extracted from the Sixth Volume of the Western Journal of Medical and Physical Sciences. Cincinnati: Printed at the Chronicle Office. E. Deming, December, 1832. 46 p.

"[Review:] Elements of Chemical Philosophy, on the Basis of Reid, Comprising the Rudiments of that Science . . . By Thomas D. Mitchell, M.D." Western Journal of the Medical and Physical Sciences (Cincinnati) 6: 394-408 (October, November, December, 1832).

"Gravitation, Affinity, Vitality. Address to the Mechanics Institute. December 8, 1832. Not delivered 'till the 15th." Opening only; continues with page 5 of revised text, "Introductory Lecture for the Second Session of the Medical College of Ohio. November, 1821." MS in Cincinnati General Hospital Library. 4 p.

"Medical College of Ohio." Western Medical Gazette 1: 13-14 (December 15, 1832). Signed "D." [?]

"Memoir of Dr. Godman." In John D. Godman, Rambles of a Naturalist. Philadelphia: Thomas T. Ashe, 1833. Pages 13-36.

Communication from Dr. Drake. To the Honorable the General Assembly of the State of Ohio. N.p., n.d. [Cincinnati, January 19, 1833]. 20 p.

"'Free People of Color.'" Cincinnati Chronicle and Literary Gazette, February 2, 1833. Signed "D." [?]

"The Medical College of Ohio." Cincinnati Chronicle and Literary Gazette, February 9, 1833.

"[Review:] A. Cornelii Celsi de Medicina Libri Octo et Recensione Leonardi Targae. The Eight Books of Aulus Cornelius Celsus on Medicine, from the Text of Leonardus Targa." Western Journal of the Medical and Physical Sciences (Cincinnati) 6: 574-86 (January, February, March, 1833).

"[Review:] A Manual of Surgery, Founded Upon the Principles and Practice Lately

Taught by Sir Astley Cooper and Joseph Henry Green . . . Ed. by Thomas Castle, F.L.S." Western Journal of the Medical and Physical Sciences (Cincinnati) 6: 586–92 (January, February, March, 1833).

"[Review:] United States Dispensatory." Western Medical Gazette (Cincinnati) 1: 109 (March 15, 1833). Signed "D." [?]

"On the Use of Ergot in Leucorrhoea." Western Medical Gazette (Cincinnati) 1: 132 (April 15, 1833). Signed "D." [?]

"Attendance on Medical Lectures." Western Medical Gazette (Cincinnati) 1: 155–56 (May 1, 1833). Signed "D." [?]

"Observations on Some of the Uses and Abuses of Purgatives." Western Journal of the Medical and Physical Sciences (Cincinnati) 7: 32–42 (April, May, June, 1833).

"Our Enterprize." Western Journal of the Medical and Physical Sciences (Cincinnati) 7: 151–53 (April, May, June, 1833).

"The Epidemic." Western Journal of the Medical and Physical Sciences (Cincinnati) 7: 153–54 (April, May, June, 1833).

"Measles and Mumps in Combination." Western Journal of the Medical and Physical Sciences (Cincinnati) 7: 154 (April, May, June, 1833).

"Convulsions of Young Infants." Western Journal of the Medical and Physical Sciences (Cincinnati) 7: 155 (April, May, June, 1833).

"Animalcules in our Hydrant Water." Western Journal of the Medical and Physical Sciences (Cincinnati) 7: 155–56 (April, May, June, 1833).

"Neglect of Vaccination—Free Schools." Western Journal of the Medical and Physical Sciences (Cincinnati) 7: 156 (April, May, June, 1833).

"Inquiry into the Condition of the Medical College of Ohio." Western Journal of the Medical and Physical Sciences (Cincinnati) 7: 156–57 (April, May, June, 1833).

"Western Graduates." Western Journal of the Medical and Physical Sciences (Cincinnati) 7: 157 (April, May, June, 1833).

"Popular Lectures on Anatomy and Surgery." Western Journal of the Medical and Physical Sciences (Cincinnati) 7: 157–58 (April, May, June, 1833).

"Lithotomy." Western Journal of the Medical and Physical Sciences (Cincinnati) 7: 158 (April, May, June, 1833).

"Medical Obituary [James Roane, M.D., Dr. Richard Pindell, Joseph Challen, M.D., Thomas Flanner, M.D., William Richards, M.D., Lewis Heermann, M.D.]." Western Journal of the Medical and Physical Sciences (Cincinnati) 7: 158–60 (April, May, June, 1833).

"Cholera, Town and Country [June 13, 1833]." Cincinnati Chronicle and Literary Gazette, June 15, 1833.

"Cholera—Town and Country [June 13, 1833]." National Intelligencer (Washington, D.C.), June 27, 1833.

"Progress of the Cholera. July 23, noon." Cincinnati Chronicle and Literary Gazette, July 27, 1833.

"The Epidemic—Again. July 24, noon." Cincinnati Chronicle and Literary Gazette, July 27, 1833.

"The Epidemic—Again. July 26, 2 P.M." Cincinnati Chronicle and Literary Gazette, July 27, 1833.

"The Science of Nutrition." Western Medical Gazette (Cincinnati) 1: 217–18 (July 1, 1833); 228–29 (July 15, 1833); 273–75 (September 1, 1833). Signed "***." [?]

"Epidemic Cholera. Miscellaneous Observations—Historical, Statistical, Aetiological and Therapeutic, on the Prevailing Epidemic." Western Journal of the Medical and Physical Sciences (Cincinnati) 7: 161-81 (July, August, September, 1833).

"Extra Professional Duties." Western Journal of the Medical and Physical Sciences (Cincinnati) 7: 316-18 (July, August, September, 1833).

"Education, Physical and Intellectual, of the Two Sexes. [Delivered before the Literary Convention of Kentucky, Lexington, November 8, 1833.]" MS in Cincinnati General Hospital Library. 23 p.

Remarks on the Importance of Promoting Literary and Social Concert, in the Valley of the Mississippi, as a Means of Elevating Its Character, and Perpetuating the Union. Delivered in the Chapel of Transylvania University, to the Literary Convention of Kentucky, November 8, 1833. Louisville: Published by Members of the Convention, 1833. 26 p.

"Cholera." American Journal of the Medical Sciences (Philadelphia) 13: 287-88 (November, 1833).

"To the Public. [Physicians' Statement on the Use of Ardent Spirits.]" Cincinnati Chronicle and Literary Gazette, November 30, 1833. Repeated, December 14, 1833.

"Editorial Note [on Science]." Western Journal of the Medical and Physical Sciences (Cincinnati) 7: 363-68 (October, November, December, 1833).

"Shooting Stars. An Account of a Remarkable Exhibition of Shooting Stars, Seen at Various and Distant Places, on the Night of Wednesday, 13th of November, 1833." Western Journal of the Medical and Physical Sciences (Cincinnati) 7: 368-84 (October, November, December, 1833).

"[Review:] The Cyclopoedia of Practical Medicine and Surgery . . . Edited by Isaac Hays, M.D." Western Journal of the Medical and Physical Sciences (Cincinnati) 7: 385-410 (October, November, December, 1833).

"Bad Effects of Animal Diet." Western Journal of the Medical and Physical Sciences (Cincinnati) 7: 466 (October, November, December, 1833).

"Poisoning of Oil of Tansey." Western Journal of the Medical and Physical Sciences (Cincinnati) 7: 469-70 (October, November, December, 1833).

"Editorial Observations [on a Suddenly Fatal Disease at Stanford, Kentucky]." Western Journal of the Medical and Physical Sciences (Cincinnati) 7: 472-73 (October, November, December, 1833).

"Catalogue of the Plants of Kentucky." Western Journal of the Medical and Physical Sciences (Cincinnati) 7: 474-75 (October, November, December, 1833).

"Cultivation of Botany in Ohio." Western Journal of the Medical and Physical Sciences (Cincinnati) 7: 476 (October, November, December, 1833).

"Cincinnati Medical Society." Western Journal of the Medical and Physical Sciences (Cincinnati) 7: 476-77 (October, November, December, 1833).

"Repeal of the Medical Law of Ohio. Qualified Approval. Suggestion about Another Law." Western Journal of the Medical and Physical Sciences (Cincinnati) 7: 479-83 (October, November, December, 1833).

"School for Instruction of the Blind." Western Journal of the Medical and Physical Sciences (Cincinnati) 7: 483-85 (October, November, December, 1833).

"Education Convention of Kentucky." Western Journal of the Medical and Physical Sciences (Cincinnati) 7: 486 (October, November, December, 1833).

"Introduction of Tropical Plants into Florida." Western Journal of the Medical and

Physical Sciences (Cincinnati) 7: 486–87 (October, November, December, 1833).

"Subsidence of the Cholera in the West. Health of Cincinnati." Western Journal of the Medical and Physical Sciences (Cincinnati) 7: 487–88 (October, November, December, 1833).

"Medical Obituary [John Wooley, M.D., Thomas Sargent, M.D., William Gardner, M.D.]." Western Journal of the Medical and Physical Sciences (Cincinnati) 7: 488 (October, November, December, 1833).

"[On the Qualities of the Buckeye Tree; Reply to a Toast at The Buckeye Dinner, December 26, 1833.]" Western Monthly Magazine (Cincinnati) 3: 151–54 (March 1834).

"[The Buckeye Tree.]" Cincinnati Chronicle and Literary Gazette, May 10, 1834.

"The Buckeye Tree [Delivered at the Cincinnati Forty-fifth Anniversary Celebration, December 26, 1833]." In Benjamin Drake, Tales and Sketches, From the Queen City. Cincinnati: E. Morgan & Co., 1838. Pages 173–80.

Extracts, from the Western Journal of the Medical and Physical Sciences . . . Addressed to the Legislators, Physicians, and Students of Medicine, of Ohio . . . Cincinnati: January 8, 1834. 7 p.

"School for Instruction of the Blind." Cincinnati Mirror and Western Gazette 3: 115–16 (January 25, 1834).

"To the Public [Concerning the Publication of Drs. R. and J. Moorhead, Relative to the Case of Mr. Brooke]." Cincinnati Daily Gazette, January 30, February 8, 12, 1834.

"[Resolutions to the Temperance Convention.]" Cincinnati Daily Gazette, February 3, 1834.

"Cincinnati Medical Society. From the Western Journal of the Medical and Physical Sciences." Cincinnati Chronicle and Literary Gazette, February 15, 1834.

"Irritation and Inflammation of the Peritoneum." Western Journal of the Medical and Physical Sciences (Cincinnati) 7: 489–531 (January, February, March, 1834).

"The Medical College of Ohio." Western Journal of the Medical and Physical Sciences (Cincinnati) 7: 623–51 (January, February, March, 1834).

"Popular Study of Anatomy and Physiology." Western Journal of the Medical and Physical Sciences (Cincinnati) 7: 651–52 (January, February, March, 1834).

"Epidemic Cholera." Western Journal of the Medical and Physical Sciences (Cincinnati) 7: 652 (January, February, March, 1834).

"National Marine Hospitals on the Mississippi and Its Tributary Streams." Western Journal of the Medical and Physical Sciences (Cincinnati) 7: 652–53 (January, February, March, 1834).

"Inquiries Concerning Tubercular Diseases." Western Journal of the Medical and Physical Sciences (Cincinnati) 7: 653–54 (January, February, March, 1834).

"College Beneficiaries." Western Medical Gazette (Cincinnati) 1: 340–41 (March 1, 1834). Signed "Medicus." [?]

"On the Importance of Hospitals to Medical Schools." Western Medical Gazette (Cincinnati) 1: 359–61 (April 1, 1834). Signed "***." [?]

"Academical and Popular Study of Anatomy and Physiology. Ch. I: General Views and Observations." Cincinnati Chronicle and Literary Gazette, May 17, 1834.

"Academical and Popular Study of Anatomy and Physiology. Ch. II: The Skeleton." Cincinnati Chronicle and Literary Gazette, May 24, 31, 1834.

"[Review:] Observations on the Abuse of Spirituous Liquors, by the European Troops in India; and on the Impolicy of Uniformly and Indiscriminately Issuing Spirit Rations to Soldiers. By Henry Marshall, Esq. [Edinburgh Medical and Surgical Journal, January, 1834]." Western Medical Gazette (Cincinnati) 2: 68–72 (June 1834). Signed "***." [?]

"A Projected School and College Book of Anatomy and Physiology, Designed for Both Sexes." Western Journal of the Medical and Physical Sciences (Cincinnati) 8: 99–104 (April, May, June, 1834).

"[Note:] Medical Lectures." Western Medical Gazette (Cincinnati) 2: 192 (August 1834). Signed "D." [?]

"National Hospitals in the Valley of the Mississippi. From the forthcoming Western Medical Journal." Cincinnati Chronicle and Literary Gazette, September 27, 1834.

"An Account of the Third Visitation of Epidemic Cholera, in Connexion with Dysentary, at Cincinnati, in the Summer of 1834." Western Journal of the Medical and Physical Sciences (Cincinnati) 8: 169–93 (July, August, September, 1834).

"A Case of Partial Amnesia, in which the Memory for Proper Names was Lost." Western Journal of the Medical and Physical Sciences (Cincinnati) 8: 209–12 (July, August, September, 1834).

"National Hospitals in the Valley of the Mississippi." Western Journal of the Medical and Physical Sciences (Cincinnati) 8: 301–305 (July, August, September, 1834).

"Lunatic Asylum of Ohio." Western Journal of the Medical and Physical Sciences (Cincinnati) 8: 305–308 (July, August, September, 1834).

"Library of the Medical College of Ohio." Western Journal of the Medical and Physical Sciences (Cincinnati) 8: 308–10 (July, August, September, 1834).

"Clinical Ticket of Admission to the Commercial Hospital and Lunatic Asylum of Ohio." Western Journal of the Medical and Physical Sciences (Cincinnati) 8: 310–12 (July, August, September, 1834).

"Cooperative Classes of the Medical Schools in Cincinnati and Lexington." Western Journal of the Medical and Physical Sciences (Cincinnati) 8: 312–14 (July, August, September, 1834).

"Clinical Proofs of Distinct Nerves for Motion and Sensation." Western Journal of the Medical and Physical Sciences (Cincinnati) 8: 316–17 (July, August, September, 1834).

"Efficacy of the Hot and Cold Bath in Alternation, Illustrated with a Case." Western Journal of the Medical and Physical Sciences (Cincinnati) 8: 317–19 (July, August, September, 1834).

"Large Urinary Calculus in a Young Female." Western Journal of the Medical and Physical Sciences (Cincinnati) 8: 320–21 (July, August, September, 1834).

"To Students of Medicine." Western Medical Gazette (Cincinnati) 2: 237–38 (September, 1834). Signed "Medicus." [?]

Discourse on the History, Character, and Prospects of the West: Delivered to the Union Literary Society of Miami University, Oxford, Ohio, at their Ninth Anniversary, September 23, 1834. Cincinnati: Truman & Smith, 1834. 53 p.

"[Extracts:] History, Character, and Prospects of the West." Cincinnati Chronicle and Literary Gazette, December 6, 1834.

"Discourse on the History, Character, and Prospects of the West: Delivered to the Union Literary Society of Miami University, Oxford, Ohio, at their Ninth Anniversary, September 23, 1834." In [Miami University,] Oxford Addresses. Hanover, Indiana: Joseph G. Monfort, 1835. Pages 193-226.

The West: Extract from a Discourse on the History, Character, and Prospects of the West: Delivered to the Union Literary Society of Oxford, Ohio, at Their Ninth Anniversary, September 23, 1334 [sic]. Connecticut Courant (Hartford), Supplement, 4: 5-8 (1835).

"The Patriotism of Western Literature [from Discourse on the History, Character, and Prospects of the West]." In William Holmes McGuffey, The Eclectic Fourth Reader. 23d ed. Cincinnati: Truman & Smith, 1842. Pages 228-30; Rev. ed. Cincinnati: Winthrop B. Smith, 1848. Pages 317-19; Newly rev. ed. Cincinnati: Winthrop B. Smith, 1853. Pages 313-15.

"Natural Ties among the Western States [from Discourse on the History, Character, and Prospects of the West]." In William Holmes McGuffey, The Eclectic Fourth Reader. 23d ed. Cincinnati: Truman & Smith, 1842. Pages 272-75.

A Discourse on the Philosophy of Discipline, in Families, Schools, and Colleges; Delivered before the Western Literary Institute, and College of Professional Teachers, In Cincinnati, on the 6th of October, 1834. Cincinnati: U. P. James, 1834. 31 p.

"Discourse on the Philosophy of Family, School, and College Discipline." In Transactions of the Fourth Annual Meeting of the Western Literary Institute, and College of Professional Teachers, Held in Cincinnati, October, 1834. Cincinnati: Josiah Drake, 1835. Pages 31-62.

"[Report on] The Necessity for Hospitals in the Valleys of the Mississippi and the Lakes [for the Medical Convention of Ohio]." Western Journal of the Medical and Physical Sciences (Cincinnati) 8: 459-64 (October, November, December, 1834).

"[Report on] A School for the Education of the Blind [for the Medical Convention of Ohio]." Western Journal of the Medical and Physical Sciences (Cincinnati) 8: 465-67 (October, November, December, 1834).

"Note on the Efficacy of the Acetate of Lead, in Several Cases of Epidemic Cholera, in Cincinnati, October, 1834." Western Journal of the Medical and Physical Sciences (Cincinnati) 8: 402-408 (October, November, December, 1834).

"Gossamer [remarks on the Medical Convention of Ohio]." Western Journal of the Medical and Physical Sciences (Cincinnati) 8: 482-86 (October, November, December, 1834).

"Gossamer. From the Western Medical Journal." Cincinnati Chronicle and Literary Gazette, February 14, 1835.

"Report of the Committee on Western Commercial Hospitals." In Journal of the Proceedings of A Convention of Physicians, of Ohio, held in the City of Columbus, on January 5, 1835. Cincinnati: Printed by Order of the Convention, 1835. Pages 8-13.

"Report of the Committee on Western Commercial Hospitals [of the Medical Convention of Ohio, January, 1835]." Western Medical Gazette (Cincinnati) 2: 440-45 (February 1835).

"National Hospitals in the West. . . . Report of the Committee on Western Commercial Hospitals." In U.S. Congress, House of Representatives, 24th Congress, 1st session, "Marine Hospitals in the West." House Document 264, May 31, 1836. 5 p.

"Report of the Committee on a School for the Blind [of the Medical Convention of Ohio, January, 1835]." Western Medical Gazette (Cincinnati) 2: 446–48 (February 1835).

"Parting Song (To the Tune of 'Auld Lang Syne')." [Written for the Medical Convention of Ohio, January, 1835.] Western Medical Gazette (Cincinnati) 2: 466–67 (February 1835).

"Report of the Committee on a School for the Education of the Blind, to be Established by the State of Ohio, near her Capitol." In Journal of the Proceedings of A Convention of Physicians, of Ohio, held in the City of Columbus, January 5, 1835. Cincinnati: Published by Order of the Convention, 1835. Pages 14–16.

"Report on the Subject of the Education of the Blind by Daniel Drake, M.D., to the Ohio Medical Convention, January 6, 1835." N.p., n.d. [Cincinnati, 1835]. 6 p.

A Reply to the Pamphlet Entitled "Facts Concisely Stated for the Information of the Legislature of Ohio. Published by the Faculty of the Medical College of Ohio." N.p., n.d. [Cincinnati, January 23, 1835.] 6 p.

A Reply to the Pamphlet Entitled "Facts Concisely Stated for the Information of the Legislature of Ohio. Published by the Faculty of the Medical College of Ohio." 2d ed. N.p., n.d. [Cincinnati, January 28, 1835.] 7 p.

A Reply to the Pamphlet Entitled "Facts Concisely Stated for the Information of the Legislature of Ohio. Published by the Faculty of the Medical College of Ohio." 3d ed. N.p., n.d. [Cincinnati, January 31, 1835.] 8 p.

"Obituary [Dr. Samuel Watson Hales]." Western Journal of the Medical and Physical Sciences (Cincinnati) 8: 646 (January, February, March, 1835).

"Change of Climate in Pulmonary Consumption." Western Journal of the Medical and Physical Sciences (Cincinnati) 8: 624–28 (January, February, March, 1835).

"[Remarks on an] Account of the Epidemic Yellow Fever which Prevailed in New Orleans, during the Autumn of 1833. By Edward H. Barton." Western Journal of the Medical and Physical Sciences (Cincinnati) 8: 628–33 (January, February, March, 1835).

"Obituary [Dr. Samuel Watson Hales]." Western Journal of the Medical and Physical Sciences (Cincinnati) 8: 646–48 (January, February, March, 1835).

"Observations on the Properties and Proximate Principles of Ergot. By a Correspondent." Western Medical Gazette (Cincinnati) 2: 484–88 (March 1835). [?]

"Cholera Fantasies." Western Journal of the Medical and Physical Sciences (Cincinnati) 9: 13–24 (April, May, June, 1835).

"Medical Society of Tennessee. [Review:] Minutes of the Proceedings of the Medical Society of Tennessee, at the Sixth Annual Meeting, Held in Nashville, May 4th, 1835." Western Journal of the Medical and Physical Sciences (Cincinnati) 9: 77–78 (April, May, June, 1835).

"Editorial and Publishing Arrangements." Western Journal of the Medical and Physical Sciences (Cincinnati) 9: 154 (April, May, June, 1835).

"[Note:] Medical Schools." Western Journal of the Medical and Physical Sciences (Cincinnati) 9: 154 (April, May, June, 1835).

"Western Academy of Natural Sciences." Western Journal of the Medical and Physical Sciences (Cincinnati) 9: 155–56 (April, May, June, 1835).

"Benefits of Practical Phrenology." Western Journal of the Medical and Physical Sciences (Cincinnati) 9: 156–57 (April, May, June, 1835).

"Death of Leaves from the Fumes of Lead." Western Journal of the Medical and Physical Sciences (Cincinnati) 9: 157 (April, May, June, 1835).

"Nitrate of Silver in Tonsilitis." Western Journal of the Medical and Physical Sciences (Cincinnati) 9: 157 (April, May, June, 1835).

"Chalybeates in Dropsy." Western Journal of the Medical and Physical Sciences (Cincinnati) 9: 158 (April, May, June, 1835).

"Greenville Epsom Springs, Harrodsburg, Kentucky." Western Journal of the Medical and Physical Sciences (Cincinnati) 9: 159 (April, May, June, 1835).

"Blue Sulphur Springs, Virginia." Western Journal of the Medical and Physical Sciences (Cincinnati) 9: 159–60 (April, May, June, 1835).

"Exemptions of Certain Parts of the Delta of the Mississippi from Autumnal Diseases." Western Journal of the Medical and Physical Sciences (Cincinnati) 9: 160–61 (April, May, June, 1835).

"Winter Retreat for Those Inclined to Pulmonary Consumption." Western Journal of the Medical and Physical Sciences (Cincinnati) 9: 161–62 (April, May, June, 1835).

"Method of Finding the Dew Point." Western Journal of the Medical and Physical Sciences (Cincinnati) 9: 162 (April, May, June, 1835).

"Hot Douching in Sprains and Bruises." Western Journal of the Medical and Physical Sciences (Cincinnati) 9: 163 (April, May, June, 1835).

"Reform of the Medical Colleges of Ohio." Western Journal of the Medical and Physical Sciences (Cincinnati) 9: 169–203 (April, May, June, 1835).

Supplement to the Western Journal of the Medical and Physical Sciences: For April, May and June, 1835. Reform of the Medical College of Ohio. [Cincinnati,] June 30, 1835. 35 p.

"[Review:] Discourse on Self-Limited Diseases . . . By Jacob Bigelow, M.D." Western Journal of the Medical and Physical Sciences (Cincinnati) 9: 285–88 (July, August, September, 1835).

"[Editorial remarks on:] On the Mode in which Colchicum Operates in the Cure of Acute Rheumatism [. . . By Dr. A. G. Thompson]." Western Journal of the Medical and Physical Sciences (Cincinnati) 9: 334 (July, August, September, 1835).

"[Editorial remarks on:] Practical and Pharmaceutical Remarks on the Veratrum Viride [Itch-weed or American Hellebore]." Western Journal of the Medical and Physical Sciences (Cincinnati) 9: 335 (July, August, September, 1835).

"[Editorial Remarks on:] The Physiology of Respiration and Chemistry of the Blood Applied to Epidemic Cholera." Western Journal of the Medical and Physical Sciences (Cincinnati) 9: 361–65 (July, August, September, 1835).

"Absorption of the Liquor Amnii—Ascent of the Gravid Uterus." Western Journal of the Medical and Physical Sciences (Cincinnati) 9: 366–67 (July, August, September, 1835).

"Death Following the Operation for Congenital Cataract." Western Journal of the Medical and Physical Sciences (Cincinnati) 9: 368 (July, August, September, 1835).

"Comparative Pain and Constitutional Affections from Conching and Extracting the Cataract." Western Journal of the Medical and Physical Sciences (Cincinnati) 9: 368 (July, August, September, 1835).

"Medical Department of Willoughby University, on Lake Erie; and Medical College of Louisiana, New Orleans." Western Journal of the Medical and Physical Sciences (Cincinnati) 9: 369 (July, August, September, 1835).

"Safety and Success of Extracting the Cataract in Old Age." Western Journal of the Medical and Physical Sciences (Cincinnati) 9: 369 (July, August, September, 1835).

"Dr. Julius, of Prussia." Western Journal of the Medical and Physical Sciences (Cincinnati) 9: 369–70 (July, August, September, 1835).

"Analysis of Minerals and Mineral Waters." Western Journal of the Medical and Physical Sciences (Cincinnati) 9: 371 (July, August, September, 1835).

"Study of General and Pathological Anatomy." Western Journal of the Medical and Physical Sciences (Cincinnati) 9: 370–71 (July, August, September, 1835).

"Professor Jameson." Western Journal of the Medical and Physical Sciences (Cincinnati) 9: 371 (July, August, September, 1835).

"Intermittent Fever in the West." Western Journal of the Medical and Physical Sciences (Cincinnati) 9: 372–73 (July, August, September, 1835).

"Geological Surveys." Western Journal of the Medical and Physical Sciences (Cincinnati) 9: 376 (July, August, September, 1835).

Rail-road from the Banks of the Ohio River to the Tide Waters of the Carolinas and Georgia. Cincinnati: James & Gazlay, [September,] 1835. 30 p.

"[Remarks] On the Education of Emigrants." In Transactions of the Fifth Annual Meeting of the Western Literary Institute, and College of Professional Teachers, Held in Cincinnati, October, 1835. Cincinnati: Published by the Executive Committee, 1836. Pages 80–84.

"Remarks on Anatomy and Physiology, as a Branch of Study in Schools." In Transactions of the Fifth Annual Meeting of the Western Literary Institute, and College of Professional Teachers, Held in Cincinnati, October, 1835. Cincinnati: Published by the Executive Committee, 1836. Pages 125–26.

"Dr. Drake's Remarks [on Common Schools in the West]." In Transactions of the Fifth Annual Meeting of the Western Literary Institute, and College of Professional Teachers, Held in Cincinnati, October, 1835. Cincinnati: Published by the Executive Committee, 1836. Pages 172–75.

"Formation of Professional Character. [Introductory Lecture to the First Session, Medical Department of Cincinnati College.] November 2, 1835." Dated October 28, 1835. MS in Cincinnati General Hospital Library. 36 p.

"Dr. Drake's Reply [to Charges in a Pamphlet Signed 'Vindex']. October 30, 1835." In Medical Students of the Cincinnati College, To the Public. N.p., n.d. [Cincinnati, October 31, 1835.] Pages 3–11.

"Dr. Drake's Reply to 'Vindex.'" Cincinnati Whig and Commercial Advertiser, November 5, 6, 1835.

"[Review:] The American Cyclopaedia of Practical Medicine and Surgery. Part IX." Western Journal of the Medical and Physical Sciences (Cincinnati) 9: 444–50 (October, November, December, 1835).

"The Late Dr. David Hosack." Western Journal of the Medical and Physical Sciences (Cincinnati) 9: 513–15 (October, November, December, 1835).

"The Radical Cure of Hernia." Western Journal of the Medical and Physical Sciences (Cincinnati) 9: 515–19 (October, November, December, 1835).

"Call of the Maryland Academy of Science." Western Journal of the Medical and

Physical Sciences (Cincinnati) 9: 519–20 (October, November, December, 1835).

"Our Periodical." Western Journal of the Medical and Physical Sciences (Cincinnati) 9: 521–24 (October, November, December, 1835).

"The Medical Department of the Cincinnati College." Western Journal of the Medical and Physical Sciences (Cincinnati) 9: 524–26 (October, November, December, 1835).

"Tribute of Respect to Professor McDowell." Western Journal of the Medical and Physical Sciences (Cincinnati) 9: 526 (October, November, December, 1835).

Rail Road to Carolina from the Ohio River. N.p., n.d. [Cincinnati, March 1, 1836]. 6 p.

"To E. S. Thomas, Esq., Editor of the Evening Post [on the origin of the Cincinnati and Charleston Rail Road project]." Daily Cincinnati Republican, and Commercial Register, May 2, 1836.

"Facts Relative to the Influence of Variations of Food on the Qualities of the Flesh and Fluids of Animals. [Delivered at Transylvania University, November, 1823, one of a series on Materia Alimentaria.]" Western Journal of the Medical and Physical Sciences (Cincinnati) 9: 533–47 (January, February, March, 1836).

"[Editorial Remarks on:] Abstract of Meteorology. Observations, for the Calendar Year 1835. By Professor [Joseph] Ray of Woodward College, Cincinnati." Western Journal of the Medical and Physical Sciences (Cincinnati) 9: 547–48 (January, February, March, 1836).

"[Editorial Remarks on:] Meteorological Notices in Indiana. By D. Dale Owen." Western Journal of the Medical and Physical Sciences (Cincinnati) 9: 550 (January, February, March, 1836).

"[Editorial Remarks on:] Extract of a Report on the Recent Progress and Present State of Meteorology. By James D. Forbes." Western Journal of the Medical and Physical Sciences (Cincinnati) 9: 552 (January, February, March, 1836).

"[Review Essay:] Western Medical Schools." Western Journal of the Medical and Physical Sciences (Cincinnati) 9: 607–18 (January, February, March, 1836).

"[Review:] Journal of Pharmacy . . . Philadelphia College of Pharmacy." Western Journal of the Medical and Physical Sciences (Cincinnati) 9: 618–22 (January, February, March, 1836).

"Medical History of the West." Western Journal of the Medical and Physical Sciences (Cincinnati) 9: 679–85 (January, February, March, 1836).

"Medical Department of the Cincinnati College—Ediface—Trustees." Western Journal of the Medical and Physical Sciences (Cincinnati) 9: 685–86 (January, February, March, 1836).

"Preface [to vol. 10 of the Journal]." Western Journal of the Medical and Physical Sciences (Cincinnati) 10: 7–8 (April, May, June, 1836).

"Traveling Memoranda." Western Journal of the Medical and Physical Sciences (Cincinnati) 10: 311–19 (July, August, September, 1836).

"The Cincinnati College." Western Journal of the Medical and Physical Sciences (Cincinnati) 10: 319–21 (July, August, September, 1836).

"The Cincinnati Hospital." Western Journal of the Medical and Physical Sciences (Cincinnati) 10: 321–22 (July, August, September, 1836).

"National Marine Hospitals in the West." Western Journal of the Medical and Physical Sciences (Cincinnati) 10: 323 (July, August, September, 1836).

"Lectures on Mental Philosophy of Students of Medicine." Western Journal of the Medical and Physical Sciences (Cincinnati) 10: 324 (July, August, September, 1836).

"On the Advantages of the Louisville, Cincinnati, and Charleston Rail-Road." Western Literary Journal, and Monthly Review (Cincinnati) 1: 249–56 (September 1836).

"Report on the Best Method of Introducing and Prosecuting the Study of Anatomy and Physiology [in Schools and Colleges]." In Transactions of the Sixth Annual Meeting of the Western Literary Institute, and College of Professional Teachers, Held in Cincinnati, October, 1836. Cincinnati: Published by the Executive Committee, 1837. Pages 235–37.

"[Review:] The Cyclopaedia of Practical Medicine . . . Edited by John Forbes, M.D., Alexander Tweedie, M.D., and John Connolly, M.D." Western Journal of the Medical and Physical Sciences (Cincinnati) 10: 385–414 (October, November, December, 1836).

"Traveling Memoranda." Western Journal of the Medical and Physical Sciences (Cincinnati) 10: 469–76 (October, November, December, 1836).

"Professional Observations for the Year 1837." Western Journal of the Medical and Physical Sciences (Cincinnati) 10: 476–77 (October, November, December, 1836).

"Convulsions of Children." Western Journal of the Medical and Physical Sciences (Cincinnati) 10: 477–78 (October, November, December, 1836).

"Scalding by Steam." Western Journal of the Medical and Physical Sciences (Cincinnati) 10: 478–79 (October, November, December, 1836).

"Cincinnati Eye Infirmary." Western Journal of the Medical and Physical Sciences (Cincinnati) 10: 479–80 (October, November, December, 1836).

"Summer Medical Instruction." Western Journal of the Medical and Physical Sciences (Cincinnati) 10: 480 (October, November, December, 1836).

"Doctors, Quack-Doctors, and Doctor-Quacks. The Newspapers of the United States, for 1837." Western Journal of the Medical and Physical Sciences (Cincinnati) 10: 613–21 (January, February, March, 1837).

"Patronage of Quacks." Western Journal of the Medical and Physical Sciences (Cincinnati) 10: 622 (January, February, March, 1837).

"The Medical Department of the College of Cincinnati." Western Journal of the Medical and Physical Sciences (Cincinnati) 10: 623–30 (January, February, March, 1837).

"Brief Notices of the Weather and Diseases in Cincinnati, in the Months of January, February and March, 1837." Western Journal of the Medical and Physical Sciences (Cincinnati) 10: 641–43 (January, February, March, 1837).

"Consultation Letters." Western Journal of the Medical and Physical Sciences (Cincinnati) 10: 643–44 (January, February, March, 1837).

"Progress of Quackery." Western Journal of the Medical and Physical Sciences (Cincinnati) 10: 644 (January, February, March, 1837).

"Absurdities in Dress." In James Hall, The Western Reader. Cincinnati: Burgess & Crane, 1837. Pages 81–82.

"Medical Graduates of the University of Pennsylvania." Western Journal of the Medical and Physical Sciences (Cincinnati) 11: 138–41 (April, May, June, 1837).

"Medical Proceedings in the State of Connecticut." Western Journal of the Medical and Physical Sciences (Cincinnati) 11: 141–44 (April, May, June, 1837).

"Medical School of Fairfield." Western Journal of the Medical and Physical Sciences (Cincinnati) 11: 144–45 (April, May, June, 1837).

"Private Anatomical School at Auburn, New York." Western Journal of the Medical and Physical Sciences (Cincinnati) 11: 145–47 (April, May, June, 1837).

"The Willoughby Medical College." Western Journal of the Medical and Physical Sciences (Cincinnati) 11: 148–50 (April, May, June, 1837).

"Medical Department of Transylvania University." Western Journal of the Medical and Physical Sciences (Cincinnati) 11: 150–51 (April, May, June, 1837).

"Acclimation—Temperance in the South—Medical College of Louisiana." Western Journal of the Medical and Physical Sciences (Cincinnati) 11: 152–55 (April, May, June, 1837).

"The American Cyclopaedia of Practical Medicine and Surgery." Western Journal of the Medical and Physical Sciences (Cincinnati) 11: 155–56 (April, May, June, 1837).

"Bell's and Dunglison's Medical Libraries." Western Journal of the Medical and Physical Sciences (Cincinnati) 11: 156 (April, May, June, 1837).

"The Philosophy of Medical Investigation." Western Journal of the Medical and Physical Sciences (Cincinnati) 11: 156–58 (April, May, June, 1837).

"New Prospectus of the Western Journal of the Medical and Physical Sciences: Owned, Published and Edited by the Medical Faculty of the Cincinnati College." Western Journal of the Medical and Physical Sciences (Cincinnati) 11: 159–60 (April, May, June, 1837).

"Re-Organization of the Medical Department of Transylvania University." Western Journal of the Medical and Physical Sciences (Cincinnati) 11: 162–63 (April, May, June, 1837).

"The College of Surgeons and Physicians, of the University of the State of New York." Western Journal of the Medical and Physical Sciences (Cincinnati) 11: 163–65 (April, May, June, 1837).

"Medical and Physical Works in the Library of Lane Theological Seminary." Western Journal of the Medical and Physical Sciences (Cincinnati) 11: 166–68 (April, May, June, 1837).

"[Review:] Physiological Observations on the Pulsation of the Heart . . . By Dr. Knox, of Edinburgh." Western Journal of the Medical and Physical Sciences (Cincinnati) 11: 320–21 (July, August, September, 1837).

"Large Doses of Sulphate of Quinine in Malignant Intermittents." Western Journal of the Medical and Physical Sciences (Cincinnati) 11: 321–22 (July, August, September, 1837).

"Microscopic Insects, Apparently Developed from Inorganic Matter, by Galvanism." Western Journal of the Medical and Physical Sciences (Cincinnati) 11: 325–27 (July, August, September, 1837).

"The Graham Journal of Health and Longevity." Western Journal of the Medical and Physical Sciences (Cincinnati) 11: 329–31 (July, August, September, 1837).

"Medical Journals and Medical Schools." Western Journal of the Medical and Physical Sciences (Cincinnati) 11: 331–32 (July, August, September, 1837).

"De Abortu." Western Journal of the Medical and Physical Sciences (Cincinnati) 11: 332–34 (July, August, September, 1837).

"Our Institution." Western Journal of the Medical and Physical Sciences (Cincinnati) 11: 335–36 (July, August, September, 1837).

"Filling Up the Louisville Medical Institute." Western Journal of the Medical and Physical Sciences (Cincinnati) 11: 336–37 (July, August, September, 1837).

"Medical Convention of Ohio." Western Journal of the Medical and Physical Sciences (Cincinnati) 11: 337–39 (July, August, September, 1837).

"Voluntary Medical Societies in Kentucky." Western Journal of the Medical and Physical Sciences (Cincinnati) 11: 339–40 (July, August, September, 1837).

"Successful Tracheotomy." Western Journal of the Medical and Physical Sciences (Cincinnati) 11: 341–42 (July, August, September, 1837).

"Cow-Pox and Measles at the Same Time, in the Same Patient." Western Journal of the Medical and Physical Sciences (Cincinnati) 11: 342–43 (July, August, September, 1837).

"Varientia Rerum. Recognitions of the Cincinnati College." Western Journal of the Medical and Physical Sciences (Cincinnati) 11: 343–44 (July, August, September, 1837).

"Introductory Lecture. Third Session, Medical Department of Cincinnati College. October 28, 1837." Dated October 13, 1837. Pages 29–32 missing. MS in Cincinnati General Hospital Library. 61 p.

"Number of Pupils." Western Journal of the Medical and Physical Sciences (Cincinnati) 11: 498 (October, November, December, 1837).

"Lectures on Animal Magnetism, Apparitions and Witchcraft." Western Journal of the Medical and Physical Sciences (Cincinnati) 11: 499 (October, November, December, 1837).

"Exhortation to Gentility of Style." Western Journal of the Medical and Physical Sciences (Cincinnati) 11: 499–500 (October, November, December, 1837).

"Obituary Notices [Dr. Physick and Mr. Edward A. Scruggs]." Western Journal of the Medical and Physical Sciences (Cincinnati) 11: 502 (January, February, March, 1838).

"Preface [Concerning the Convention's Support of Scientific Investigation]." In Journal of the Proceedings of the Medical Convention of Ohio . . . held in . . . Columbus, on January 1, 1838 . . . Cincinnati: Pugh & Dodd, 1838. Page [2].

"Report of the Committee on the Causes which Contribute to Depress the Science, Dignity and Usefulness of the Medical Profession in the State of Ohio." In Journal of the Proceedings of the Medical Convention of Ohio . . . held in . . . Columbus, on January 1, 1838 . . . Cincinnati: Pugh & Dodd, 1838. Pages 11–14.

"Report of the Committee on Future Meetings of the Convention." In Journal of the Proceedings of the Medical Convention of Ohio . . . held in . . . Columbus, on January 1, 1838 . . . Cincinnati: Pugh & Dodd, 1838. Pages 15–16.

"Report of the Committee on the Defects in the Organization and Administration of the Medical Schools of the United States." In Journal of the Proceedings of the Medical Convention of Ohio . . . held in . . . Columbus, on January 1, 1838 . . . Cincinnati: Pugh & Dodd, 1838. Page 17.

"Resolution on the Establishment of [Common] Schools in Ohio." In Journal of the Proceedings of the Medical Convention of Ohio . . . held in . . . Columbus, on January 1, 1838 . . . Cincinnati: Pugh & Dodd, 1838. Page 18.

"[Request for Information on the] Medical and Economical Botany of Ohio." In Journal of the Proceedings of the Medical Convention of Ohio . . . held in . . . Columbus, on January 1, 1838 . . . Cincinnati: Pugh & Dodd, 1838. Page [31].

"[Request for Information on the] Medical Topography, Climate and Diseases of Ohio." In Journal of the Proceedings of the Medical Convention of Ohio . . . held in . . . Columbus, on January 1, 1838 . . . Cincinnati: Pugh & Dodd, 1838. Pages [31-32].

"Animal Magnetism." Western Journal of the Medical and Physical Sciences (Cincinnati) 11: 407-40 (January, February, March, 1838).

"Editorial Gossip." Western Journal of the Medical and Physical Sciences (Cincinnati) 11: 497-501 (January, February, March, 1838).

"Editorial Remarks [on Death from Eating Fruit]." Western Journal of the Medical and Physical Sciences (Cincinnati) 11: 564-65 (January, February, March, 1838).

"Mineral Springs of Virginia." Western Journal of the Medical and Physical Sciences (Cincinnati) 11: 565-70 (January, February, March, 1838).

"To Our Subscribers." Western Journal of the Medical and Physical Sciences (Cincinnati) 11: 642A-642B (January, February, March, 1838).

"The Cincinnati College." Western Journal of the Medical and Physical Sciences (Cincinnati) 11: 642B (January, February, March, 1838).

"Elements of Pathology." Western Journal of the Medical and Physical Sciences (Cincinnati) 11: 642B-43 (January, February, March, 1838).

"The Medical Convention of Ohio." Western Journal of the Medical and Physical Sciences (Cincinnati) 11: 643-44 (January, February, March, 1838).

"Obituary Notices [Dr. McIntosh, M. Alibert, Dr. Ives, Dr. Eberle]." Western Journal of the Medical and Physical Sciences (Cincinnati) 11: 648-50 (January, February, March, 1838).

"Compensation of Physicians." Western Journal of the Medical and Physical Sciences (Cincinnati) 11: 651 (January, February, March, 1838).

"Varicose Veins and Ulcers of the Legs." Western Journal of the Medical and Physical Sciences (Cincinnati) 11: 652-59 (January, February, March, 1838).

"Geological Surveys of Ohio, Indiana and Michigan." Western Journal of the Medical and Physical Sciences (Cincinnati) 11: 659-62 (January, February, March, 1838).

"To the Patrons of the Western Journal." Western Journal of the Medical and Physical Sciences (Cincinnati) 11: 664 (January, February, March, 1838).

"Observations on the Compensation of Physicians, Especially in the Western States." Western Journal of the Medical and Physical Sciences (Cincinnati) 12: 49-56 (April, May, June, 1838).

"The Cincinnati College." Western Journal of the Medical and Physical Sciences (Cincinnati) 12: 153-56 (April, May, June, 1838).

"The East and the West." Western Journal of the Medical and Physical Sciences (Cincinnati) 12: 156-62 (April, May, June, 1838).

"Western Medical Appointments." Western Journal of the Medical and Physical Sciences (Cincinnati) 12: 162 (April, May, June, 1838).

"Medical Journals in the West." Western Journal of the Medical and Physical Sciences (Cincinnati) 12: 163 (April, May, June, 1838).

"The Black Art." Western Journal of the Medical and Physical Sciences (Cincinnati) 12: 164–66 (April, May, June, 1838).

"New Medical School in Virginia." Western Journal of the Medical and Physical Sciences (Cincinnati) 12: 166–67 (April, May, June, 1838).

"Hope [A Poem in Five Stanzas]." Family Magazine or Monthly Abstract of General Knowledge (Cincinnati) 3: 399 ([September] 1838).

"To Doctor John Locke. [Citizens' Letter concerning his Report on Steamboat Explosions.]" Cincinnati Daily Gazette, September 24, 1838.

"Memoir of the Miami Country, 1779–1794. [Delivered at the Cincinnati Semicentennial Celebration, December 26, 1838.]" Edited by Beverley W. Bond, Jr. Quarterly Publication of the Historical and Philosophical Society of Ohio (Cincinnati) 18: 39–93 (April–September 1923).

"To Professor Mussey. [August 15, 1837.]" In Cincinnati Advertiser and Western Journal—Extra. February 1, 1839. Handbill. 1 p.

"To Professor Mussey. [August 15, 1837.]" Daily Advertiser and Journal (Cincinnati), February 2, 1839.

The War of Extermination. N.p. [Cincinnati,] March 25, 1839. 15 p.

"To Mr. Joseph Bonsall [March 28, 1839]." Cincinnati Daily Republican, March 29, 1839.

"To Mr. Joseph Bonsall [March 28, 1839]." In [Joseph Bonsall,] Controversy in Relation to the Medical Schools of Cincinnati and Commercial Hospital and Lunatic Asylum of Ohio. Cincinnati: Isaac Hefley & Co., 1839. Pages 9–10.

"The War Not Exterminated." Handbill, Cincinnati, March 29, 1839. 1 p.

"The War Not Exterminated [March 29, 1839]." In [Joseph Bonsall,] Controversy in Relation to the Medical Schools of Cincinnati and Commercial Hospital and Lunatic Asylum of Ohio. Cincinnati: Isaac Hefley & Co., 1839. Pages 11–17.

"[Introductory Lecture to the] Medical Institute of the City of Louisville. November 9, 1839." MS in Cincinnati General Hospital Library. 2 p.

"[Review:] On Scarlatina . . . By William Ingalls, M.D." Western Journal of Medicine and Surgery (Louisville) 1: 55–56 (January 1840).

"Our Enterprise." Western Journal of Medicine and Surgery (Louisville) 1: 79–80 (January 1840).

"Yellow Fever in the South-West." Western Journal of Medicine and Surgery (Louisville) 1: 80–81 (January 1840).

"Milk Sickness, *alias* Sick Stomach." Western Journal of Medicine and Surgery (Louisville) 1: 84 (January 1840).

"Translation of Trosseau and Bellog on Laryngitis." Western Journal of Medicine and Surgery (Louisville) 1: 85 (January 1840).

"Gross' Pathological Anatomy." Western Journal of Medicine and Surgery (Louisville) 1: 85 (January 1840).

"Scarlatina Simplex." Western Journal of Medicine and Surgery (Louisville) 1: 85 (January 1840).

"[Review:] Statistics on the Medical Colleges of the United States. By T. Romney Beck." Western Journal of Medicine and Surgery (Louisville) 1: 273–78 (February 1840).

"[Review:] A Catalogue of Plants . . . in the Vicinity of Columbus, Ohio. By

William S. Sullivant." Western Journal of Medicine and Surgery (Louisville) 1: 278–80 (February 1840).

"Our Delay." Western Journal of Medicine and Surgery (Louisville) 1: 359 (February 1840).

"Editorial Department." Western Journal of Medicine and Surgery (Louisville) 1: 360 (February 1840).

"Variations in the Quantity of Rain–Malaria." Western Journal of Medicine and Surgery (Louisville) 1: 360-61 (February 1840).

"Death from Salivation in a Child." Western Journal of Medicine and Surgery (Louisville) 1: 361-62 (February 1840).

"Commercial Hospitals in the West." Western Journal of Medicine and Surgery (Louisville) 1: 362-63 (February 1840).

"Proposed Remedy for Spina Bifida." Western Journal of Medicine and Surgery (Louisville) 1: 363-64 (February 1840).

"Production of Castor Oil in Illinois." Western Journal of Medicine and Surgery (Louisville) 1: 364-65 (February 1840).

"Pathological Museum." Western Journal of Medicine and Surgery (Louisville) 1: 365 (February 1840).

"Meetings of the Physicians of North-Eastern Kentucky." Western Journal of Medicine and Surgery (Louisville) 1: 365-66 (February 1840).

"National Medical Convention." Western Journal of Medicine and Surgery (Louisville) 1: 366-68 (February 1840).

"Baltimore College of Dental Surgery." Western Journal of Medicine and Surgery (Louisville) 1: 369-70 (February 1840).

"Clarendon Springs, Vermont." Western Journal of Medicine and Surgery (Louisville) 1: 370 (February 1840).

"Winter Retreat for Consumptives." Western Journal of Medicine and Surgery (Louisville) 1: 370-71 (February 1840).

"Medical Society of Stark County, Ohio." Western Journal of Medicine and Surgery (Louisville) 1: 371-72 (February 1840).

"Cure of a Deaf Mute, by Puncturing the Membrana Tympani." Western Journal of Medicine and Surgery (Louisville) 1: 372-73 (February 1840).

"Education of the Deaf and Dumb." Western Journal of Medicine and Surgery (Louisville) 1: 373-75 (February 1840).

"Revival of Medical Instruction in Egypt." Western Journal of Medicine and Surgery (Louisville) 1: 375-76 (February 1840).

"Professor Espy's Lectures on Meteorology." Western Journal of Medicine and Surgery (Louisville) 1: 377-78 (February 1840).

"Hail Storm." Western Journal of Medicine and Surgery (Louisville) 1: 378-79 (February 1840).

"Harrodsburg Springs." Western Journal of Medicine and Surgery (Louisville) 1: 379 (February 1840).

"Subterranean Retreat for Invalids." Western Journal of Medicine and Surgery (Louisville) 1: 379-80 (February 1840).

"Lunatic Asylums." Western Journal of Medicine and Surgery (Louisville) 1: 380-84 (February 1840).

"Candidates for Admission into the Medical Staff of the Army. Medical Schools."

Western Journal of Medicine and Surgery (Louisville) 1: 384–86 (February 1840).

"Death of Dr. Joseph Parrish." Western Journal of Medicine and Surgery (Louisville) 1: 387–88 (February 1840).

"Animal Magnetism." Western Journal of Medicine and Surgery (Louisville) 1: 388 (February 1840).

"Valedictory Address to the Graduates of the Louisville Medical Institute, Third Session. March 10, 1840." MS in Cincinnati General Hospital Library. 42 p.

"Doctorate Address to the Louisville Medical Institute [March 10, 1840]." In Louisville Medical Institute [Catalogue, 1839–40]. N.p., n.d. [Louisville 1840]. Pages 11–14.

"Case of Artificial Anus Cured." Western Journal of Medicine and Surgery (Louisville) 1: 481 (June 1840).

"Permanent Cure of Hernia from Fracture of the Upper Extremity of the Thigh Bone." Western Journal of Medicine and Surgery (Louisville) 1: 481 (June 1840).

"Treatment of Dysmenorrhoea with the Pessary." Western Journal of Medicine and Surgery (Louisville) 1: 482 (June 1840).

"Surgical Anatomy and Operative Surgery." Western Journal of Medicine and Surgery (Louisville) 1: 484 (June 1840).

"Saturated Alcoholic Tincture in Eupatorium Perfoliatum." Western Journal of Medicine and Surgery (Louisville) 2: 79 (July 1840).

"Congenital Fungus Haematodes." Western Journal of Medicine and Surgery (Louisville) 2: 79–80 (July 1840).

"Milk-Sickness." Western Journal of Medicine and Surgery (Louisville) 2: 80–81 (July 1840).

"Cases of Partial Paralysis in Children." Western Journal of Medicine and Surgery (Louisville) 2: 82–84 (July 1840).

"Address to the Philomathesian Society of Kenyon College, to be Delivered on Commencement Day, August 5, 1840." Pages numbered 6–67. Page 6 headed "Introductory to Chapter III." Also delivered Chillicothe, July 26, 1840; Louisville, March 1, 1841; St. Louis, March 1841; Lane Seminary, June 1841; Athens, August 21, 1841; Chillicothe, September 1841. MS in Cincinnati General Hospital Library. 61 p.

"[Review:] Elements of Pathological Anatomy . . . By Samuel D. Gross." Western Journal of Medicine and Surgery (Louisville) 2: 131–56 (August 1840); 187–96 (September 1840); 289–304 (October 1840); 367–82 (November 1840).

"The Natchez Tornado." Western Journal of Medicine and Surgery (Louisville) 2: 237 (September 1840).

"Nasal Polypus Cured with Sanguinara Canadensis." Western Journal of Medicine and Surgery (Louisville) 2: 237–38 (September 1840).

"Family Cataract." Western Journal of Medicine and Surgery (Louisville) 2: 238–40 (September 1840).

"Lancaster Medical Institute." Western Journal of Medicine and Surgery (Louisville) 2: 315–16 (October 1840).

"Dover's Powder Modified." Western Journal of Medicine and Surgery (Louisville) 2: 316–17 (October 1840).

"Temporary Hemiplegia and Aphonia, Apparently from Worms." Western Journal of Medicine and Surgery (Louisville) 2: 317–18 (October 1840).

"Progress of Temperance Reform." Western Journal of Medicine and Surgery (Louisville) 2: 318–19 (October 1840).

"Death of Dr. Perrine." Western Journal of Medicine and Surgery (Louisville) 2: 321–23 (October 1840).

"[Valedictory, on Moral Improvement.] October 8, 1840." MS in Cincinnati Historical Society. 4 p.

"Intermittent Fever, Hepatitis, Pneumonitis, Laryngitis and Intestinal Mucous Irritation." Western Journal of Medicine and Surgery (Louisville) 2: 345–52 (November 1840).

"Summer and Autumn Diseases of 1840." Western Journal of Medicine and Surgery (Louisville) 2: 399–400 (November 1840).

"Scarlet Fever." Western Journal of Medicine and Surgery (Louisville) 2: 400 (November 1840).

"Cow-Pox." Western Journal of Medicine and Surgery (Louisville) 2: 400–401 (November 1840).

"Licking County Medical and Philosophical Society." Western Journal of Medicine and Surgery (Louisville) 2: 401 (November 1840).

"Green County Medical Society." Western Journal of Medicine and Surgery (Louisville) 2: 402 (November 1840).

An Introductory Discourse to a Course of Lectures in Clinical Medicine and Pathological Anatomy. Delivered at the Opening of the New Clinical Amphitheatre of the Louisville Marine Hospital, November 5, 1840. Louisville: Prentice & Weissinger, 1840. 16 p.

"Address to the Louisville Medical Society. November 27, 1840." Opening only; continues with page 3 of "Gravitation, Affinity, Vitality. Address to the Mechanics Institute, December 8, 1832." MS in Cincinnati General Hospital Library. 2 p.

"Peroration to a Lecture on the Ear and the Joys of Hearing, the Fourth in a Course of Popular Physiology, delivering every Monday night, in the Louisville Medical Institute. Written from 6 to 12 P.M., Saturday, November 28, 1840." MS in Cincinnati General Hospital Library. 24 p.

"[Notes:] Popular Lecture on the Ear, Hearing, Deafness, &c. &c. For Monday night, November 30, [1840]. Louisville Medical Institute." MS in Cincinnati General Hospital Library. 2 p.

"To Readers and Correspondents." Western Journal of Medicine and Surgery (Louisville) 2: 479–80 (December 1840).

"Medical Convention of Kentucky." Western Journal of Medicine and Surgery (Louisville) 2: 480–81 (December 1840).

"Medical Obituary [Dudley Woodbridge Rhodes, M.D., John A. Turner, M.D.]." Western Journal of Medicine and Surgery (Louisville) 2: 481–84 (December 1840).

"[Notes:] Physiological Lecture. Smell and Taste. December 7, 1840." MS in Cincinnati General Hospital Library. 4 p.

"Rachitis." Western Journal of Medicine and Surgery (Louisville) 3: 74–75 (January 1841).

A Memoir on the Disease Called by the People the Trembles, the Sick Stomach or

Milksickness, as they Appear in the Virginia Military District in the State of Ohio. Read before the Medical Convention of Kentucky, January 12, 1841. Louisville: Maxwell, 1841. 57 p.

"A Memoir on the Disease Called by the People 'Trembles,' and the 'Sick Stomach' or 'Milk-Sickness;' as they have Occurred in the Counties of Fayette, Madison, Clark, and Green in the State of Ohio." Western Journal of Medicine and Surgery (Louisville) 3: 161–226 (March 1841).

"Medical Convention of Kentucky." Western Journal of Medicine and Surgery (Louisville) 3: 157–59 (February 1841).

"Death of President Harrison." Western Journal of Medicine and Surgery (Louisville) 3: 398–99 (May 1841).

"Medical Hoax." Western Journal of Medicine and Surgery (Louisville) 3: 399 (May 1841).

"Geological Voyage." Western Journal of Medicine and Surgery (Louisville) 3: 399–400 (May 1841).

"The Spring." Western Journal of Medicine and Surgery (Louisville) 3: 478 (June 1841).

"Medical Society of Tennessee." Western Journal of Medicine and Surgery (Louisville) 3: 478–79 (June 1841).

"Contribution to the Pathological Museum of the Medical Institute of Louisville." Western Journal of Medicine and Surgery (Louisville) 4: 73–74 (July 1841).

"Contributions to the Journal by Societies." Western Journal of Medicine and Surgery (Louisville) 4: 75 (July 1841).

"The Leviathan Discovered at Last." Western Journal of Medicine and Surgery (Louisville) 4: 149–56 (August 1841).

"Medical College of the State of North Carolina." Western Journal of Medicine and Surgery (Louisville) 4: 156–57 (August 1841).

"Address to the Athens Literary Society [on the Proper Objects of Study]. Athens [Ohio], August 2, 1841." Originally 30 p., pages 11–21 missing. MS in Cincinnati General Hospital Library. 19 p.

"Medical Institute of Louisville." Western Journal of Medicine and Surgery (Louisville) 4: 228–29 (September 1841).

"Unpublished Contributions." Western Journal of Medicine and Surgery (Louisville) 4: 230–32 (September 1841).

"Worthy of Imitation." Western Journal of Medicine and Surgery (Louisville) 4: 232–33 (September 1841).

"Morbid Anatomy of Milksickness." Western Journal of Medicine and Surgery (Louisville) 4: 234–35 (September 1841).

"Clerical Encouragement of Quackery." Western Journal of Medicine and Surgery (Louisville) 4: 235–36 (September 1841).

"[Review:] Statistical Researches relative to the Etiology of Pulmonary and Rheumatic Diseases . . . By Samuel Forry." Western Journal of Medicine and Surgery (Louisville) 4: 271–94 (October 1841).

"Endemic Dysentery." Western Journal of Medicine and Surgery (Louisville) 4: 315 (October 1841).

"Board of Health of New Orleans." Western Journal of Medicine and Surgery (Louisville) 4: 315–16 (October 1841).

"State Medical Society of Kentucky." Western Journal of Medicine and Surgery (Louisville) 4: 316-17 (October 1841).

"Coup de Soleil, or Stroke of the Sun." Western Journal of Medicine and Surgery (Louisville) 4: 317 (October 1841).

"Absence of Lacteal Secretion after Parturition." Western Journal of Medicine and Surgery (Louisville) 4: 317 (October 1841).

"[Review:] Physiology and Animal Mechanism." Western Journal of Medicine and Surgery (Louisville) 4: 371 (November 1841).

"[Review:] Bulletin of the Proceedings of the National Institution for the Promotion of Science." Western Journal of Medicine and Surgery (Louisville) 4: 372 (November 1841).

"Progress of Science." Western Journal of Medicine and Surgery (Louisville) 4: 400-402 (November 1841).

"Sulphate of Quinine in Congestive Fever." Western Journal of Medicine and Surgery (Louisville) 4: 402-403 (November 1841).

"Uticaria." Western Journal of Medicine and Surgery (Louisville) 4: 403-404 (November 1841).

"Consultations through the Journal." Western Journal of Medicine and Surgery (Louisville) 4: 404 (November 1841).

"Arsenic Sold for Antimony.—Empoisoning." Western Journal of Medicine and Surgery (Louisville) 4: 405-407 (November 1841).

"Murrain of Cattle." Western Journal of Medicine and Surgery (Louisville) 4: 407-408 (November 1841).

"Alleged Discharge of Hairs from the Hand." Western Journal of Medicine and Surgery (Louisville) 4: 483-85 (December 1841).

"New Work on the 'Climate of the United States and Its Endemic Influences.'" Western Journal of Medicine and Surgery (Louisville) 4: 487 (December 1841).

"Spinal Counter Stimulation in Congestive Fever." Western Journal of Medicine and Surgery (Louisville) 4: 487-88 (December 1841).

"Address to Boys and Young Men [on the Formation of Sound and Respectable Character. First of a Series of Popular Lectures on Monday Evenings at the Louisville Medical Institute.] December 4, 1841." Originally 50 p., including text of "Address to the Athens Literary Society, August 2, 1841" as pages 21-50. Pages 31-41 missing. MS in Cincinnati General Hospital Library. 39 p.

"Breweries vs. Foundries." Dated Christmas, 1841. Delivered, Chillicothe, July 4, 1842. MS in Cincinnati Historical Society. 28 p.

"State Medical Society of Kentucky." Western Journal of Medicine and Surgery (Louisville) 5: 79 (January 1842).

"Lithotomy." Western Journal of Medicine and Surgery (Louisville) 5: 80 (January 1842).

"Physiological Temperance Society." Western Journal of Medicine and Surgery (Louisville) 5: 80 (January 1842).

"[Motives for Refraining from Drink, and Joining the Physiological Temperance Society.] January 1, 1842." MS in Cincinnati Historical Society. 10 p.

"[Notes:] Necessity and Value of Cooperative Effort. [For the Physiological Temperance Society.] January 1, 1842." MS in Cincinnati Historical Society. 2 p.

"Report of the Committee to Memorialize [Congress] for Repeal of that Part of the Patent Law which Authorizes Granting Patents for Compounds of Medicine." In Proceedings of the Medical Convention of Ohio, held at Cincinnati, January 16-20, 1842 . . . Cincinnati: R. P. Brooks, 1842. Page 10.

"[Request for Information on] Snake Bites [and their Treatment]." In Proceedings of the Medical Convention of Ohio, held at Cincinnati, January 16-20, 1842 . . . Cincinnati: R. P. Brooks, 1842. Pages 10-11.

"[Resolution to Recommend the Promotion of Temperance by Physicians 'as a Means of Averting Many Formidable Diseases of Body, and Disorders of Mind.']" In Proceedings of the Medical Convention of Ohio, held at Cincinnati, January 16-20, 1842 . . . Cincinnati: R. P. Brooks, 1842. Page 11.

"Lithotomy—Again." Western Journal of Medicine and Surgery (Louisville) 5: 159 (February 1842).

"The Trembles and Milk Sickness." Western Journal of Medicine and Surgery (Louisville) 5: 235-36 (March 1842).

"Arsenic and the Trembles." Western Journal of Medicine and Surgery (Louisville) 5: 236-37 (March 1842).

"Kentucky School for the Instruction of the Blind." Western Journal of Medicine and Surgery (Louisville) 5: 237 (March 1842).

"Ohio Lunatic Asylum." Western Journal of Medicine and Surgery (Louisville) 5: 237-38 (March 1842).

"Kentucky Lunatic Asylum." Western Journal of Medicine and Surgery (Louisville) 5: 238-40 (March 1842).

"Ohio School for the Blind." Western Journal of Medicine and Surgery (Louisville) 5: 240 (March 1842).

"The Past Winter and Present Spring." Western Journal of Medicine and Surgery (Louisville) 5: 393-94 (May 1842).

"Progress of Total Abstinence." Western Journal of Medicine and Surgery (Louisville) 5: 394 (May 1842).

"The Sanicula Marilandica a Cure for the Bite of the Snake." Western Journal of Medicine and Surgery (Louisville) 5: 394-95 (May 1842).

"Operation for Club Foot." Western Journal of Medicine and Surgery (Louisville) 5: 395-97 (May 1842).

"French Measles." Western Journal of Medicine and Surgery (Louisville) 5: 397 (May 1842).

"Dry Gangrene." Western Journal of Medicine and Surgery (Louisville) 5: 397-98 (May 1842).

"Flowers of Sulphur and Cream of Tartar and Magnesia in Dysentary." Western Journal of Medicine and Surgery (Louisville) 5: 398 (May 1842).

"Cold Dash in the Asphyxia of Infants." Western Journal of Medicine and Surgery (Louisville) 5: 398-99 (May 1842).

"Leaden Canula in the Treatment of Fistula in Ano." Western Journal of Medicine and Surgery (Louisville) 5: 399 (May 1842).

"Medical Convention of Ohio." Western Journal of Medicine and Surgery (Louisville) 5: 469-72 (June 1842).

"Western Periodicals." Western Journal of Medicine and Surgery (Louisville) 5: 472-73 (June 1842).

"Medical Schools of the West." Western Journal of Medicine and Surgery (Louisville) 5: 473–74 (June 1842).

"Death of Dr. Blythe." Western Journal of Medicine and Surgery (Louisville) 5: 474–76 (June 1842).

"[Address before a society for the promotion of general knowledge. Cincinnati,] June 6, 1842." Pages 1–3 missing. MS in Cincinnati General Hospital Library. 9 p.

"Two Evenings' Experiments on Mesmerism." Western Journal of Medicine and Surgery (Louisville) 6: 71–79 (July 1842).

"The College of Professional Teachers." Western Journal of Medicine and Surgery (Louisville) 6: 79–80 (July 1842).

"Notes for a Temperance Address in the Missionary Presbyterian Church: Mackinac, July 31, 1842." MS in Cincinnati Historical Society. 19 p.

"Causes and Consequences of Temperate Drinking." In Rev. James Young, ed., The Lights of Temperance. Louisville: Morton & Griswold, 1851; Hull & Brother, 1853. Pages 163–92.

"Notes for an Address on Intemperance, to be Delivered to the Troops in Fort Mackinac, Commanded by Captain Scott, August 1, 1842." MS in Cincinnati Historical Society. 8 p.

"Private Medical Instruction." Western Journal of Medicine and Surgery (Louisville) 6: 239 (September 1842).

"To the Editor of the Western Lancet. [Requesting Information on Chronic Laryngitis of Clergymen and Other Public Speakers.] November 5, 1842." Western Lancet (Cincinnati) 1: 328 (November 1842).

"[Address at the] Opening of the Physiological Temperance Society of the Medical Institute of Louisville. November 12, 1842." Repeated, 1844. MS in Cincinnati Historical Society. 12 p.

"The Northern Lakes a Summer Residence for Invalids of the South." Western Journal of Medicine and Surgery (Louisville) 6: 401–26 (December 1842).

The Northern Lakes a Summer Residence for Invalids of the South. Louisville: J. Maxwell, Jr., 1842. 29 p.

"Northern Lakes a Summer Residence for Invalids of the South [Extract]." Western Lancet (Cincinnati) 1: 407–409 (January 1843).

The Northern Lakes a Summer Residence for Invalids of the South. N.p., n.d. 13 p.

"Intemperance Discourse: Second Presbyterian Church, Louisville. December 20, 1842." Dated December 17, 1842. MS in Cincinnati Historical Society. 62 p.

"Western Meteorology." Western Journal of Medicine and Surgery (Louisville) 7: 63–64 (January 1843).

"Temperature and Diseases of the Year 1842." Western Journal of Medicine and Surgery (Louisville) 7: 64–65 (January 1843).

"Diseases of the West [extract from the Western Journal of Medicine and Surgery]." Boston Medical and Surgical Journal 28: 308 (May 17, 1843).

"Pneumonia of the Winter of 1841–42." Western Journal of Medicine and Surgery (Louisville) 7: 65 (January 1843).

"Northern Limits of Milk-Sickness." Western Journal of Medicine and Surgery (Louisville) 7: 65 (January 1843).

"Snake-Bites." Western Journal of Medicine and Surgery (Louisville) 7: 65–66 (January 1843).

"Throat Disease of Public Speakers." Western Journal of Medicine and Surgery (Louisville) 7: 66–68 (January 1843).

"Abolition of a Rule in Medical Graduation—Willoughby University of Lake Erie." Western Journal of Medicine and Surgery (Louisville) 7: 68–72 (January 1843).

"Saturated Solution of Corrosive Sublimate in Erysipelas." Western Journal of Medicine and Surgery (Louisville) 7: 72–73 (January 1843).

"A Discovery of the True Cause of the Disease Called by the People Trembles or Milk Sickness." Western Journal of Medicine and Surgery (Louisville) 7: 73–77 (January 1843).

"School for the Blind." Western Journal of Medicine and Surgery (Louisville) 7: 77–78 (January 1843).

"The Mammoth Cave a Winter Retreat for Invalids." Western Journal of Medicine and Surgery (Louisville) 7: 78 (January 1843).

"Lithotomy on a Man Seventy-Eight Years of Age." Western Journal of Medicine and Surgery (Louisville) 7: 78 (January 1843).

"Progress of Temperance." Western Journal of Medicine and Surgery (Louisville) 7: 79 (January 1843).

"To Professional Correspondents." Western Journal of Medicine and Surgery (Louisville) 7: 79 (January 1843).

"A Lecture on the Morbid Effects of Intemperance on the Mind; Delivered to the Physiological Temperance Society of the Medical Institute of Louisville [extract]." Western and Southern Medical Recorder (Lexington) 2: 91 (February 1843).

"Travelling Editorials [from New Orleans]." Western Journal of Medicine and Surgery (Louisville) 7: 235–40 (March 1843).

"Travelling Editorials [from New Orleans]." Western Journal of Medicine and Surgery (Louisville) 7: 313–15 (April 1843).

"Travelling Editorials [from Pensacola Bay]." Western Journal of Medicine and Surgery (Louisville) 7: 393–400 (May 1843).

"Travelling Editorials [from Greensboro, Alabama]." Western Journal of Medicine and Surgery (Louisville) 7: 469–75 (June 1843).

"Travelling Editorials [from Tombeckbee River]." Western Journal of Medicine and Surgery (Louisville) 8: 73–78 (July 1843).

"Travelling Editorials. Voyage up the Mississippi [from Sioto Valley, Ohio]." Western Journal of Medicine and Surgery (Louisville) 8: 153–59 (August 1843).

"Travelling Editorials [from Portsmouth, Ohio]." Western Journal of Medicine and Surgery (Louisville) 8: 236–39 (September 1843).

"For the Opening of the Physiological Temperance Society, Third Session. November 17, 1843." Repeated, 1845. MS in Cincinnati Historical Society. 15 p.

"The Western Journal." Western Journal of Medicine and Surgery (Louisville) 8: 473–75 (December 1843).

"[Notes:] Lecture on Drunkenness. Physiological Temperance Society of the Medical Institute. December 9, 1843." MS in Cincinnati Historical Society. 4 p.

"[Address to the Physiological Temperance Society. December 16, 1843.]" MS in Cincinnati Historical Society. 5 p.

"[Notes:] Diseases from the Habitual Use and Abuse of Alcoholic Drinks. December

30, 1843." For the Physiological Temperance Society. MS in Cincinnati Historical Society. 4 p.

"The Western Journal." Western Journal of Medicine and Surgery (Louisville) n.s. 1: 82–83 (January 1844).

"Lecture before the Physiological Temperance Society of the Medical Institute on the Uses and Abuses of Opium. February 17, 1844." Dated January 20, 1844. Also delivered, Mobile, Alabama, April 1844. MS in Cincinnati Historical Society. 26 p.

Analytical Report of a Series of Experiments in Mesmeric Somniloquism, Performed by An Association of Gentlemen: With Speculations on the Production of Its Phenomena. Louisville: F. W. Prescott & Co., [February 1,] 1844. 56 p.

"Report of a Series of Experiments on the Alleged Mental Sympathy of a Person Said to be in a State of Mesmeric Somniloquism." Western Journal of Medicine and Surgery (Louisville) n.s. 1: 189–219 (March 1844).

"Speculations on the Facts Presented in the Analytical Report on Mesmeric Somniloquism Published in the Last Number of This Journal." Western Journal of Medicine and Surgery (Louisville) n.s. 1: 285–313 (April 1844).

"[Review:] The Principles and Practice of Medicine. By John Elliotson." Western Journal of Medicine and Surgery (Louisville) n.s. 1: 340–51 (April 1844).

"Notes for a Lecture on the Diseases Produced by Intemperance, to be Delivered in the Methodist Church at Baton Rouge, May 25, 1844." MS in Cincinnati Historical Society. 2 p.

"Dr. Drake's Letter [Concerning his Research in the South]." New Orleans Medical Journal 1: 69–70 (May 9, 1844).

"Dr. Drake's Letter [to the Editors of the New Orleans Medical Journal]." Western Journal of Medicine and Surgery (Louisville) n.s. 2: 159–61 (August 1844).

"[Review:] A Memoir on the Means of Modifying the Forms of Living Bodies by Regimen. By M. Hippolyte Roger-Collard." Western Journal of Medicine and Surgery (Louisville) n.s. 2: 141–47 (August 1844).

"Travelling Letters from the Senior Editor. No. I [from Mobile]." Western Journal of Medicine and Surgery (Louisville) n.s. 1: 546–54 (June 1844).

"Travelling Letters from the Senior Editor. No. II [from New Orleans]." Western Journal of Medicine and Surgery (Louisville) n.s. 2: 163–74 (August 1844).

"Travelling Letters from the Senior Editor. No. III [from New Orleans]." Western Journal of Medicine and Surgery (Louisville) n.s. 2: 174–79 (August 1844).

"Travelling Letters from the Senior Editor. No. IV [from Kanzas River, Indian Territory]." Western Journal of Medicine and Surgery (Louisville) n.s. 2: 270–79 (September 1844).

"[Notes:] Some Maternal and Social Duties of Woman. Bloomington [Illinois], September 14, 1844." MS in Cincinnati Historical Society. 2 p.

"Travelling Letters from the Senior Editor. No. V [from Missouri River]." Western Journal of Medicine and Surgery (Louisville) n.s. 2: 354–60 (October 1844).

"Travelling Letters from the Senior Editor. No. VI [from St. Louis]." Western Journal of Medicine and Surgery (Louisville) n.s. 2: 360–66 (October 1844).

"Travelling Letters from the Senior Editor. No. VII [from Peoria, Illinois]." Western Journal of Medicine and Surgery (Louisville) n.s. 2: 445–51 (November 1844).

"Travelling Letters from the Senior Editor. No. VIII [from Chicago, Illinois]."

Western Journal of Medicine and Surgery (Louisville) n.s. 2: 456–62 (November 1844).

"Diseases of the Negro Population. In a Letter to Rev. Mr. Pinney, November 15, 1844." New Orleans Medical Journal 1: 583–84 (May 1845).

"Diseases of the Negro Population." Western Journal of Medicine and Surgery (Louisville) n.s. 3: 164–66 (February 1845).

An Introductory Lecture, on the Means of Promoting The Intellectual Improvement of the Students and Physicians, of the Valley of the Mississippi. Delivered in the Medical Institute of Louisville, November 4th, 1844. Published by the Class. Louisville: Prentice & Weissinger, 1844. 2 eds.: 24 p.; 24 p.

"Office Furniture of the Physician at the West. [Extract from An Introductory Lecture, on the Means of Promoting the Intellectual Improvement of the Students and Physicians, of the Valley of the Mississippi.]" Boston Medical and Surgical Journal 31: 445 (January 1, 1845).

"[Notes:] Physiological Temperance Society: Opening Lecture, November 9, 1844. On the Necessity and Consequences of Temperate Drinking." MS in Cincinnati Historical Society. 2 p.

"Travelling Letters from the Senior Editor. No. IX [from Galena, Illinois]." Western Journal of Medicine and Surgery (Louisville) n.s. 2: 537–44 (December 1844).

"Travelling Letters from the Senior Editor. No. X [from the Upper Mississippi]." Western Journal of Medicine and Surgery (Louisville) n.s. 2: 545–48 (December 1844).

"[Review:] Ethics of Medicine . . . By Thomas M. Logan. The Claims of Religion on Medical Men . . . By B. A. Boardman." Western Journal of Medicine and Surgery (Louisville) n.s. 3: 47–58 (January 1845).

"Medical Institute of Louisville." Western Journal of Medicine and Surgery (Louisville) n.s. 3: 85–87 (January 1845).

"Epidemic Erysipelas." Western Journal of Medicine and Surgery (Louisville) n.s. 3: 180–81 (February 1845).

"Rate of Mortality Among the Students of the Medical Institute of Louisville." Western Journal of Medicine and Surgery (Louisville) n.s. 3: 181–82 (February 1845).

"Sixth Annual Report of the Ohio Lunatic Asylum." Western Journal of Medicine and Surgery (Louisville) n.s. 3: 183–84 (February 1845).

"Valedictory [to the Physiological Temperance Society]. February 20, 1845." MS in Cincinnati Historical Society. 9 p.

"Unofficial Editorial Notice." Western Journal of Medicine and Surgery (Louisville) n.s. 3: 269 (March 1845).

"Study of Botany by Young Physicians." Western Journal of Medicine and Surgery (Louisville) n.s. 3: 269–73 (March 1845).

"Kentucky Institution for the Education of the Blind." Western Journal of Medicine and Surgery (Louisville) n.s. 3: 274–75 (March 1845).

"From the Senior Editor [from Cincinnati, June 23, 1845]." Western Journal of Medicine and Surgery (Louisville) n.s. 4: 176–82 (August 1845).

"To the Medical Convention of Ohio [requesting information on the physical condition, climate, and diseases of the Mississippi Valley]." In Proceedings of the Ohio Medical Convention, Held at Columbus, May 20, 1845. Columbus: Charles Scott & Co., 1845. Pages 8–9.

"National Convention in the City of New York." Western Journal of Medicine and Surgery (Louisville) n.s. 4: 451–52 (November 1845).

"Southern and Western Convention at Memphis." Western Journal of Medicine and Surgery (Louisville) n.s. 4: 452–53 (November 1845).

"Mississippi Valley Association of Dental Surgeons, First Annual Meeting." Western Journal of Medicine and Surgery (Louisville) n.s. 4: 454–55 (November 1845).

"To the Physicians and Meteorological Observers of the Valley of the Mississippi and the Lakes." Western Journal of Medicine and Surgery (Louisville) n.s. 4: 541–42 (December 1845).

"To Our Correspondents." Western Journal of Medicine and Surgery (Louisville) n.s. 4: 545–46 (December 1845).

"Reminiscential Letters of Daniel Drake, M.D., to His Children [December 1845 to January 1848]." MSS in Cincinnati General Hospital Library.

Pioneer Life in Kentucky: A Series of Reminiscential Letters from Daniel Drake, M.D., of Cincinnati, to his Children. Edited with Notes and a Biographical Sketch by his Son, Charles D. Drake. Cincinnati: Robert Clarke & Co., 1870. xlvi + 263 p.

"Letter to the President of the Mississippi Valley Association of Dental Surgeons [September 2, 1846. Requesting information on the causes of decay and diseases of the mouth]." In Mississippi Valley Association of Dental Surgeons, Proceedings of the Second Annual Meeting. Cincinnati: Printed for the Association, 1846. Pages 10–11.

"Editorial Letter [from Cincinnati, September 15, 1846]." Western Journal of Medicine and Surgery (Louisville) n.s. 6: 357–64 (October 1846).

"Opening Address. To the Medical and Law Students of the University of Louisville, before its Physiological Temperance Society. November 7, 1846." MS in Cincinnati Historical Society. 12 p.

"Letters from a Western Layman, No. 1. January 1, 1847." Episcopal Recorder (Philadelphia) 24: 182 (1847).

"Letters from a Western Layman, No. 2." Episcopal Recorder (Philadelphia) 24: 186 (1847).

"Letters from a Western Layman, No. 3. February 18, 1847." Episcopal Recorder (Philadelphia) 25: 2 (1847).

"Letters from a Western Layman, No. 4." Episcopal Recorder (Philadelphia) 25: 10 (1847).

"Letters from a Western Layman, No. 5. April 4, 1847." Episcopal Recorder (Philadelphia) 25: 22 (1847).

"Letters from a Western Layman, No. 6." Episcopal Recorder (Philadelphia) 25:30 (1847).

"Letters from a Western Layman, No. 7. May 9, 1847." Episcopal Recorder (Philadelphia) 25: 46 (1847).

"Letters from a Western Layman, No. 8." Episcopal Recorder (Philadelphia) 25: 58 (1847).

"[Letter from Cincinnati, May 22, 1847, on the Medical Convention of Ohio.]" Western Journal of Medicine and Surgery (Louisville) n.s. 7: 547–48 (June 1847).

"Travelling Letters from the Senior Editor. Number One [from Mountains of Western Virginia, July 9, 1847]." Western Journal of Medicine and Surgery (Louisville) n.s. 8: 263–70 (September 1847).

"Travelling Letters from the Senior Editor. Number Two [from Mountains of Pennsylvania, July 28, 1847]." Western Journal of Medicine and Surgery (Louisville) n.s. 8: 270–74 (September 1847).

"The Irish Immigrants' Fever. [August 27, 1847.]" Boston Medical and Surgical Journal 37: 149–57 (September 22, 1847).

"Travelling Letters from the Senior Editor. Number Three [from Chautauque Lake, N.Y., August 1, 1847]." Western Journal of Medicine and Surgery (Louisville) n.s. 8: 351–57 (October 1847).

"Travelling Letters from the Senior Editor. Number Four [from Gros Isle, 33 miles below Quebec, August 27]." Western Journal of Medicine and Surgery (Louisville) n.s. 8: 358–67 (October 1847).

"Travelling Letters from the Senior Editor. Number Five [from Toronto, September 13, 1847]." Western Journal of Medicine and Surgery (Louisville) n.s. 8: 455–61 (November 1847).

Strictures on Some of the Defects and Infirmities of Intellectual and Moral Character, in Students of Medicine: An Introductory Lecture, Delivered in the University of Louisville, November 1st, 1847. Published by the Class. Louisville: Prentice & Weissinger, 1847. 16 p.

"Cure of Cancer." Western Journal of Medicine and Surgery (Louisville) 3d series, 2: 455–56 (November 1848).

"Unsuccessful Attempt at Poisoning with Pounded Glass." Western Journal of Medicine and Surgery (Louisville) 3d series, 2: 456–57 (November 1848).

"Part of Valedictory [to the Physiological Temperance Society]. February 15, 1849." An Addition to the Valedictory Address delivered February 20, 1845. MS in Cincinnati Historical Society. 3 p.

An Introductory Lecture at the Opening of the Thirtieth Session of the Medical College of Ohio. Delivered at the Request of the Faculty, November 5, 1849. Published by the Class. Cincinnati: Morgan & Overend, 1849. 16 p.

"[Lectures on Pathology and the Practice of Medicine.]" In W. W. Dawson, "Notes Taken at the Medical College of Ohio, 1849–50," passim. MS in Cincinnati Historical Society.

A Systematic Treatise, Historical, Etiological, and Practical, on the Principal Diseases of the Interior Valley of North America, as They Appear in the Caucasion, African, Indian, and Esquimaux Varieties of Its Population. Cincinnati: Winthrop B. Smith & Co.; Philadelphia: Grigg, Elliot & Co.; New York: Mason & Law, 1850. 878 p.

"Communication from Dr. Drake to the Medico-Chirurgical Society of Cincinnati." Western Lancet (Cincinnati) 11: 557–69 (September 1850).

"Letters on Slavery, to Dr. John C. Warren, of Boston. [December 26, 31, 1850; January 4, 1851.]" National Intelligencer (Washington, D.C.), April 3, 5, 7, 1851.

"[Memorial] To the President of the [Ohio Constitutional] Convention [on slavery and colonization]." Ohio State Journal (Columbus), January 4, 1851.

"[Letter on] A City Park." Daily Cincinnati Gazette, July 29, 1851.

Discourses Delivered by Appointment, before the Cincinnati Medical Library Association, January 9th and 10th, 1852. Cincinnati: Published for the Association, by Moore & Anderson, 1852. 93 p.

A Systematic Treatise, Historical, Etiological, and Practical, on the Principal Diseases of the Interior Valley of North America, as They Appear in the Caucasion, African, Indian, and Esquimaux Varieties of Its Population. Second Series. Edited by S. Hanbury Smith, M.D., and Francis G. Smith, M.D. Philadelphia: Lippincott, Grambo & Co., 1854. 985 p.

"[Peroration to an Address on the Preventives of Intemperance.]" Undated. MS in Cincinnati Historical Society. 17 p.

"[Opportunities and Obligations of Men of Wealth and Learning.]" Undated. Pages numbered 11–21 and 31–41. MS in Cincinnati Historical Society. 10 p.

"[On the Need to Cultivate a Spirit of Piety, Patience, and Promptitude.]" For the Louisville Medical Institute? 1840? MS in Cincinnati General Hospital Library. 4 p.

"[Notes:] Education According to the Propensities, Sentiments, and Faculties." MS in Cincinnati General Hospital Library. 3 p.

"Of Gambling." MS in Cincinnati General Hospital Library. 14 p.

"[Address to the] Representatives of the People and City of Louisville—Founders of the Hospital and Institute." No date, ca. 1845? Pages numbered 29–35. MS in Cincinnati General Hospital Library. 7 p.

"[Notes:] Effects of Intemperance on the Intellectual and Moral Faculties." Undated. For the Physiological Temperance Society? MS in Cincinnati Historical Society. 3 p.

"[That Which Should be Cherished for a Wife.]" For the Louisville Medical Institute? MS in Cincinnati General Hospital Library. 3 p.

"[Duties of Students and Young Physicians.]" For the Louisville Medical Institute? MS in Cincinnati General Hospital Library. 5 p.